"The writers, two senior journalists at *The Boston Globe*, have thoroughly trawled through Romney's history, and indeed that of his zealous Mormon ancestors. The book charts the various stages of his polymorphic life in impressive detail. . . . All this is well done. The analysis of Romney's time at Bain is balanced and fair."
—*The Economist*

"Balanced and informative. . . . A well-written and useful resource for Romneyana great and small." —Louis Menand, *The New Yorker*

"Kranish and Helman have done a good job of finding and interviewing people who knew Romney at various early stages of his life. . . . They deserve credit for a well-balanced and thorough biography."
—*The Washington Independent Review of Books*

"A comprehensive and eminently fair-minded biography."
—*The New Republic*

THE REAL ROMNEY

THE REAL ROMNEY

Michael Kranish *and* Scott Helman

HARPER

NEW YORK • LONDON • TORONTO • SYDNEY

HARPER

FIRST HARPER PAPERBACK PUBLISHED 2012.

The Library of Congress has catalogued the hardcover edition as follows:

Kranish, Michael.
    The real Romney / by Michael Kranish and Scott Helman.—FIRST EDITION.
      pages cm
    Includes bibliographical references.
    ISBN 978-0-06-212327-5
1. Romney, Mitt. 2. Governors—Biography. 3. Presidential candidates—United States—Biography. I. Helman, Scott. II. Title.
    E901.1.R66K73   2012
    974.4'044092—dc23
    [B]
                                                                2011045483

ISBN 978-0-06-212328-2 (pbk.)

12 13 14 15 16   OV/RRD   10 9 8 7 6 5 4 3 2 1

Michael Kranish:

To the journalists of *The Boston Globe*.

Scott Helman:

To Jonas and Eli, and the rich lives that await them.

———————————

CONTENTS

About This Book   *ix*

ABOUT THIS BOOK

The Real Romney was written by Scott Helman and Michael Kranish of *The Boston Globe* staff, and edited by Mark S. Morrow, the deputy managing editor who oversees the Sunday *Globe* and major news projects. It builds on the work of the many *Globe* staffers who have tracked Mitt Romney's life and career over the decades, including those who contributed to a landmark seven-part biographical series published as he launched his first presidential run. Among those critical contributors: Brian C. Mooney, a veteran investigative reporter, whose deep reporting on Romney's years as governor and his push for state health care reform was a critical resource; Beth Healy, a business reporter who brought expert reporting and vital sources to our analysis of Romney's years as a leveraged buyout specialist; Bob Hohler, a sports reporter who trained a sharp eye on Romney's turnaround of the Utah Olympics; Michael Paulson, the *Globe*'s former religion reporter and now an editor at *The New York Times*, who was the first to trace Romney's missionary years in detail; Neil Swidey, a Sunday magazine staff writer who played a crucial writing and editing role in the 2007 series; and Peter S. Canellos, who edited the series as the paper's Washington bureau chief and now leads its editorial page. Other key contributions to that series came from *Globe* staffers Stephanie Ebbert, Robert Gavin, and Sacha Pfeiffer (now an on-air host for WBUR radio). The authors are deeply indebted, and grateful, to them all.

THE REAL ROMNEY

PROLOGUE

W hen he finally enters the auditorium, they jump to their feet, the murmur crescendoing to a spirited ovation. Through a gauntlet of whoops, whistles, fist pumps, and camera flashes, he emerges, microphone in hand, his charcoal hair speckled with gray, wearing an easy smile as the warm reception washes over him. "Let's go, Mitt!" someone screams. He mildly protests the adulation, imploring everyone to sit down, to please sit down. On this balmy autumn day in New Hampshire, in a campaign being fought very much on his turf, Willard Mitt Romney is confident, serenely so. He's an old hand at this now. He's ditched the stiff suits for cuffed khakis, scrapped the elaborate stagecraft for a minimalist presentation—just a guy holding court at a modest town hall meeting, ready for anything from voters who feel entitled to this ritual of political intimacy. He opens with a patriotic riff, promising "a campaign of American greatness." He wants everyone to know: Mitt Romney loves America, and he believes in its people.

It is a simple, Reaganesque anthem, served up to a Republican audience hungry for a credible, electable leader who will deny President Barack Obama a second term. Romney's whole demeanor, here at the center of this theater-in-the-round at Saint Anselm College in Manchester, New Hampshire, is meant to convey, beyond any doubt, that he is ready. If he could take over now, he would. Just give him the keys to the White House. Why wait? After all, he's been preparing for this moment—his moment—all his life. Many politicians say that. Or have it said about them. In Mitt Romney's case, to a remarkable degree, it happens to be true.

He is, as he likes to say in debates, in speeches, and on the stump, a turnaround specialist running to lead a nation that desperately needs one. In his narrative, President Barack Obama has steered the country into a ditch, and Mitt Romney is the only one capable of yanking it out. Mr. Fix-it, reporting for duty. He's already fixed his approach as a candidate, self-assured and savvy where he was often slipshod and self-defeating in 2008. He is, as one local politician tells him point-blank at Saint Anselm, a far stronger contender than he was four years ago, much more at home in a campaign centered on the economy. Last time, Romney looked like an actor playing a presidential candidate. This time, he seems like the real thing.

All around him on this day, mounted high on the walls of the college's New Hampshire Institute of Politics, are imposing photos of men who have achieved what he has long dreamed about. There's George H. W. Bush in a red jacket, standing with supporters near the ocean in Maine. There's Bill Clinton on the tarmac beside Air Force One. There's Jimmy Carter with Bill and Jeanne Shaheen, one of New Hampshire's most famous political couples. But on these walls also hang reminders of the brutal selectivity of presidential politics, of the men whose reach came up short. There's Bill Bradley shooting a basketball in a tweed coat and red tie. There's the young Al Gore, in blue jeans and a barn jacket, on the doorstep of a man who is evidently not thrilled at the visit. And then, down a hallway in a quiet corner, is an even more poignant reminder that politics is a fickle business.

There, at eye level, in a simple black frame against a wood panel, hangs a campaign poster from the 1968 presidential bid of his father, George Romney, a Republican moderate who bowed out of the race Richard Nixon would go on to win. The poster, in keeping with the era, has a psychedelic feel, with the words I WANT ROMNEY IN '68 printed in a fun-house font that wouldn't be out of place on a Jimi Hendrix concert flyer. George is smiling broadly in the gray-and-white rendering of his handsome face, his teeth impossibly white and his hair helmetlike in its perfection. It is an image strangely in tune with the moment, an artifact of both inspiration and warning.

This book is the first complete, independent biography of Mitt Romney, a man whose journey to national political fame is at once remarkable and thoroughly unsurprising. It would have been unthinkable to his ancestors just a few generations ago, yet countless people whose lives intersected with Romney's over the past seven decades have drawn the same conclusion: this man might just be president someday. *The Real Romney*, which draws on our many years tracking the man and his career for *The Boston Globe*, is an attempt to capture him in whole, to plumb the many chapters of his life for insight into his character, his worldview, his drive, and his contradictions.

It is the story of a man guided by his faith and firmly grounded in family. It is the story of a once marginal and feared religion, Mormonism, a brand of Christianity homemade in America that he and his forebears helped move to the mainstream. It is the story of the counterculture movement of the 1960s and the proudly square young man who found it all appalling. It is the story of the wildly lucrative world of private equity and leveraged buyouts, a world largely opaque to outsiders, in which wealth is built and concentrated in novel ways, sometimes at others' expense. It is the story of an uneasy relationship between conviction and vaulting ambition and how political dreams can die when tactics outrun beliefs. And it is the story, of course, of a father and a son, George and Mitt Romney, and the ways in which their lives aligned and diverged.

George Romney remains, more than four decades after his own adventure in presidential politics and sixteen years after his death, a constant spiritual presence at his son's side. When a young girl asks Mitt Romney during the New Hampshire gathering what he would tell her class to make them want to be politicians, he deadpans at first, saying "The answer is: nothing. Don't do it. Run as far as you can." But when he turns serious, he invokes the advice he says his father offered years ago: "He said, 'Don't get into politics as your profession. . . . Get into the world of the real economy. And if someday you're able to make a contribution, do it.'" This is the essence of Romney's pitch, and it has been ever since his days as a deal maker in the 1980s and 1990s. He's made his money—a mountain of it, in fact—and believes, as his father did, that he now owes a debt to the country that made a place for him.

Once upon a time, Mitt Romney was just a kid with a famous last name. He was born into a family of pioneers—early adopters of Mormonism who set out on brave journeys to spread their faith and preserve their traditions; a father who rose from dusty exile in Mexico to the gleaming boardrooms of business and politics. Through those strong men and women, the genes of leadership accumulated, adapted, and reassembled in a boy whose birth, the doctors said, should have been impossible, the youngest of four in a Detroit family whose fortunes, like so many in postwar America, were hitched to the automobile. Thanks to his father's drive and hard work, Mitt Romney grew up differently from the Romneys before him. He enjoyed a largely privileged childhood of private schools and wide suburban streets, the product of a close-knit Mormon community that had gradually gained acceptance in a diversifying society. He was, more or less, an all-American baby-boom kid—a success story written in a land once so hostile to his ancestors.

If early Romneys got by on grit and frontier pluck, Mitt Romney would develop a different skill set. From early childhood, he formed his views through observation and analysis, by hanging back as the world unfolded before him. He watched his siblings grow up before him. He watched his parents build a life of family and faith that he would endeavor to follow. In college, he watched the social unrest of the 1960s convulse the country and chose to go another way, his obedient nature inclined to side with authority. He watched his father win three elections for governor of Michigan and then lose a bid for president, taking away critical insights from the highs and lows alike.

In his own career, armed with exceptional intelligence and analytical skills honed in one of the nation's elite graduate programs, Romney studied and dissected companies around the world, eventually trading on his data mining to score incredibly sweet investment returns. When he did make pioneering breakthroughs, whether expanding the boundaries of private equity or enacting a novel universal health care plan as governor of Massachusetts, he followed a trusty formula: pursue data aggressively, analyze rigorously, test constantly, and observe always. Having grown up around engines, Romney adopted a kind of car hobbyist's mind-set. Almost anything, he believed, could be taken apart, studied, and reengineered. Strategy informed nearly everything he did. This is a man who saw the millions of dollars one could make in selling paper clips. The man who once explained to a reporter in an airport that he preferred eating only the tops of muffins, so as to avoid the butter that melted and sank during baking.

Yet as successful as his strategic impulse made him in the private sector, Romney would find it unreliable in politics, where intangibles often reign. Not everything, he learned, would be so easy to quantify. Strategy can take a candidate only so far and sometimes to the wrong place entirely. Romney has, to date, lost more political races than he's won, and his failure to see past the limits of his strategic outlook is one reason why.

The proximate cautionary tale, of course, is his campaign of 2008, which thrust him to the fore of national politics but exposed a tendency to assume whatever political profile he thought would best help him win. At the outset of that race, two household names, John McCain and Rudy Giuliani, occupied the political center of the Republican field. The vacuum was on the right, and Romney set out to fill it, launching a prolonged and aggressive courtship of social conservatives, fiscal conservatives, religious conservatives—any conservatives he could find. The trouble was that it looked too much like opportunism—or, worse, insincerity, given his long record of syncing his political views with the party's moderate wing. "Everything could always be tweaked, reshaped, fixed, addressed," said one former aide, describing Romney's outlook. "It was foreign to him on policy issues that core principles mattered—that somebody would go back and say, 'Well, three years ago you said this.'" The perception of expedience, along with lingering bigotry against Mormonism, helped bury his hopes.

Those challenges have not exactly subsided with time, despite Romney's best efforts to overcome them. To this day, he remains an enigmatic presence to people outside his closest circle, a puzzle whose pieces don't neatly fit. Many see in him what they want to see: a centrist or a conservative, an economic wizard or a rapacious capitalist, an adaptable leader or a calculating politician who will do anything to get elected. He acknowledges that he has changed on some key issues, casting his shifts as evidence of his nimble and flexible mind. He tells voters in New Hampshire to read his latest book if they want to know where he stands. "I'm very happy with where I am and the things I believe," he says. Even the phrasing, though, implies that conviction is a destination. However tortuous the path, he's saying, he is comfortable where he ended up.

It's a journey his father might have had difficulty understanding. Alike in so many ways, George and Mitt Romney had distinct views of politics. George was famously headstrong and outspoken, willing to follow his gut where it took him. He was, in that way, more idealistic than pragmatic. He charged out of the 1964 Republican National Convention over the party's foot-dragging on civil rights. He ensured the end of his presidential hopes in 1968 with an honest outburst about Vietnam. And he infuriated Richard Nixon while serving in Nixon's cabinet, by pushing hard on behalf of racially integrated housing, a cause dear to his heart. George did what he felt was right, and if the torpedoes came, the torpedoes came. "There is no leader who can provide sound leadership on the basis of unsound principles," he once said. "Principles are more important than men."

If George Romney shot from the hip, his son, before he shoots at all, carefully studies the target, lines up the barrel just right, and might even fire a few practice rounds. Mitt Romney, who saw the shortcomings of his father's approach, has often been more inclined to identify the consequences he wants, then figure out how to get there. In politics, those methods have varied depending on the race. In the current campaign, Romney is hoping that his undeniable economic expertise and record of accomplishment in business and government will trump questions about the striking changes, over time, in his political persona. And it may well be that the virtue of constancy will count for less when the unemployment rate is so stubbornly high and economic uncertainty so pervasive.

Romney is also, by now, a pro at this, knowing he will take his knocks and much better at parrying them. It's a trade-off he would gladly make in exchange for avenging his father's loss and becoming the country's forty-fifth president. "Politics is like washing diapers," Romney's mother, Lenore, once said. "You want the baby so much, you don't mind washing his diapers."

And Mitt Romney wants the baby.

A wall. A shell. A mask. There are many names for it, but many who have known or worked with Romney say the same thing: he carries himself as a man apart, a man who sometimes seems to be looking not into your eyes but past them. This detachment, in political settings, can make him seem too programmed, self-aware, bottled up. Even some of Romney's closest friends don't always recognize the man they see from afar.

This is a vexing rap to those in his inner circle—his wife, his family, and his closest confidants. They see a very different Mitt Romney and can't imagine why anyone wouldn't want him to lead their country. The man they know is warm. He's human. He's silly. He's funny, though sometimes his attempts at humor drift into corniness or just pure oddness. He's deeply generous with both his time and his money when people need a lift. It seems that everyone who has known him has a tale of his altruism, whether it's quietly funding a charitable cause or helping build a playground to honor the late son of a friend and neighbor. And he is an authentically devoted husband and father, commitments often honored by politicians more in the breach than in fact.

Romney's challenge is to narrow the gulf between these dueling perspectives, of the outsiders and the insiders. Especially in a race bursting with outsize personalities, he must show that he can connect to people and their lives; that, his expansive bank account aside, he's just like one of them; that he can get voters excited about, or at least comfortable with, the prospect of him as president.

The stripped-down campaign apparatus has surely helped; four years ago his entourage trailed him like a wedding party, signaling an off-putting self-importance. So has the decision by Romney and his advisers not to be so tethered to a script, a tack that pays off at the town hall meeting in New Hampshire, where the first question is a doozy. "I see you've fallen for the fallacy that raising taxes stifles growth," the man begins, then lectures Romney about running a business, and about how Wall Street traders and bank presidents should pay higher taxes. Through it all, Romney nods and smirks like a dinner guest awaiting his moment for a pithy retort. Then he pounces: "Have you thought about running in the Democratic primary for president?" Romney asks the man, and laughs fill the Saint Anselm auditorium. "You might consider it." It's the kind of challenge that might have ruffled Romney four years ago, but he swats it away effortlessly.

On the stump, Romney loves dichotomies, and he frames the 2012 race using some of his favorites: strong versus weak, stagnation versus prosperity, leadership versus drift. Political campaigns always magnify the villain, and Romney is happy to play along, even when the facts don't quite support his critique. But his is a largely sober message, carefully calibrated to appeal to the center. So even as he attacks environmental regulators for killing jobs, he's willing to say he believes that people contribute to global warming. As he celebrates the United States' strong military tradition, he's willing to say that parts of the defense budget deserve cutting, too. And as he rails against Democrats and their liberal policies, he's willing to say that Democrats also love America. These shades of centrism are always a bit of a gamble in a primary fight, but Romney's hope is that they will appeal to independent voters in a general election. His core message is deceptively simple. "I know how business works," he says. "I know why jobs come and why they go."

Bob Sutton is sitting in the front row in a long-sleeved blue-and-white-striped shirt, a collection of pens resting in a breast pocket. Like many eager supporters, he wears a blue Romney sticker, proud to show his allegiance to the man standing before him, promising to fix his ailing nation. When the floor opens, he seizes the chance.

"I'm from Michigan," he announces after grabbing the microphone. "I voted for your father three times. And now I'm going to vote for you." Sutton says he doesn't believe those who argue that Romney's not like his father, that he lacks core convictions. "How do I fight these people?"

"Let me give you some brass knuckles—that'll help," Romney jokes. "Where are you from in Michigan?"

"Muskegon."

"Muskegon, Michigan. Beautiful place," Romney says. "My wife's family had a cottage just north of there."

Sutton explains afterward that he is seventy-seven years old and now the owner of a New Hampshire software company that's taken a beating in the sluggish economy. Romney's the only one Sutton thinks has the stuff to change that. And he knows the stock from which Romney comes. "His father was a good man—a family man, and a leader," he says. "You could trust him."

Romney, answering Sutton's question before the audience, says he doesn't whine about the attacks on him. He's always known they were part of the deal. He's seen it all before. "I watched my dad run, all right?" he says. "I watched my dad run three times. He ran for president as well. Most of you don't remember George Romney, but he was governor of Michigan." He is interrupted by applause. "You remember? A couple of folks—good! And it's a great honor to represent a state or to represent the nation." The important thing, he says, is to stand up for what you believe in.

Then, after a few closing flourishes, he's done. He's made his case. The cheers die down. He is rushed by well-wishers anxious for grip-and-grin photos, for autographed brochures, for firm handshakes, or just for a few fleeting seconds of personal connection. The media crush surrounds him, forming a bubble of cameras, microphones, tape recorders, and notebooks. Romney is right at the center, where he loves to be, with a challenge ahead, a problem to solve, eager to complete one last turnaround and close his biggest deal yet.

ONE

PRAYING FOR A MIRACLE

I grew up idolizing him. I thought everything
he said was interesting.

—MITT ROMNEY ON HIS FATHER

George and Lenore Romney had always wanted a large family. This was God's will, the devout Mormons believed, not just personal longing. They had two girls and a boy, but then, for five long years, no more. Now, in 1946, the time for another child seemed especially right. Peace had returned, the guns of World War II at last silenced. Prosperity was sweeping through Detroit's auto industry, whose assembly lines were putting out family sedans again instead of tanks and jeeps and where George was a rising star. The Romneys lived in a spacious home and were leading citizens of Michigan and their church. Everything seemed to be going perfectly except for what felt most important—that Lenore might not survive another pregnancy and probably could not get pregnant at all.

Her doctor was adamant. Her health "would not permit . . . another child," he told her. She needed a major operation as soon as possible. Finally accepting this devastating news but still determined to expand their family, George and Lenore assembled papers to adopt a child in Switzerland. Yet all the time, they clung to hope.

The first sign that the doctor might be wrong came as the Romneys were on vacation, boating in the Dakotas. Lenore sensed that she might be pregnant, and the concerns for her health that had first been raised at the doctor's office came rushing back. "I remember my father's face, the worry and concern," one of their children, Jane, would say years later. "I had never seen that face before."

George was, after all, an inveterate optimist, a pillar of a man, tall and ruddy, movie-star handsome, with neatly parted black hair and steely eyes. Born in Mexico in a Mormon colony, he had come to the United States at the age of five and built himself a fine life. And much of that life had been centered on the woman whose health was now at risk.

From the time George was a young teenager in Utah, he had been in pursuit of Lenore. She was smart, independent-minded, and beautiful, with porcelain skin and a winning personality, assets she put to use in pursuit of her dream of being a Hollywood actress. George, meanwhile, moved inexorably up the career ladder. He attended George Washington University and, while still an undergraduate, got a job as a typist and then a policy aide with U.S. Senator David Walsh, a Massachusetts Democrat. He then used his Senate connections to get hired as an Aluminum Company of America salesman. Lenore, after graduating from George Washington in 1929, received an offer to work for a Hollywood film studio. George was aghast. He and Lenore had "long and heated arguments before Lenore overrode all of George's arguments and did what she wanted to do," according to one of George's biographers. George was so smitten that he dropped out of school, convinced the Aluminum Company of America to let him go west, and followed Lenore to California. But the couple nearly split for good once he got there. Lenore objected to George's overbearing effort to control her life, declaring that she would "never marry him," but the couple soon made up. Thus began a cycle of affection and argument that would define their long lives together.

Lenore's career took off quickly. She won parts in movies that starred film legends such as Greta Garbo and Jean Harlow. She entered a social whirl that included stars such as Clark Gable, leaving the lovestruck George even more anxious about losing her. At a time when the Great

Depression gripped the nation, Lenore was better educated and better off than George, a relatively independent woman at a time when that was hardly typical. A movie studio issued a promotional clip of her that showed a striking young woman wearing a short-sleeved dress and an elegant beribboned hat, sharing equal billing with a dog. The screen filled with a caption: "Miss Lenore LaFount, Metro-Goldwyn-Mayer actress, proves the futility of trying to force Buster, the dog star, to do tricks." It was something of a celluloid wink to the audience, as Lenore hugged the dog and looked up with an alluring smile. It had nothing to do with dogs, of course, and everything to do with building Lenore into a star of the future. She was on her way.

George soon learned that Metro-Goldwyn-Mayer had offered Lenore a three-year, $50,000 contract. He then made his own take-it-or-leave-it offer: give up Hollywood and seal herself with him for eternity in a traditional Mormon marriage. George would later call it "the biggest sale I ever made in my life." Lenore was reluctant at first to give up her career, but years later she insisted that after three or four months of playing bit parts, she was ready to get married and "never had any regrets about giving up movies." The couple married in 1931 and moved to Washington when the Aluminum Company of America promoted George to be one of the company's lobbyists, tasked with making the industry's case to members of Congress and the administration. "This was the legend we grew up on," said one of the Romney children, Jane. "Dad comes on his white horse, gets a job transfer from Washington to L.A., talks her out of [the] contract and into marrying [him]. This was storybook stuff."

Just as Lenore had given up acting, George also gave up two of his dreams. He would not finish college or go to Harvard's Graduate School of Business Administration. But he was, in fact, already doing well without such golden credentials. He left his lobbyist job to become head of the Detroit office of the Automobile Manufacturers Association, and took charge of organizing the auto industry's contribution to the massive industrial buildup during the World War II years.

By 1946, with the war ended and the auto business booming, the

Romneys seemed to have it all. George and Lenore had moved into a rambling three-story home in an exclusive, leafy section of Detroit and were raising three children: Margo Lynn, Jane, and Scott, the last born in 1941. For the following five years, the Romneys prayed for another child that they were told would never come. And then, on a boat in the Dakotas, Lenore disclosed to George that she believed, contrary to all expectations, that she was pregnant.

The Romneys returned home and, excited but fearful, went again to see the doctors. Lenore was promptly hospitalized. They concluded that "the baby should be taken and a major operation performed," George wrote later, a reference to the need for a cesarean birth and a related procedure. The prospect frightened the Romneys even more. Lenore's sister, Elsie, had had an operation two years earlier and "nearly lost her life," George wrote. But there was no alternative. The operation was performed, leaving Lenore "suffering." After it was over, the doctor told George, "I don't see how she became pregnant, or how she carried the child." But carry the child she did, and she delivered a six-pound son in good health. He had "strong features," George wrote, and even on this first day of life, the child's "dark hair" was notable. The rejoicing among the family seemed nonstop.

George took out a piece of stationery and wrote a letter to his friends and colleagues.

"Dear Folks," Romney wrote on March 13, 1947, on the letterhead of the Automobile Manufacturers Association. "Well, by now most of you have had the really big news, but for those who haven't, Willard Mitt Romney arrived at Ten AM March 12." As George detailed how precarious his wife's pregnancy had been, it became clear that there was a special level of wonderment embedded in this announcement, this birth. He told the family's friends how it had happened: Lenore "had a lot of faith." George wrote, "We consider it a blessing for which we must thank the Creator of all." From then on, Lenore referred to Mitt as her miracle baby.

His sisters, Margo Lynn and Jane, were nearly twelve and nine at the time, his brother, Scott, almost six, and they didn't wait long to begin the debate over whether to call the baby "Bill" or "Mitt." The

Willard was in honor of J. Willard Marriott, a family friend, fellow Mormon, and future hotel magnate; Mitt was a nod to Milton "Mitt" Romney, a cousin of George and a former Chicago Bears quarterback.

From birth, Mitt enjoyed a starring role in the "family bulletins" George mailed out. When Mitt was not yet two years old and making his first visit to see Santa Claus, George wrote with pride, "He walked right up like a man and shook hands!" In the same letter, he noted that Mitt was "bold and inclined to be a bit reckless—loves to climb up on high chairs and say, 'Careful, careful, careful!' " Throughout his childhood, Mitt logged lots of time sitting on his father's lap, watching him read the paper. As George flipped through the pages, the passing headlines prompted him to share with his son his insights about the wider world.

In 1953, when Mitt was about six years old, the Romneys moved from their home in Detroit to the suburb of Bloomfield Hills, one of the nation's wealthiest enclaves. The Romneys lived in a contemporary home adjacent to the Bloomfield Hills Country Club. The city of several thousand people, a domain of sprawling homes, emerald lawns, and elite private schools was a world away from nearby Detroit. George became president of the regional Mormon "stake," overseeing a number of wards, and a meetinghouse was constructed in Bloomfield Hills. A *Time* magazine reporter who visited the home during Mitt's childhood described a typical morning, with George taking a prebreakfast jog around the links or perhaps a fast round of golf.

Shortly after Mitt's birth, George had been hired as an executive with the Nash-Kelvinator Corporation, which later merged with Hudson Motor Car Company to become American Motors Corporation, where George served as executive vice president. In 1954, after the head of AMC died, George became chairman and CEO. He took over when the company was on the brink of bankruptcy. As Mitt would later recall it, his father took the money from the sale of their Detroit house and used it to buy company stock. "He literally risked his net worth on his ability to turn things around," Mitt wrote years later. Mitt recalled walking the factory floor with his dad, who would say, "We're going to make this company great." One of the keys to George's continued success was his ability to fight off a corporate raider who bought up shares of American Motors stock and seemed poised to try to sell off company assets and take a profit.

In perhaps the most important business decision of his life, George embraced a proposal to focus American Motors on producing a fuel-efficient car, predicting that Americans would choose it over the "gas-guzzling dinosaurs" produced by larger automakers as well as by his own company. Even Mitt questioned the strategy.

"If Ramblers are such great cars, why doesn't everybody have one?" he asked his father one day.

"People don't always recognize what was best," George responded, according to a recollection by Mitt's brother, Scott.

George was ahead of his time. In the mid-1950s, Romney's Rambler was touted as getting thirty miles per gallon. It would be several decades before the federal government required the average passenger car fuel efficiency to be nearly that high. American consumers living in a time of 30-cents-per-gallon gas—$2.41 in 2010 dollars—were beginning to realize the economic attractiveness of a compact car. George's embrace of the Rambler put American Motors on the road to profitability and helped make Romney wealthy and successful. It was a decision that had a lasting impact on Mitt, whose favorite pastime was discussing cars and business with his father. When Mitt announced his first presidential campaign, he arranged for an antique Rambler to be put on the stage behind him, underscoring what he called his family's commitment to innovation.

Listening to his father, Mitt wrote later, didn't seem like listening to a businessman. "It was more like he was on a great mission with American Motors to build innovative cars so that people could save money and fuel, and have better lives. Work was never just a way to make a buck to my dad. There was a calling and purpose to it. It was about making life better for people."

Mixed with the talk about cars and business was an ongoing emphasis on living the Mormon life. The Romneys were one of the Mormon faith's leading families. George had followed family tradition and regularly delivered talks at the local Mormon church as well as at the dinner table. Unlike Utah, where George had spent most of his childhood, Michigan had relatively few Mormons, and members of the faith were widely considered to be outsiders. So, in a religion that relied upon proselytizing to grow its ranks, George urged his children to set an example of upholding the traditions of the Church of Jesus Christ of Latter-day Saints. Lenore also had deep roots in the faith; she came from a well-to-do Mormon family in Utah. But as a woman she was not allowed to hold a leadership role in the church.

The Mormon faith infused Mitt's life, and he was expected to be a leader from the start, immersing himself in theology and delivering sermons in his teens. When Mitt was fourteen years old, he and his parents were featured in a story in the *Detroit Free Press* about Mormons in Michigan. A photo showed a Mormon teacher at the Romney home, instructing an intense-looking Mitt seated next to George and Lenore. The story noted that Mormons were "one of the smallest and least understood faiths" in the Detroit area, with only four thousand Mormons living in southeast Michigan and very few elsewhere in the state. The story, which ran in the months before George sought the governorship, was the first introduction of Mitt to a wide audience through the lens of his religion.

George emphasized to his children how one Romney after another had excelled by following the tenets of their faith. To underscore his point, George instructed Mitt to follow what he called a "three point formula for joyous achievement." It was one of his favorite pieces of advice, and it came directly from Mormon doctrine: "Search diligently, pray always and be believing, and all things shall work together for your good." When George died, Mitt and the rest of the family would inscribe those words on his tombstone.

"Dream and dream big, and if you'll dream big and if you'll work hard and if you pray always, your dreams will come true," George also advised his son. The advice from his father, and the example that George set in business, leadership, and faith, would impact Mitt for a lifetime as he sought to live up to his father's expectations. "I grew up idolizing him," Mitt wrote about his father. "I thought everything he said was interesting." As a boy, he had seen his father's square-jawed face on the cover of national magazines such as *Time* and heard George lionized as the industrial "man of the year." Despite his celebrity, George managed to make more time for Mitt than he did for his older children. "Dad was more settled by then," said Mitt's sister Jane.

Throughout Mitt's life, the striking parallels between him and his father would be widely noted, from their meticulously groomed hair to their business and political careers. But Mitt was devoted to his mother as well, and he inherited her tact and even temperament—qualities that were often absent in her blunt, intense husband. The Romney children relished the story of how their mother had been ahead of her time as an independent woman and a college graduate, briefly worked as a Hollywood actress, and then devoted herself to her family and religious duties, all while retaining her own outspoken identity. When George was elected governor, Lenore gave the public a sense of her liberated style, telling *Time* magazine she planned to use her role as first lady of Michigan to make "a real breakthrough in human relations by bringing people together as people—just like George has enunciated. Women have a very interesting role in this, and I don't expect to be a society leader holding a series of meaningless teas."

Mitt's sister Jane would speak years later of the indelible impression Lenore left with her children for being a "gutsy" woman, "invincible and courageous as heck, setting the tone of our home, her joy, her love of life."

The tone at home was both loving and jarring. George tried to bring Lenore a single rose every day, a true sign of their affection. But they also argued so much that their grandchildren would call them "the Bickersons." Luckily for George, Lenore was well matched to a man like him—unintimidated, indeed drawn, by his powerful persona. She once said she had rejected other suitors because she was so attracted to George's "forceful personality." George was the sort of person who tried to set people straight by starting his argument saying, "Look," and didn't hesitate to bluntly challenge those who disagreed with him. *Life* magazine memorably called him "a loner who is really close to one person, Lenore." That closeness enabled the couple to settle their arguments amicably, with Lenore calming her fiery husband. Still, such spats were something Mitt would resolve to banish from his own life, adopting his mother's diplomatic style. He tried to avoid arguments at all costs in his family life, if not always in business and government. His wife, Ann, would later say that they had argued only once, without raising voices, and that they had never argued after they were married. Mitt wanted no repeat showing of the Bickersons.

Lenore's impact on Mitt was clear, according to his sister Jane. "Mitt's more like my mother," she said. "Mitt's the diplomat. Mitt knows how to smooth things." Still, when it came to his career aspirations, Mitt modeled himself on his force-of-nature father, and their relationship would be a central axis in his life. Indeed, *Life* magazine may have been off in its assessment; George the loner expanded his circle of confidants to include Mitt, and the impact of that decision would last a lifetime. The young Mitt aspired to run a car company, just like Dad, and later would try to follow his father's political path. The bond between the two could be seen on summer weekends, when George would join the rest of the family at their cottage on Lake Huron in Ontario. Mitt and his best friend, Tom McCaffrey, would sneak into his father's briefcase for a first look at photos of the cars planned for the new model year.

On the paddle tennis court in front of the cottage, George would compete with his children in matches played to the death. Mitt's older brother, Scott, was the fiercest challenger, sharing their father's competitive streak and exceeding him in athletic ability. Like his father, Mitt was never much of an athlete, but even that seemed to work in his favor. Although Scott went through the common adolescent phase of occasionally competing with his father, Mitt always maintained an easy rapport. Scott would marvel at his little brother's confidence in talking with their father almost as a peer. When George held "family councils" to discuss big decisions he was contemplating, Scott and his sisters would say, "Gee, that sounds fabulous," while Mitt would pipe up with "Well, have you thought about this?"

In the seventh grade, Mitt enrolled at the elite Cranbrook School, a boarding school in his hometown of Bloomfield Hills. The 315-acre campus has been described as "one of the most enchanted architectural settings in America." Boys attended classes in elegant buildings modeled after an English boarding school. Girls went to separate facilities designed in the Arts and Crafts style. Amid this prim and preppy landscape was a world-class art school, surrounded by wondrous sculptures by Carl Milles, including a replica of his masterwork, the *Orpheus* fountain, which features nine bronze figures thrusting skyward from a pool of water.

Surrounded by other sons of privilege, Mitt wasn't a standout. "He was in many ways the antithesis of what he's portrayed as today," classmate Jim Bailey said. "He was tall, skinny, gawky, had a bad complexion." A quarterly report card told the story of a bright boy who had yet to feel the urge to apply himself fully. "He can do a lot better," his English teacher wrote. "He tended to let up on time spent studying and had a low mark on his last test. He wastes much time in class," his algebra teacher wrote. "He had some trouble with Genetics," his biology teacher wrote. His worst grade was in French; he got a C. Following the pattern of other classes, Mitt had started off well, but he had then fallen into a "lull." The French teacher predicted Mitt could improve, and Mitt indeed later became fluent while living in France. A teacher added a handwritten notation at the bottom of a 1961 report card: "Mitt is doing well. He is a more responsible citizen this year." In six years at Cranbrook, he never showed himself to be a leader; Bailey, not Mitt, went on to be president of their class.

Mitt was a day student the first three years, going home after classes. But when his father became governor and his parents spent much of their time at the capital of Lansing, Mitt boarded at the school. As a result, his last three years were a time of increasing independence, and Mitt thrust himself more deeply into the prep school world. He became known as an inveterate prankster. On one occasion, he staged an elaborate formal dinner in the median strip of a busy thoroughfare. For another prank, Mitt came close to stepping over the line, if not crossing it. He dressed up in a uniform similar to that worn by a police officer, put a flashing red "cherry top" on his car, and raced after a vehicle carrying two of his male friends and their dates. By prearrangement, the friends had stashed beer bottles in the trunk and knew that Romney would pretend to be an officer chasing them. The dates had no idea of the plot.

As planned, Romney pulled up to the car belonging to his friends. "He came up to the car and pretended to be a cop. . . . He asked me to get out of the car and open the trunk, at which point we found the beer," recalled Graham McDonald, one of the friends in the car. Romney told McDonald and the other friend to come with him, "and we drove off," McDonald recalled. "It was a terrible thing to do. We came back shortly. We didn't leave damsels in distress." In retrospect, the idea of the governor's son impersonating a police officer is startling. But McDonald's point in telling the story is that "I am surprised when I read about him being stiff and humorless. That is the opposite image I have of him. He was almost slapstick to a fault."

Despite the school's rarefied air, it was still a high school, so the jocks tended to be the most popular. Mitt took on the role of cheerleader, putting on his school sweater, working the sidelines at football games, and shouting traditional cheers. The Cranbrook teams were called the Cranes, but no one could find a crane to be a mascot. Instead, the cheerleading squad acquired a duck. Mitt volunteered to take care of it because his parents' house had a pond. But when Mitt went to retrieve the bird a few days later, all he found was a pile of feathers. A local fox had apparently dined well. Mitt "was genuinely distressed because he was the guardian of the duck," said Gregg Dearth, a fellow cheerleader. "You can imagine, Mitt had a responsibility for the duck and the duck no longer existed. He felt he had let us all down."

When Mitt tried to transition into being an athlete, it proved even more embarrassing. He joined the cross-country team for a 2.5-mile race traditionally held during a football game, setting off with the rest of the runners at the start of halftime. But in his eagerness to compete, he failed to pace himself. Everyone except Mitt returned before the second half began. Finally, the several hundred spectators noticed Mitt making an agonizingly slow approach to the cinder track. "Mitt kept falling and getting up, falling and getting up, and eventually he just crawled across the line," McDonald recalled. It could have been one of the most humiliating moments of his young life. But then the crowd began to rise to its feet, giving Mitt a standing ovation for his effort. "It was definitely looked upon as a show of character. Other people would have quit," another classmate, Sidney Barthwell, Jr., said. Mitt's fellow runners realized what had happened: he had run too fast at the start and, unprepared for the distance, had cramped up.

Dearth, who also ran the race, is convinced that this moment provided a lesson for Mitt, which he would use in his later political life. Mitt had started an endurance race as if it were a dash. "It is something that has stayed with him the rest of his life—to pace yourself and to run the whole race, and to temper your enthusiasm with judgment," Dearth said.

Given the family history, it likely prompted a round of discussion at home. George relished telling the story of how his own high school coach had never seen someone with so little athletic ability "try so hard." In later years, George jogged most mornings and, even when he was elderly, walked at such a fast pace that many companions practically had to run to keep up. Mitt, too, would work to improve his athletic performance and, when he ran for president in 2008, would run an advertisement that showed him running through the New Hampshire woods.

While Mitt's world revolved around the relatively pristine preserve of Cranbrook and Bloomfield Hills, the urban and racial problems of Detroit were just a half-hour drive away. George often brought troubling details about that other world home with him. Detroit's leaders had urged George to help lead an effort to improve the city schools. George agreed to work with what became known as the Citizens Advisory Committee on School Needs, helping to pass a $60 million education bond, and he then helped lead an effort to resuscitate the economy of Michigan, which was suffering with the rest of the country through a pair of recessions that had hit in the late 1950s and 1960. George soon concluded that the best way to help the state was to run for governor in 1962—if his family supported the idea. One day, he came down the stairs, entered the dining room, and asked Mitt and the rest of the family a question: "You know, I think I'm going to run for governor. Should I run as a Republican or a Democrat?" George decided that he would run as a Republican, and Mitt was among the most enthusiastic about the race.

Fifteen-year-old Mitt was increasingly interested in politics and became involved in his father's campaign from the start. He appeared at George's side at the announcement speech, spoke at county fairs, and traveled throughout the state in a campaign van. He worked the campaign switchboard and set up "Romney for Governor" booths. "I would introduce myself and shout out to people walking past, 'You should vote for my father for governor. He's a truly great person. You've got to support him. He's going to make things better,'" he wrote years later. And, he said, "I really believed it. We all did. It was true."

George ran as a liberal-to-moderate politician, playing down his Republican Party affiliation, and worked hard to win the votes of groups that didn't always support the GOP, including the black and labor votes. A month before election day, President John F. Kennedy arrived to campaign with the Democratic candidate, the incumbent governor, John Swainson. Kennedy and Swainson went on a twelve-mile tour of Detroit and the surrounding area, with the presidential motorcade greeted by crowds that may have reached a hundred thousand people. Without mentioning Romney's name, Kennedy mocked GOP candidates, who he said were attempting to win without declaring that they were Republicans. "You can't find the word 'Republican' on their literature," he said. The presidential appearance almost certainly made an impact on Mitt, who would have been aware of the possibility that Kennedy's intervention might doom his father's chances. But even President Kennedy could not stop George's momentum, and he won narrowly.

Watching his father, Mitt learned that George had a disarming way of rebutting arguments against him. Sometimes George did so by distancing himself from his party and speaking sympathetically about an opponent's viewpoint. For example, George had been attacked by labor groups, one of which published a pamphlet titled "Who Is the Real George Romney?" The booklet accused Romney of doublespeak, such as when he said that Michigan needed a hundred thousand new jobs but then declared that he had never promised he could deliver them. But George, who had mixed relations with the unions at American Motors, sometimes sounded sympathetic to his labor antagonists, as when he lamented that the Republican Party was identified "too much as a business party." The blunt way was the George Romney way. It worked, at least at the time. He was easily reelected and eventually gained enough clout to win passage of the state's first income tax. For Mitt, these were early lessons in how a Republican could win in a Democratic state.

On the Cranbrook campus, Mitt downplayed his family's fame, though others showed less restraint. *The Detroit News*'s report on a small fire at Cranbrook carried this headline: "Romney Son Helps Fight School Fire." Deep in the article, it turned out that Mitt's heroism had consisted of opening the building's front door and directing the firefighters toward the small blaze.

More important to Mitt was sharing his father's front-row seat on government as an intern in the governor's office. When Mitt was sixteen years old, in 1963, he joined his father late one night at the capitol in Lansing. As Governor Romney sought passage of a bill before a midnight deadline, an Associated Press reporter observed the "tall, slim boy" delivering advice to his father about how to deal with recalcitrant legislators.

"Dad, go in there and talk to them," Mitt told his father.

"I don't think that will get it done," George told his son.

But it did, and a resolution was finally reached at 1:30 a.m. At that point, the AP reporter checked on Mitt and his father. "It was lots more fun before they let the reporters in," Mitt said, perhaps foreshadowing his dour view of the press when he was a candidate himself. It was laid out in a front-page story headlined "Mitt Romney Keeps Vigil as Clock Is Running Out; Hears Dad Argue, Win Point with Solons."

It was a formative moment, that December evening in Lansing.

Three weeks after the assassination of President John F. Kennedy, with the country in turmoil and searching for leadership, his father was fast emerging as a shining light of the Republican Party. The bond between father and son, as well as between Mitt and the political world, was only growing stronger.

Dick Milliman, who served as Romney's press secretary, was struck by how much the governor delighted in having his teenage son around. "They would hug upon meeting, and not just any hug," he recalled. "He would give Mitt a big bear hug and a kiss." To Milliman, it was clearly not just a father-son bond but almost a "partner relationship." Around the office, just as around the family home, Mitt seldom held back. "He would chime in, 'Have you thought about this?'" Milliman said, admitting, "Sometimes you'd think, 'That kid oughta shut up!' But he was always nice to be around."

Mitt's sister Jane likened their upbringing amid the swirl of politics to "living in a drama." It was, she said, a fascinating time, with interesting people always parading through, reporters often at the doorstep, the issues of the day deliberated at the dinner table. George Romney's success in Michigan prompted talk of him as a presidential candidate in 1964. That didn't happen, but he arrived at the Republican National Convention in San Francisco that summer as a star, inviting Mitt to come orbit around him. The elder Romney would make headlines by walking out on nominee Barry Goldwater because of Goldwater's opposition to civil rights legislation. In a subsequent letter to Goldwater, Romney wrote, "The rights of some must not be enjoyed by denying the rights of others." Romney refused to endorse Goldwater's candidacy, embittering conservatives within the party and solidifying Romney's reputation as a more liberal iconoclast.

When Goldwater complained about Romney's failure to endorse him, Romney responded in a blistering letter—soon leaked to *The New York Times*—in which he took sharp aim at Goldwater's right-wing philosophy. In what amounted to a Romney Manifesto, the governor wrote, "Dogmatic ideological parties tend to splinter the political and social fabric of a nation, lead to governmental crises and deadlocks, and stymie the compromises so often necessary to preserve freedom and achieve progress." Years later, Mitt would run as a moderate and win the governorship of Massachusetts and as a conservative in seeking the presidency. Throughout his career, Mitt would be accused of changing his positions according to the politics of the moment. He would reject charges of opportunism, but his father's blunt forthrightness, his adamancy on the divisive civil rights issue and others, was a trait he would not so fully emulate. George more easily, even proudly, embraced what some saw as philosophical conflicts. When Brigham Young University named an institute at its School of Management after George, it erected a plaque that boasted of the wide embrace of his spirit: "A liberal in his treatment of his fellow humans, a conservative with other people's money." George often said he hated political labels, knowing how they could typecast a politician and put off potential supporters. He was presumed to be conservative in his actions as a business and religious leader, but he would perhaps be best remembered as a liberal for his views of racial equality and social justice.

George's progressive stance on race earned him critics not only in the right wing of his party but at the highest levels of his church. The church policy at the time was that blacks could join as members but not become members of the priesthood. In 1964, a top Mormon official wrote to Romney, calling a civil rights bill "vicious legislation" and warning Romney that it was not man's job to remove what he termed the Lord's "curse upon the Negro." Romney refused to back down. Mitt would be particularly proud of his father's willingness to take on their own church when it came to the Mormons' treatment of blacks. When Mitt was asked later in his life about the church's refusal until 1978 to let blacks fully participate in Mormon rituals, he cited his father's work on civil rights as evidence of how far the family had distanced itself from the church on the issue.

Still, there weren't many blacks in Mitt's exclusive neighborhood of Bloomfield Hills. Mitt's primary exposure to black people was his family's beloved housekeeper, Birdie Nailing, and a fellow student, Sidney Barthwell, Jr., whose father had worked with George Romney in revising the state's constitution. Coming from Detroit to attend

Cranbrook, Barthwell had entered a different world. "It was primarily white WASP. I was the only African-American student in our graduating class," Barthwell said. Though Mitt and Sidney were not close, they were friendly and went through all six years at the school together. It made an impression on Barthwell that the Romney family opposed the Mormon prohibition against blacks holding the priesthood. His recollection of Mitt is that he seemed like a "very nice guy" although "not a standout student." During one of his father's gubernatorial campaigns, Mitt and his brother, Scott, were sent to talk to African Americans in a Detroit neighborhood, and they came away pleased that their father was respected by blacks, even if they didn't receive assurances of votes.

In his final years at Cranbrook, Mitt emerged a more serious student and a good-looking teen. Adding to the package was his great head of hair. Mitt had grown up hearing people comment on his father's sweep of slicked-back black hair, white at the temples. But since his early teens, Mitt had patterned his own hairstyle after a man named Edwin Jones, who served as his father's top aide in running the Detroit operations of the Mormon church. "He sat up front, to the side at a desk, keeping records," Mitt would recall years later. "I remember that he had very dark hair, that it was quite shiny, and that you could see it in distinct comb lines from front to back. Have you looked at my hair? Yep, it's just like his was some forty years ago."

When graduation arrived, the speaker was none other than George Romney. He hit upon a surprising theme. Girlfriends, the governor told the seventy-six graduating boys, "will have more to do with shaping your life than probably anybody else. . . . If the girl you're interested in doesn't inspire you to greater effort than you would undertake without knowing her, then you'd better look around and get another."

Mitt had looked around and, just like his father, had found at a relatively early age the girl he wanted to marry. Her name was Ann Davies, and she was beautiful, smart, and independent-minded. The parallels to Mitt's mother were unmistakable. But there was one difference—and a major problem: Ann was not a Mormon; she came from a mainline Protestant family.

Mitt had first met Ann when they were both in elementary school in Bloomfield Hills. He was dressed in his Cub Scout uniform and saw Ann riding her horse over the railroad tracks. He picked up some stones and threw them at her, he recalled years later. They lost track of each other over the years. Then they were at neighboring prep schools. Ann attended Cranbrook's sister school, Kingswood, on the other side of campus. Mitt had just turned eighteen and Ann was fifteen, almost exactly the same ages his parents had been when they met. One day, Mitt went to a friend's birthday party. Across the room, he spied Ann. "Wow, has she changed," Mitt said he thought to himself. He went over to Ann's date and offered to drive Ann home.

Cranbrook in the 1960s still adhered to a strict separation of the sexes. The girls were allowed to see the boys for athletic events, dance lessons, and a weekly movie night in the gym. Beyond that, their inter-action largely was confined to letters, which the Kingswood girls lined up to receive daily. Shortly after they ran into each other at the party, Mitt asked Ann out for a date. It was March 21, 1965, and they saw the movie *The Sound of Music.* "I caught his eye and he never let me go," Ann recalled years later. "I mean, he hotly pursued me." They fell "deeply in love," she said, but "we didn't tell anyone, because no one would have believed it." Mitt later said, "I fell in love with her the second I saw her."

When Ann arrived at the Kingswood School, she wasn't much in-terested in academics; that would come later. She was more into riding horses and playing field hockey, lacrosse, basketball, and tennis. Mitt learned to keep up. They strode across the sprawling Cranbrook cam-pus, past the lake, amid the emerald landscape with its bubbling pools and surfeit of sculptures. Mitt taught Ann how to water-ski, and they went almost every day when the weather allowed. She taught him to snow-ski. She found Mitt funny and fun to be with, and "no matter where he was, there was a lot of action." Other boys pursued her, and she would date them in Mitt's absence, but she said Mitt "stole my heart from the very first."

Ann, like Mitt, had grown up in Bloomfield Hills. Her father, Edward R. Davies, was the city's former mayor and had become the wealthy president of Jered Industries, which made maritime machinery. He was also something of an inventor. One time, her father got mad at her and her two brothers for not closing a sliding door. So he built a pulley system that automatically closed the door. Ann considered him a "creative genius." What Mitt didn't know about Ann was that she had been brought up in a home with a father who had no use for religion, and that she had been on a spiritual search since a young age. Her father had grown up in a coal-mining family in Wales, and Ann's brothers say he associated the religion of his childhood—a Welsh Congregational church he found as dreary as the climate of Wales—with drudgery and hogwash. Before their dad married their mom, Lois Davies, he insisted that she give up organized religion. "Dad," said Ann's older brother, Roderick Davies, "considered people who were religious to be weak in the knees."

But like Mitt, Ann had a special relationship with her father. So he occasionally indulged his only daughter's requests that the family attend services at one Protestant church or another. He remained unswayed by the pulpit and believed his daughter would eventually come to her senses. As for her romance, Ann's father knew that Mitt was heading to California for college while Ann still had two years of high school left. So how serious could they be?

They were serious. One night, Mitt went to pick up Ann for the prom, driving his "goofy-looking" AMC Marlin, a two-door fastback with a sloping roof. After they had been at the festivities for a while, he nervously took his sixteen-year-old girlfriend aside and asked—informally—if she would one day marry him. Yes, Ann said. It was a tentative yes, with the couple knowing they would soon part ways as Mitt headed to college. Years later they would remember the moment not just for the romance but for also the hilarity. Mitt, the car guy, had forgotten to gas up. He blamed it on nerves. The Marlin puttered to a halt as he drove Ann home. Somehow, Mitt and the exquisitely outfitted woman he had just asked to be his wife made it home.

From Mitt's perspective, their path was set. But there had been a lingering, critical question. On one of their earliest dates, Mitt had leaned in for a kiss, but Ann had other ideas.

"What," she asked, "do Mormons believe?"

Mitt was suddenly uneasy. He knew his religion made him something of an outsider. He didn't want Ann to consider changing religions just because of him. It had to be from her heart. Now here he was on a date with one of the prettiest girls on campus, someone he knew came from mainline Protestant stock, and she was asking for a tutorial on the Mormon Church?

"I was not in the mood to talk about religion," he would say later. "I was much more interested in physical expressions of love."

Mitt looked Ann in the eyes and tried to answer her question. He turned to the church's "Articles of Faith," propounded by church founder Joseph Smith and typically memorized by followers. Mitt began by quoting the first article. "We believe in God, the Eternal Father, and in His Son, Jesus Christ, and in the Holy Ghost." When he finished, he noticed that Ann had started to cry.

There was, of course, much more to explain, much more to the story of this little-known faith. It was a story in which the Romney family itself was deeply intertwined, from the early days of the religion's founding to the modern era, when the Romneys held leadership positions within the church. It was a story of scripture and revelation but also of polygamy, an early Mormon practice in which Romneys had played a pivotal role, and of a harrowing westward journey that tested the hardiest settlers. The story of the faith, it turned out, was in many ways the story of the Romneys. It had begun in an English village more than a century earlier, when Mitt's great-great-grandfather heard an astonishing tale from an American visitor who insisted he was a saint.

TWO

FOLLOWING THE CALL

Brother Miles, I want you to take another wife.

—BRIGHAM YOUNG TO MITT ROMNEY'S
GREAT-GRANDFATHER IN 1867

"I believe in my Mormon faith, and I endeavor to live by it. My faith is the faith of my fathers," Romney said at a moment of grave political peril during his 2008 presidential campaign, a time when he was caught between the need to cultivate the support of Christian evangelicals and the reality that many people viewed Mormonism with skepticism or hostility. "I will be true to them and to my beliefs," Romney continued. "Some believe that such a confession of my faith will sink my candidacy. If they are right, so be it."

Forty-two years after Romney's earnest attempt to explain his faith to the girl he wanted to marry, and almost two hundred years after the religion he professed was born, it was still necessary to explain and justify Mormonism to Americans, or at least to some Americans. Romney, when he entered public life, had long—probably too long—avoided the need to make that case before bowing to the necessity. And the case he made was as general as it was sincere, as if to dip into the core controversy over Mormon beliefs and early practices would only be kindling for further bigotry—as indeed, it might well be. To many people the Mormon story still sounds strange or is simply unknown; even less well known outside the church is the central place of the Romney family in that story.

One reason for that disconnect is that Mitt rarely talked about the special legacy of his ancestors. It is something he has held close, in a deeply private place. But the pride in his standing as one of Mormonism's first families is plainly there, a fact that was obvious to any visitor to Romney's home in Belmont, Massachusetts. In the foyer he had mounted framed portraits of five leading Mormon men, all Romneys, all figures who bear introduction if one is to reach into the elusive core of Mitt Romney.

The first face on the wall—long, lean, high-browed—was Miles A. Romney, the man who brought the name to America, the earliest among them to hear the call.

The boomtown of Nauvoo, Illinois, rose along a horseshoe bend on the Mississippi River, spreading across the muddy flats and up onto the bluffs, just as its founder had prophesied. By 1841, wagons overburdened with immigrants and their worldly goods rattled down the bustling streets, a great white temple was being constructed on a grassy hillock, and ships filled with still more newcomers pulled into the river dock. The settlers had come with a common goal: to build a haven for believers in a growing new religion, with new saints, new gospels, and a new story of Jesus, called Mormonism. Onto this scene arrived a nearly destitute family with the name of Romney. They came from the quiet village of Lower Penwortham, near Liverpool, England. For years, Romneys had been moving out and on; one named George Romney had gone to London and become a celebrated eighteenth-century portrait painter. But most were of modest means. Such was the case with a carpenter named Miles Archibald Romney and his wife, Elizabeth. One day in 1837, the couple heard a group of Mormon missionaries preach at a town square. The Romneys were so taken with the message that they became one of the first families in Great Britain to convert.

Four years after that first encounter, Miles and Elizabeth had become such fervent believers that they risked everything, emigrated to the United States, and made their way, as instructed, to Nauvoo with as many as five children. They had little money and only a vague idea of what awaited them there, following a prophet's call. They soon learned of a small stone house that stood in the north end of Nauvoo and, according to family lore, took possession of it in exchange for a paisley shawl.

The Mormon leaders, meanwhile, were overjoyed when they learned that Miles was an expert in carpentry and construction, and they gave him one of the most important jobs in the city, naming him "master mechanic" in the building of the temple. Soon another Romney child was on the way. He would be named Miles Park Romney, and it is this son, known to all as "Miles P.," who would play a pivotal role in the faith—and later be known as the great-grandfather of Mitt Romney. The journey of the Romneys in America, and the trials of their faith, would unfold largely through his eyes. His was the second portrait in the line on Mitt Romney's wall.

Nauvoo was in great tumult when, two years after the Romneys arrived, Miles was born on August 18, 1843. The founder of Mormonism, Joseph Smith, had chosen Nauvoo four years earlier as a haven for the faithful after they had been expelled from other areas. But even as the faith grew, so did attacks against it. The story of Mormonism's beginning came from Smith, who had been born in Vermont and had moved to New York, where he had worked on farms and searched for treasure. He said he had been praying in the woods in 1820 when he saw "a pillar of light exactly over my head." He said that two "personages" had appeared before him, telling him that other sects "were all wrong" and putting him on a path toward restoring what he regarded as the true Christian faith. Smith said an angel named Moroni had later led him to find golden plates in western New York that he translated into the Book of Mormon, describing how Christ had come to America and how Native Americans were descended from the lost tribe of Israel. This, Smith told his followers, was the true and restored Christian faith. While living in Nauvoo, Smith took multiple wives, a practice some in the faith said was a "divine pronouncement" and a

restoration of a practice common in early biblical times. Some histori-
ans would later calculate that Smith had taken at least several dozen
wives.

Smith's declaration that other religions were wrong and the fears of
some that he would lead a regime of theocratic "despotism" had fueled
anger among non-Mormons in the region. His call for a new religious
order had led to a chain reaction of violence and exodus; his flock had
been kicked out of Ohio and Missouri; and he now talked of making
Nauvoo a quasi-independent state. He railed against traditional Christian faiths, Illinois authorities, and the federal government and even
declared he was seeking the U.S. presidency. Smith's gift for oratory
had attracted thousands but also repelled nonbelievers. Mobs gathered
regularly to attack Mormons.

Then a local newspaper, the *Nauvoo Expositor*, printed accusations
that Smith was a tyrant and polygamist. "We are earnestly seeking to
explode the vicious principles of Joseph Smith, and those who practice
the same abominations and whoredoms," the newspaper said. Tensions
rose, and a series of complaints were made against Smith, who was also
the mayor of Nauvoo. The Nauvoo City Council, acting at Smith's be-
hest, voted that the newspaper's printing press be destroyed. The town
marshal and his men dragged the *Expositor*'s press into the street and
pounded it with sledgehammers. Smith was arrested on charges that
he had incited a riot. Imprisoned in the summer of 1844 in the nearby
town of Carthage, Smith was shot and killed by a mob that stormed
the jail; he was dead at age thirty-eight, two decades after receiving
what he declared to be a mandate from God.

Members of the faith began to flee Nauvoo, but the Romney family stayed behind, partly to enable the senior Romney to finish his
work on the temple. The temple was barely completed when mobs
forced Mormons from the city at the "point of a bayonet." Vandals soon
set the temple on fire, destroying the elder Romney's years of patient
craftsmanship. Nauvoo was a sanctuary no longer, and some twenty
thousand Mormons deserted a city that had briefly rivaled Chicago
in population. The Romney family was too poor to follow the main
group of settlers to what Mormon leaders said was the new promised

land, later known as Utah. Instead, the Romneys fled to Burlington, Iowa, and then to St. Louis, Missouri. They became part of a wayward band, moving from camp to camp, with few provisions and many sick and malnourished Mormons in need of care. It would be four years before the family had enough money to load an ox-drawn wagon for the westward trek.

Miles P. Romney was seven years old as he made the harrowing journey of 1,300 miles. He traveled over rough trails in often frigid weather under the threat of attack by anti-Mormon mobs, Indians, and wild animals. The Romneys passed through Iowa, Nebraska, and Wyoming, traveling by pioneer trail landmarks such as the hulking mass of Independence Rock and the massive pillars of Devil's Gate on the Sweetwater River. Finally, the Romneys descended through Echo Canyon and Emigration Canyon and beheld the Great Salt Lake Valley. In the shadow of the jagged peaks of the Wasatch Range, the valley spread for miles, vast stretches of arid land intersected by creeks and streams, reaching to the 1,700 square miles of the Great Salt Lake. After this journey across the plains, mountains, canyons, and basins, the Romneys settled in Salt Lake City and initially lived in a wagon just like the one in which they had traveled. And the elder Romney soon had a new task for his tools and hands: he began work on the city's new temple. At last the family began to scrape together a decent living.

Ringed by mountains and remote from other population centers, Salt Lake City seemed like a sanctuary where the outcast Mormons could securely thrive. But as Miles entered his teenage years, the Mormons faced a new threat. In 1856, the Republican Party platform denounced polygamy and slavery as the "twin relics of barbarism." The following year, amid concerns in Washington that Mormons put their church doctrine above loyalty to the federal government, the U.S. Army was sent to Utah to quell Mormons and ensure federal control of the territory. Miles, fourteen years old, scrambled to join the fight, but he was ordered to remain in Salt Lake City while his older brother George joined the Mormon brigade of perhaps two thousand men in nearby Echo Canyon. The conflict turned into a series of standoffs, and the federal troops eventually retreated, losing a number of soldiers to desertion.

The principal flash point was the Mormon practice of polygamy, or plural marriage, as it was also known. Of all the facets of the history of his faith, this is the one Mitt Romney, in his speeches and writings, has said the least about—and he has reason. It is a practice three generations behind him in his own family and long illegal in the United States. He is a devoted family man and finds the whole idea unspeakable and repugnant. "They were trying to build a generation out there in the desert, and so he [Miles P. Romney] took additional wives as he was told to do. And I must admit I can't imagine anything more awful than polygamy," Romney has said.

It was a firm repudiation, but again, as in the description of the "faith of his fathers" in his campaign speech, remarkably colorless. Understanding where Mitt Romney fits into the long line of his ancestors requires going deeper than that. For Miles P. Romney's story is interlaced, to an astonishing degree, with the history of polygamy in the church and in the church's transition away from it. He was a trusted leader who lived through, and would barely survive, this bloody cataclysm in the early Mormon way.

Miles had grown into one of the fiercest defenders of his faith, as well as one of the most engaging members of the Romney family. One family member described him as "emotionally high strung" and prone to issuing "scathing denunciations" against those who disagreed with him. Blustery and imposing, Miles bore the family trademark of a long face and high forehead, and he would be remembered for working off his energy by taking to the ballroom floor and acting in theatrical productions, with a preference for playing Hamlet.

When Miles turned eighteen, the Mormon leader, Brigham Young, asked to see him. Young had succeeded Joseph Smith as president of the church, and he was known as the Mormon Moses for having led the exodus of his followers from Nauvoo to Salt Lake City.

Are you married? Young asked Miles.

No, Miles replied.

You must marry as soon as possible, Young said.

Miles was an obedient believer and, as it happened, in love. An "attractive Scotch lass" named Hannah Hood Hill had caught Miles's eye. She was a nineteen-year-old born near Toronto, Canada, who had arrived as a one-year-old in Nauvoo around the time Miles was born there. In contrast to Miles's fiery temperament, Hannah was hard to ruffle, a calm yet firm influence.

So it was that on May 10, 1862, at Salt Lake City's Endowment House, where Mormon rituals were conducted, Miles P. Romney married Hannah. She would bear him ten children and one day be known as the great-grandmother of Mitt Romney. The couple had a month together before the church sent them on their separate ways. Hannah was pregnant when she received the news. Young had ordered that Miles leave for a missionary trip to England, which would keep him apart from his new wife for three years. As Hannah later recalled it, "We had no display at our marriage, nor a very long honeymoon, but our love was for each other and [we] were happy in each other's society." When they separated, she felt as though her "only friend" had left her. She supported herself during Miles's absence by washing "all day from sunup to sundown for a dollar."

Two months after the marriage, on July 8, 1862, President Abraham Lincoln signed the Morrill Anti-bigamy Act, which was intended to prohibit polygamy in Utah and the other territories where Mormons had settled. But this was one year into the Civil War, and Lincoln's true priority was to ensure that Mormons stayed out of that conflict. The president delivered a message via a Mormon courier: "You go back and tell Brigham Young that if he will let me alone I will let him alone." The antibigamy act was rarely enforced. Miles, meanwhile, was committed to his mission in England to bring converts to the United States, laying out his defense of his religion in a fiery article titled "Persecution."

"Many, now, wonder why it is that we are so despised," Romney wrote. "Many likewise, will argue . . . that if we had the Truth we would not be so despised by the great majority of mankind." But Romney stood by his faith, writing that "from the earliest ages of the history of man, Truth and those who strictly adhere to its principles have been unpopular."

Romney returned to Utah in October 1865, meeting his two-and-a-half-year-old daughter, Isabella, for the first time, and becoming reacquainted with his wife. He was an excellent provider in many ways, an expert in preaching his faith, and a skilled carpenter, but he was not the most practical pioneer. One of his sons later wrote that Miles "never milked a cow, cut a stick of wood, or cut a chicken's head off." By contrast, it was Hannah who was "practical, resourceful," and more accomplished at adapting to pioneer life. Miles and Hannah began life together with few possessions, most notably a small stove, a bed, three chairs, and a table. Miles eventually bought some land, and the Romneys moved into a two-room wooden house. Hannah became pregnant again, and a second daughter was born. "We were happy," Hannah recalled years later. "We had two sweet little girls to bless our home."

It was then, in 1867, that Miles had another fateful meeting with Brigham Young. "Brother Miles, I want you to take another wife," Young said. Miles faced the choice of obeying U.S. law or the head of the Church of Jesus Christ of Latter-day Saints. He decided to commit himself to polygamy. A family biography put it this way: "Nothing short of a firm belief in the divine origin of the Revelation of plural marriage could have induced Miles P. Romney to take a second wife." He followed Young's order.

Hannah was distraught, and she would tell of her despair in a memoir written privately for her family when she was eighty years old. The memoir, later published in an obscure collection of pioneer tales, provides one of the most affecting and revealing accounts of the emotional toll that polygamy took on Mormon women. "I felt that was more than I could endure, to have him divide his time and affections," Hannah wrote. "I used to walk the floor and shed tears of sorrow. . . . If anything will make a woman's heart ache, it is for her husband to take another wife, but I put my trust in my Heavenly Father and prayed and pled with him to give me strength to bear this trial." Then Hannah performed her duty: she prepared a room for her husband's new wife. The practice would become familiar as her husband became among the most dedicated to the practice of taking "plural wives." Despite her despair, Hannah wrote, "I was able to live in the principle of polygamy and give my husband many wives."

So three months after Hannah gave birth to a daughter, Elizabeth, Miles began a new branch of his family. He married his second wife, Caroline Lambourne, a "very beautiful" twenty-year-old who had emigrated from Scotland. Brigham Young then gave Romney and his two wives a new mission: sell your home and move to the southern Utah town of St. George. Miles was chosen for a set of skills that would be known as Romney trademarks: he was unquestionably loyal to the faith and its leaders; he was a skilled carpenter; and he had experience building a community from the ground up. He had become a leader himself. The new settlement, about three hundred miles south of Salt Lake City, was a mostly treeless expanse in a vast desert, surrounded by red-toned ridges in a region where temperatures often reached above 100 degrees in the summer.

Yet Young envisioned a beauteous future. He had prophesied that "There will yet be built between these volcanic ridges, a city, with spires and towers and steeples, with homes containing many inhabitants." The Romneys sold their Salt Lake City home at a loss and moved to St. George, where Miles, his daughters, and his two wives lived "in a little shanty, a small board room, and a wagon box," Hannah wrote. Miles was joined in St. George by his father, who became the architect and builder of the steepled red-brick Mormon tabernacle there. The elder Romney had retained the trade that he learned in England, and his work included building two elaborate wooden spiral staircases. Later, father and son helped construct St. George's grand white Mormon temple, larger than the tabernacle and used for the most sacred rites. The Romneys were also hired by Young to help build his winter home in St. George. Father and son took on the task with zeal, constructing one of the era's most lavish residences in Utah, an adobe and sandstone dwelling with high-ceilinged rooms and an elaborate porch painted red and green.

But though Miles prospered as a builder, he had increasing trouble handling two wives. Hannah wrote that Caroline "was very jealous of me. She wanted all my husband's attention. When she couldn't get it there was always a fuss in the house. [Miles], being a just man, didn't give way to her tantrums." Miles and Caroline, meanwhile, had two children, whom Hannah agreed to care for. But Caroline was despondent due to the "trials met with in plural marriage" and the hardship of living in remote St. George, one of Miles's sons later wrote. She asked Brigham Young for permission to return to her parents in Salt Lake. Young agreed, and Miles and Caroline eventually divorced. She was the only one of Miles's wives who would leave him.

As Miles rose in the church's hierarchy, he was given a new responsibility: to help defeat a new congressional effort to enforce antipolygamy laws. Many Mormons feared that the legislation would disenfranchise them. The measure had passed the House of Representatives and was under consideration by the Senate. Romney and four other Mormon leaders took responsibility for trying to defeat the measure. They signed a letter in 1870 stating that "the Anti-polygamy bill is an act of . . . ostracism, never before heard of in a republican government and in its parallel hardly to be found in the most absolute despotisms, disfranchising and criminating, as it does, 200,000 free and loyal citizens, because of a particular tenet in their religious faith." The lobbying paid off, and the bill died in the Senate

For a brief time, with Caroline having left, Miles and Hannah were once again in a single-wife marriage. It was then, in 1871, that Hannah gave birth to Gaskell, the grandfather of Mitt Romney. "He was a fine specimen of a child, always healthy and happy," Hannah wrote.

Two years after Gaskell's birth, Miles met a woman described as the "prettiest girl in St. George," the fair-skinned Catharine Cottam, who had flowing hair and a serene smile. Catharine's father was well known for building furniture, including the chairs in Brigham Young's home. Catharine wrote a letter at the time that provides a revealing glimpse of Miles. Telling her parents how much she loved him, she assured them that he had "reformed" considerably and "he firmly believes that with my love and influence and assistance, he will become a better man." A family historian who is a descendant of Catharine and Miles wrote that this referred to Miles's "weakness for wine." In any case, the assurances of self-improvement seemed to take hold, and Miles married Catharine in Salt Lake City on September 15, 1873.

For the second time, Hannah steeled herself to share her husband, though, seven months pregnant, she did not attend the wedding. Instead, she prepared a room for Catharine, whom she called "a girl of good principles and a good Latter-day Saint." She had the house plastered, painted, and papered and sewed rags into a carpet. She felt she had to "do my duty" even "if my heart did ache." All the while, Hannah worried that this new wife would be as difficult as the departed Caroline. But Catharine was different. She was "considerate of my feelings and good to the children," and the two wives got along well for the rest of their lives. Yet it was still difficult to accept another wife in the house. When Miles thanked Hannah for welcoming Catharine into their home, it only partly took away her despair. "Many nights I would cry myself to sleep," she wrote.

For a brief time, Miles settled into a routine, living with his wives, Hannah and Catharine. He continued working with his father on church-related construction projects. Then tragedy struck anew. It was at 5:30 a.m., on May 3, 1877, that the seventy-year-old Romney, working on a high window of the Tabernacle of St. George, plummeted to the ground. Thirty-six years after leaving England, the patriarch of the Romney family was dead. The *Deseret News* paid tribute to him as "one of the noblest works of God, an honest man." Now it was left to his namesake, Miles P., to carry on as a family leader.

Four months later, Miles decided it was time to take another wife, choosing Annie Maria Woodbury, a schoolteacher. He now lived with three wives and thought they would all remain for years in St. George, which had grown into a city of about 1,800 people. Then the church issued a new call. Church leaders in Salt Lake City had devised a plan to plant Mormon communities in an arc throughout the West. Miles was told in 1881 to uproot his family and help settle the small town of St. Johns, Arizona, where an initial wave of Mormon settlers was having difficulty getting along with local Mexicans, Indians, and non-Mormon migrants. The Romneys were required to leave their home and head some four hundred miles across the wilderness to a sparsely settled territory. Hannah laid down one condition for the trip: she would go only if they could take their prized organ, which the Romney children were learning to play. "I said I would take it if I had to walk," Hannah wrote, "so we took it."

———————

The Romneys journeyed southeast, taking their wagon trains on a harrowing trail that skirted the northern rim of the Grand Canyon, a barrier that walled off their passage for two hundred miles until an opening appeared at a ferry landing in Glen Canyon. After traversing mountains and riding beneath the Vermilion Cliffs, the Romneys paid the ferry tender $2 per wagon to cross the Colorado River. After the horses and wagons were ferried across the river, the Romneys stood beneath a series of red-rock, triangular-shaped cliffs. Just up the river, the cliffs gave way to a rocky formation known as Lee's Backbone, which resembled a skyward ramp toward the looming plateau. "Here you can see the river hundreds of feet below you winding its way between perpendicular banks of solid rock without a tree to be seen and devoid of vegetation," Catharine wrote.

Finally, after a journey of about four hundred miles, the Romneys arrived in St. Johns, Arizona, a Wild West amalgamation of gun-toting immigrant farmers, Native Americans, and Mexicans, many of whom despised the Mormons. There were complaints not only against the newcomers' religious practices but also about whether they had paid for the title to their land. The Romney family moved into a small wooden house, tents, and a wagon and survived on bread, beans, and gravy. "With the noise and confusion I used to feel as if I would go crazy, but still I kept my senses," Hannah wrote. She became pregnant with twins but miscarried them. With the arrival of winter, the mountains closed them in. It was, Hannah wrote, a "hard, cold country."

The Romneys eventually found more comfortable lodging, moving into a two-story farmhouse with a steeply pitched roof and center stairway. But they were constantly harassed. A new antipolygamy law, much tougher than the old one, which had rarely been acted upon, was enacted in 1882, and federal marshals were under orders to enforce it.

The marshals swept through areas with heavy Mormon populations, and the Romney family soon became a target. A local newspaper, the *Apache Chief*, urged in 1884 that townspeople use "the shotgun and rope" to get rid of Mormon settlers. During that time, Miles worked closely with the top Mormon official in the area, Bishop David Udall. "Hang a few of their polygamist leaders such as . . . Udall [and] Romney . . . and a stop will be put to it," the newspaper said. The newspaper's editor, George A. McCarter, viciously attacked Romney as "a mass of putrid pus and rotten goose pimples; a skunk, with the face of a baboon, the character of a louse, the breath of a buzzard and the record of a perjurer and common drunkard." This mass of accusations reeks of the kind of bar stool name-calling that was common in the Wild West. Still, although Miles had promised seven years earlier to reform his taste for wine, it is possible he had not succeeded.

Romney defended the Mormons in a newspaper he edited called the *Orion Era*, which had about five hundred subscribers. But that only further incensed the townspeople, who vented their "hate upon Romney," according to a Udall family history. The tension rose as local authorities sought to jail Romney on charges of polygamy. The threat was real. Around that time, Miles's older brother George, who had been a member of the Salt Lake City Council, was arrested under the antipolygamy law and jailed for six months. Other Mormon leaders were arrested and sent to confinement in a land that some in the church considered the "American Siberia": Detroit, Michigan.

In an effort to avoid conviction on polygamy charges, Romney sent two of his three wives, Catharine and Annie, into hiding. At first they hid in cornfields during the day and stayed with some neighbors at night, and later they hid in the mountains of New Mexico. As a result, Romney evaded the charges of polygamy "due to lack of evidence," his son Thomas later wrote. Had Miles been convicted, he would have been sent to the jail for polygamists in Detroit and forbidden "against speaking above a whisper at any time during the term of imprisonment."

But authorities brought new charges, alleging that Romney had lied about having title to his land. One night, a marshal arrived at

Miles's home after midnight, demanding his surrender. "The marshal had a gun in one hand and handcuffs in the other," Hannah wrote. "I told him Mr. Romney was not at home. He said he had better give himself up to save the country expense and himself more trouble." Romney had already fled on a wagon, hiding under a quilt in case marshals stopped the entourage. As Miles searched for a new safe haven, church leaders proposed a solution. They told him to go to Mexico to build a Mormon colony in which polygamy would be allowed to flourish. Miles agreed to the plan. With marshals in pursuit, Miles decided it was safest to go with only one of his wives, Annie, "disguising himself so completely that even members of his own family did not recognize him." He left behind Hannah and Catharine and their children, hoping they would reunite in the coming months.

Mitt Romney, in his autobiography, *Turnaround*, cited the story of his great-grandfather's dramatic journey but left out most of the details, with no mention of Miles's multiple wives or his perilous assignment to create a sanctuary for polygamy across the border. Instead, he wrote vaguely, "Eventually Miles was called upon to settle in northern Mexico."

───────────

After travel by train and wagon, about ninety miles south of the U.S. border, Miles reached a stunning overlook in the Mexican mountains. A valley extended for miles on the banks of the Piedras Verdes River, with mesquite and cactus carpeting the flatlands and stands of scrub oak shading the riverbanks. The valley floor was five thousand feet above sea level, providing a climate cool enough to support peach and apple trees. Beyond brown hills, the towering, pine-covered peaks of the Sierra Madre curtained the valley, catching the winter snows that would provide ample water for irrigation. As the sun crested the hilltops, desert grasses turned a golden hue, listing in the wind like lariats made of straw. This would be the colony of Juárez: Colonia Juárez.

Miles was ill prepared to start this new life. Desperately poor and responsible for an enormous family, he lived out of a wagon and then a crude hut. On December 27, 1885, shortly after helping establish the colony, Miles despaired of his plight. He feared that U.S. marshals might come to Mexico to arrest him. He was uncertain about the fate of his wives Hannah and Catharine. "I sometimes think that I am only an injury now to both my family and my friends," he wrote to a family member. "I have borrowed my friends' money, and my family receives no support from me, and the prospect ahead seems as black as midnight darkness." Hannah, meanwhile, was still trying to escape from Arizona. With eight children in tow and in fear of Indian attacks, she led a dangerous mission through the Arizona mountains, passing through snowstorms in Nutrioso, Arizona, where the Romneys had built several homes, and driving the wagon trains through the northern mountains of New Mexico.

Months later, Hannah finally arrived in Mexico. She found Miles living in a makeshift wooden house with a dirt roof. "When it rained we had mud and water coming down," Hannah wrote. " . . . We had two boxes put together for a table and some round logs sawed for chairs and a dirt floor. That was a very crude home, different from what I had been used to, but I was thankful for it as my dear children and I would be with their father and we could live in peace, with no marshals to molest us or separate us again." Then, more than a year after Romney arrived in Mexico, Catharine joined them. A festive reunion followed, with Miles, his three wives, and their children: "21 of us all together had a splendid dinner," Catharine wrote to her parents. "I think I have one of the best and kindest of husbands."

The town, meanwhile, began to take shape, due in significant part to Gaskell, the future grandfather of Mitt Romney and the third of those portraits on the wall. Descended from a family that included many tall, strikingly handsome men, Gaskell was only five feet, five inches tall. But he had the rugged look of a settler, with his piercing eyes and a face leathered by years in the Arizona and Mexican sun. Gaskell, at fifteen years old, helped build the canal that would irrigate the fields, and later played a key role in building a family farm known as Cliff Ranch in the mountains overlooking Colonia Juárez.

By 1890, the settlement was thriving. The Romneys built many of the town's handsome brick homes, along with train stations for a railroad that would provide settlers with more opportunities to sell their produce. Then the Romneys' world came crashing down again. Back in Utah, some of the same Mormon leaders who had urged Miles to create a refuge for polygamy at great personal hazard now turned against the practice. On September 24, 1890, church president Wilford Woodruff, under pressure from the U.S. government, issued what was called the Manifesto: "I now publicly declare that my advice to the Latter-day Saints is to refrain from contracting any marriages forbidden by the law of the land." The careful wording of the manifesto might have given some solace to the Romneys. They may have believed that Woodruff was referring to the law in the United States, not Mexico. They continued their polygamous practice even more isolated than before. In 1897, seven years after the Manifesto, fifty-three-year-old Miles married a wealthy widow named Emily Henrietta Eyring Snow. "There was a great closeness between the sons and daughters of the four wives," a family biography says. "All told there were thirty children."

More and more, responsibilities shifted to the industrious and head-strong Gaskell. He had become his father's "mainstay," as a family biography put it, but one day in his teens he announced to his father that he was leaving to go to school in Salt Lake. "Yes, and go to hell like your brother," Miles said, referring to another son who had gone on a church mission to France and returned with "fancy habits." But Gaskell went ahead. In 1895, after completing his education in Salt Lake City, Gaskell returned to Mexico and married Anna Amelia Pratt, who would become Mitt Romney's grandmother. Anna was descended from one of the most important families in the Mormon faith. Her grandfather, Parley Pratt, had twelve wives and had been chosen by the faith's founder, Joseph Smith, as one of the early apostles.

By the turn of the twentieth century, the Romneys' years of hardship were followed by years of plenty in Mexico. The family "accumulated a great deal of means" in those "lovely times," Hannah wrote. The Romneys owned many cattle and chickens, farmed vast lands, and built sturdy houses in a town near the original settlement. The children went to a newly built redbrick school and joined a baseball team that sometimes traveled 150 miles northeast to play in El Paso, Texas.

One day in early March 1904, sixty-year-old Miles looked out at his farm and commented how beautiful it looked. That night, as he read a newspaper, he called to Hannah and requested the gathering of all of his wives, and their children and grandchildren. Hannah held his hand for a moment and then he "breathed his last," she recalled years later. Miles's life had spanned the saga of the church from his birth in Nauvoo to the exodus over the Mormon trail to Utah, across the border to Mexico to maintain polygamy, and then into the more modern era, when the church sought broader societal acceptance.

Now the burden of family leadership shifted even more to Gaskell. As much as he had admired his father, Gaskell represented the church's new outlook, and he and his wife were married only to each other. With the foundation laid by Miles, Gaskell accumulated land and businesses and became "very prosperous," running a cattle farm and a door factory. His wealth enabled him to build a two-story red-brick home that was considered one of the nicest in his community. It was in that home in 1907 that Gaskell's wife, Anna, gave birth to George Wilcken Romney. George would go on to display many of the distinctive family traits; he was industrious, smart, and indefatigably hardworking, but also a blustery, imposing, and outspoken figure like Miles. George would tower over his short father, and his long frame would be passed along to his son, Mitt Romney. For five years, George Romney lived an idyllic life in Mexico. The wealth of his family, and the stability and sturdiness of their home, were in stark contrast to the wanderings and poverty experienced by Miles. But prosperity would again prove fleeting.

Gaskell was aware of constant talk of revolution among the local Mexican population. Factions within the country were battling one another, and the Mormon colony tried to remain neutral. At one point, little George heard gunfire as he sat on the porch of their home. In July 1912, the Romneys learned that hundreds of revolutionaries were nearby. The rebels ignored the Mormons' insistence that they were neutral. They demanded that the Romneys turn over their guns and horses. Gaskell's half brother Junius declared that he "would die before ordering our people to give up their arms." Vastly outnumbered by the rebels, the Romneys, including five-year-old George, packed their belongings and joined other Mormons at a nearby railroad station, waiting hours before boarding a packed train to El Paso, Texas. Over three days, 2,300 of 4,000 Mormons evacuated.

Twenty-seven years after Miles fled from U.S. government agents and took refuge in Mexico, the Romney family was back in the United States. In the course of a few days, Gaskell's family had gone from owning a large Mexican ranch to being nearly penniless. George would later say that his family was among "the first displaced persons of the twentieth century." George would forever be bitter about his unceremonious exile. "I was kicked out of Mexico when I was five years old because the Mexicans were envious of the fact that my people . . . became prosperous," George said years later. He also noted that "the Mexicans thought if they could just take it away from the Mormon settlers, it would be paradise. It just didn't work that way, of course."

Fortunately for the Romneys, the U.S. government, which had once chased Miles to Mexico due to his polygamy, now welcomed the Romneys and other Mormons to the United States. Congress established a $100,000 relief fund that enabled the Romneys and other Mormon exiles to receive food and lodging. Initially, the Romneys' stay on U.S. soil was to be temporary. The *El Paso Herald* reported on October 25, 1912, that Gaskell Romney and his family, including little George, had gone to Los Angeles "until it is safe for his family to return to the colonies" in Mexico. But Gaskell's family would never return to live there and made only a sentimental trip years later. Had they returned for good, Mitt Romney might never have been in a position to run for president.

The Romneys moved from house to house, from California to Idaho to Utah, as they rebuilt their lives. Gaskell once again became prosperous, constructing some of the finest homes in Salt Lake City and becoming bishop of the church's wealthiest ward, where he might have overseen five hundred members. But during the Great Depression, he "lost all he had and more," according to a family biography. The Romneys left their three-story home and moved into a rented bungalow. "Even though Father was driven out of Mexico penniless . . . he didn't make me feel poor," George wrote about Gaskell. "He never took out bankruptcy, which he could have done several times."

Gaskell regained his financial footing with help from an unlikely source: Mexico. He had never given up trying to obtain financial compensation from the Mexican government for losing his family property. Twenty-six years after the Romneys were forced from Mexico, the case of *Gaskell Romney v. United States of Mexico* was finalized in Salt Lake City in 1938. Gaskell requested $26,753 in damages. He was awarded $9,163, court records show—a sizable amount in the post-Depression years. The records say that Gaskell was to give half of the award to his son George, helping to set the family on firmer financial footing in the United States.

The Romneys had come an extraordinary distance from the day in 1841 when Miles Archibald Romney, convinced of the truth of Mormonism, had set sail for America. His son Miles Park had devoted his life to his faith and family and religious salvation and ended his days in Mexico, but in a roundabout way he had enabled succeeding generations of his family to have their chance at the American dream. He could hardly have imagined that a grandson would be governor of Michigan and run for the presidency, or that a great-grandson would be governor of Massachusetts and also seek the presidency. But the generational line passed along much: not just the angular physical characteristics, not just the fidelity to Mormon faith, but also a world-view grounded in the family's ancestral story of flight and persecution and rebuilding. The family would cycle through utter poverty and unimaginable wealth, but the Romneys would say over the years that what they held in common was clear, that they were builders all, from the carpenters to the politicians, each son trying to accomplish what the father had left undone.

———————

Today, about forty Romneys remain in Colonia Juárez, many of them living in the brick houses built by their Romney ancestors and attending school in the Academia Juárez, funded by the Mormon Church. They are descendants of some family members who, after fleeing to Texas during the revolution, did return to Mexico. Schoolchildren bound through the hallways and across a soccer field in the shadow of the same mountains that Miles P. Romney first eyed many years ago. A Mormon temple, perched on a hilltop, is brilliantly lit at night and topped by the gold-leafed figure of the angel Moroni.

Amid all this lives a man named Mike Romney, whose life has striking parallels with Mitt Romney's. Mike and Mitt Romney are both great-grandsons of Miles P. and Hannah Romney. Their grandfathers were brothers. Mitt is one year older than Mike. It seems only a twist of fate that Mike Romney lives today in Mexico and has worked as a widely respected school administrator while Mitt Romney lives in the United States and twice has sought the presidency. Mike reveres his family's history, and he takes special pride in having recovered and restored the well-worn organ that Hannah had insisted be taken on the family's journeys across the American Southwest and Mexico.

Mike has followed Mitt Romney's career and thinks he would make a great president, but as of mid-2011 he had never met or spoken to his cousin and can only hope that Mitt takes pride in the family's remarkable history—and especially its distant patriarch. "Miles Park was a pioneer in every sense of the word. He helped form part of the United States; he helped form part of Mexico. He was faithful to his church, he was faithful to his God, he was honest in his dealings. He was a good man, and I don't know what more I could ask."

The fourth and fifth portraits that were mounted on the wall in Belmont were, of course, of George and his treasured son Mitt. Mitt Romney rarely discusses the details of his family ancestry, but when he has discussed his faith at length, he has left no doubt of the importance of his family legacy, even as he has stressed that he would never let church leaders influence him if he became president. He has rejected suggestions by some that he distance himself from his religion.

It is a faith that has deepened year by year. By the time Mitt left the insular world of Bloomfield Hills, leaving Ann behind and heading to college, he was still discovering how being a Mormon put him outside the mainstream of American life. Unlike many Mormons, he did not instinctively head to Brigham Young University in Utah or another institution affiliated with the faith. He felt he had much to explore and discover, so he enrolled at Stanford University in California, near the counterculture haven of San Francisco. As Mitt left for this new journey, much of what his faith and upbringing had taught him would be tested anew.

THREE

OUTSIDE THE FRAY

I was not planning on signing up for the military.
It was not my desire to go off and serve in Vietnam.

—MITT ROMNEY ON HIS COLLEGE YEARS

In the fall of 1965, Mitt Romney moved into the third floor of a Mission Revival–style freshmen dormitory on the sprawling campus of Stanford University. All seemed serene on the grounds known fondly as The Farm. Soaring palm trees lined the pathways, and an orderly group of sandstone buildings topped with red-tile roofs clustered around the 285-foot-tall Hoover Tower, named for the former president—a Stanford alumnus—and topped with a forty-eight-bell carillon. The university had begun heavily recruiting the children of the eastern establishment to join the California-heavy student body, and Mitt, in his sporty blazer and narrow tie, seemed to fit right in. He had grown taller, his face was more angular and handsome, and he walked with the stride of a student who expected great things for himself.

The initial calm would prove deceptive. The freshmen had begun their year as if closed in a bubble, but that wouldn't last long. "The campus was quite isolated from the real world," said Wayne Brazil, who lived in Romney's dormitory. Day by day, Brazil said, "the air started leaking out of that bubble." On the dorm's first floor, one of the resident advisers, David Harris, started talking angrily about the United States' escalation of the war in Vietnam and began organizing protests. Students went to Harris's room or attended his speeches and got an earful about what was wrong with U.S. policy. The discontent began to smolder.

Mitt's third-floor room in the Rinconada dormitory seemed a haven from all that, at least at first. He placed a picture of his father on his desk, hung up his camel-hair overcoat, and shelved his books. His roommate completed the all-American picture. Mark Marquess had grown up in a lower-middle-class family in nearby Stockton and was the first in his family to go to college. He had made his way to Stanford as one of the greatest athletes of his day, the quarterback of the football team (until future Heisman Trophy winner Jim Plunkett showed up), and first baseman and outfielder on the baseball team. Marquess, a straitlaced Catholic, soon learned that Romney followed a religion called Mormonism, of which Marquess knew nothing, and that Mitt had a girl named Ann. Mitt and Mark, the son of a governor and the quarterback, each in his own way fit the big man on campus script.

Mitt was no athlete, but he made it his mission to be part of Marquess's world. As Marquess recalled it, Mitt was always running one organization or another. His most serious commitment appears to have been his role on the "Axe-Com," or the Axe Committee, charged with protecting a cherished campus tradition. In the week before the football game between Stanford and University of California–Berkeley, the material for a bonfire was gathered in a dry lake bed on campus. It was a massive setup, with telephone poles stacked like logs, ready to be lit just before the big game. But Cal students also had a tradition: they would try to sneak in and set the bonfire ablaze days before kickoff. They would also try to steal the ceremonial axe—a broad red blade mounted on a plaque that went to the winner of the game; hence the name of Romney's committee. While Marquess sent the team through its paces, Mitt took on the job of protecting the bonfire site and the axe, patrolling the grounds day and night.

When Mitt heard about a rally planned at Berkeley, he figured the axe heist might be discussed and decided to go undercover. Ditching his coat and tie, he dressed up like an antiwar protester in the hope of going unnoticed in the Berkeley crowd. In faded Levi's jeans, a heavy wool work jacket, and well-worn moccasins, Mitt infiltrated the rival campus. One classmate recalled that Romney had borrowed David Harris's clothing, although Harris has no recollection of lending an outfit. The two seemed worlds apart. Harris was protesting a war and saw himself on a mission to prevent the United States from disaster, and Romney was protecting an axe in a campus tradition. But to Romney at the time, it was a serious job.

"It sounds silly now," said Mike Roake, a classmate who accompanied Romney part of the way to Berkeley that night, "but it was the great crusade in that time of sweet innocence."

Marquess was impressed with Romney's dedication. "I don't think that sucker slept for four days," Marquess recalled, using the word "sucker" in a nice way. The bonfire and axe were protected, and an impression had been made. "You wanted Mitt on whatever committee or group you were doing. He would take charge or lead it."

Romney and Marquess did what Stanford freshmen do: they studied and they talked a lot about girls, although Mitt made it clear he would not date anyone besides Ann. They went to parties, where Mitt refrained from smoking or drinking. But Marquess, who would tire of the parties quickly, learned that Mitt would stay for hours, engaging students on whatever was the topic of the day. Oftentimes, Marquess would be worn out from practice or a game, and Mitt would leave the room to go to the dorm lobby and talk for hours more. "He was conservative and willing to express himself," said one classmate, James Baxter. Mitt also left for long stretches to attend church functions. Over the course of the year, the roommates grew to understand each other and grew close. "He didn't put on any airs about anything," Marquess said. "That's what I liked about him." On several occasions, Marquess drove with Romney to his home in Stockton, ninety miles from campus, where the two boys would have a home-cooked meal courtesy of Marquess's mother. They went to a local gym and played basketball, and slept in the modest three-bedroom home. It was a world away from Mitt's upbringing in Bloomfield Hills but a slice of normalcy that Romney seemed to embrace.

Back at the dorm, Romney became close to a number of boys whose fathers also happened to be Republican leaders. Alan Abbott had come from El Paso, Texas, where months earlier he had picked up Ronald Reagan at the train station, thanks to his father's role as county chairman of the Republican Party. On the first day of the semester, Abbott had wandered into Romney's room and found it overflowing with students. Noticing the picture of George Romney on Mitt's desk, Abbott announced that he was a fan of the Michigan governor. Everyone broke out laughing because Abbott didn't know he had entered the room of George's son. Mitt embraced Abbott as a soul mate from the start.

Also on Mitt's floor was Robert Mardian, Jr., the son of the man who had managed Barry Goldwater's 1964 presidential campaign in four western states. George Romney had walked out on Goldwater's nomination. The two sons nonetheless became close, attending football games together and spending much of the year playing practical jokes on each other. Mardian loved to tease Romney about how he had gotten into Stanford. A standing joke was that Romney, who'd had some mediocre grades on his prep school transcript, had been admitted in exchange for cars from American Motors, where George had been chairman. On other occasions, Mardian said, "I would say publicly, 'He's just not qualified. I'm getting tired of writing his papers.'" It was untrue, of course, but Romney always played along. "He took it always in a joking manner," Mardian said. "That is the part that I remember. He was a fun guy." When the two did talk seriously, it was about Republican politics and their shared dislike of radicals.

As freshman year progressed, the political demarcation lines on campus sharpened dramatically. Romney and his Republican buddies were representative of the traditional, conservative side of Stanford's ecosystem. But David Harris made surprising headway with his antiestablishment advocacy, attracting an increasing number of followers as he educated his peers about racial unrest in the South and the growing U.S. involvement in Vietnam. More air was leaking from the Stanford bubble.

It was one thing for the nearby campus of the University of California–Berkeley to have been radicalized by the antiauthority Free Speech Movement and for San Francisco's Haight-Ashbury to emerge as a mecca for free-loving hippies in peasant skirts and dashikis. But at buttoned-down Stanford this creeping radicalism was something new and unsettling. Now, as Romney returned from his mandatory freshman classes in English and Western civilization, the scent of marijuana wafted across the pathways and strains of psychedelic rock blared from the windows. Some students set up a site to take blood donations for the North Vietnamese Communist fighters. Drug-infused "acid tests" were held in courtyards, taking their name from parties popularized by a local resident, Ken Kesey, the author of *One Flew Over the Cuckoo's Nest*, who had honed his literary skills in Stanford's creative writing program. "There was this cultural current coming from the whole Bay area," Harris said. "I would assume that, coming from [Romney's] background, there were certainly things on campus that made him uncomfortable."

By the time football season got under way, Stanford was a split-personality campus of tradition and revolution, and Mitt increasingly found himself caught in a world of which he knew little, about as far from his cloistered Cranbrook School and strict Mormon upbringing as he could get. Months earlier, he had been on a date with Ann watching *The Sound of Music* back home in Michigan. Now he was at large in the land of the Grateful Dead. One fall day, a stream of protesters—most of them dressed in coat and tie or skirt and blouse—headed down a wide palm-lined pathway carrying a banner that said, VIETNAM—MATTER OF CONSCIENCE. An older, short woman carried a huge sign that nearly dwarfed her. PEACE, it said, with a giant hand-drawn peace symbol. A young man held a placard that said, FRENCH KILLED ONE MILLION. HOW MANY SHALL WE? The protests grew exponentially. More than a hundred Stanford students joined a group of at least six thousand protesters who marched between Berkeley and

Oakland. At night, Harris and other protesters would meet in a dorm room, where they smoked marijuana and listened to songs by Joan Baez, the antiwar singer whom Harris would later marry.

The protests were having an impact across the nation and on the Romney family. Three weeks after the antiwar march at Stanford, Mitt's father, George, headed to Vietnam to see for himself what was happening. The Michigan governor was comforted by what he heard, filled with assurances from U.S. generals that the conflict would turn out in the United States' favor. The issues in Vietnam were "the same that brought our country into existence," George said after meeting with the generals, adding that "the American presence in Vietnam is necessary, if the world is to maintain liberty and freedom." Nothing could have been more at odds with what the protesters at Stanford were saying—and George would soon learn about that as well. Returning from Vietnam, George stayed overnight at a San Francisco Hilton on November 12, 1965. It was during that stopover, according to two of Mitt's classmates, that George went to Stanford to visit Mitt. The campus buzzed about the visit, and a dinner was arranged for some of Mitt's dormitory friends. Peter Davenport, Mitt's classmate, recalled that the elder Romney dined with Mitt and a group of students at the dorm. "He spoke about his trip to Vietnam," Davenport said. "It was rather subdued."

George also had a more narrow parental concern. He was worried about his son's personal life. Mitt had secretly been flying back home on many weekends to visit his sixteen-year-old girlfriend, Ann. On another occasion, he drove nonstop from California to Michigan, showed up at Ann's home a sweaty mess, and dived fully clothed into her pool. Mitt planned the visits to Bloomfield Hills like a covert operation, aiming for times when his parents were staying in the state capital of Lansing due to George's gubernatorial duties. "He didn't want his parents to know," Ann recalled years later. "They had no idea he was coming home weekends." One time, Mitt and Ann were at a party and were shocked to see that Mitt's parents were there. "As soon as we saw them, we made a U-turn and left," Ann said.

At some point, however, George learned about the liaisons, and he worried that the frequent trips would affect Mitt's grades. Mitt's older brother, Scott, had attended Stanford for a year but had had trouble keeping up and had transferred to Michigan State. George did not want Mitt to encounter the same kind of troubles. So during his visit to Stanford, George sought out Mitt's friend Alan Abbott. Would you watch out for my son? he asked Abbott. Abbott, somewhat awed that one of his Republican heroes was asking for such a favor, said he would. Then George confided his anxiety about Mitt. "He said he was concerned about the time Mitt was spending traveling back to Michigan on the weekends," Abbott recalled. George planned to "cut back" on Mitt's allowance in the hope that Mitt would spend more time on campus and his studies. But Mitt was so smitten, and so determined to outwit his father, that he came up with a brazen idea. He announced to his friends that he was holding an auction. Nearly all of his clothing was for sale. Abbott arrived at Mitt's dorm room and was astonished that Mitt was selling off even his treasured camel-hair overcoat. "He auctioned off his clothing and bought a ticket to see Ann," Abbott recalled. At a time when most Stanford guys were dating an array of girls, the depth of Romney's devotion to his girl back home would make a lasting impression.

Something else stood out to Mitt's peers: his bond with his father, who remained his hero and confidant notwithstanding the friction over Mitt's home visits. Classmates could see the closeness between them. "It was especially interesting," Romney's classmate Mike Roake said, "because we were freshmen and therefore in the process of divorcing ourselves from our parents."

But even if Mitt was frequently away from campus, he could not escape the escalating turmoil fueled by Harris and other antiwar activists. A campus rally was headlined by Stokely Carmichael, the new national chairman of the influential Student Non-violent Coordinating Committee, who said that U.S. actions in Vietnam were comparable to the lynching of southern blacks. During the course of Romney's freshman year, the antiwar movement at Stanford had grown from furtive dorm room conversations to massive rallies. Harris became so popular that he was urged to run for student body president, and he launched a long-shot campaign. His growing profile was increasingly disconcerting to the university's leaders and to students such as Romney. Harris figured there was no chance he would be elected on the conservative campus; he just wanted to get his message out. "You don't stand a chance of winning," a friend told him. "This is Stanford." Wearing a wrinkled shirt, tan moccasins, and blue jeans and sporting a mustache, Harris competed against six opponents who wore suits and ties and worked their connections to the establishment and fraternities. For one rally, Harris traded an ounce of marijuana to members of an emerging band from San Francisco called the Jefferson Airplane for the use of the band's sound equipment. At another event, he was asked about his attitude toward fraternities. "I think fraternities are a crock of shit," Harris responded.

Harris's platform called for ending the university's cooperation with the war effort, abolishing its board of trustees, and legalizing marijuana. He was against the draft, but also argued that if there was a draft it should apply to everyone, including university students. Why should the war be delegated to the poor, who couldn't get student deferments? he asked. To the shock of the establishment and even more to Harris himself, he won the election with 56 percent of the vote in a runoff against a fraternity man. "The President Elect—Voice of Radicalism," reported the *Stanford Daily*. The election made national news; if supposedly conservative Stanford students elected a radical as president, what did that mean for the country? Alumni contributions plummeted. It was a repudiation not only of the establishment but also of politicians such as George Romney and, by extension, of resolute straight arrows like his son. Romney had become a leading member of the university's Republican Club. But many of his classmates were headed in the opposite direction.

The disconnect between Mitt and many of his fellow students grew increasingly pronounced. A group calling itself the Stanford Committee for a Free University sponsored a series of events called "The

New Student—Pot and/or Politics," with sessions on sex, psychedelic drugs, radicals, and "Politics Without Ideology?" And then there was growing anger over the war and anxiety about the draft. Shortly after Harris's election, Selective Service officials were quoted in the *Stanford Daily* as saying that draft boards would closely examine student deferments. If a student was at the bottom of his class, the officials suggested, he risked losing the deferment. Across the campus, fear spread that the coveted deferment might be undermined. The issue came to a head when it was announced that 850 students would have to take a Selective Service test that could affect their status. The mere presence of the Selective Service on campus prompted a new uproar. Harris and several hundred other students held a protest on White Plaza and then walked to the office of the university president, Wallace Sterling. Two dozen students (Harris not among them) occupied the office overnight, in what became the first sit-in at Stanford.

Mitt was incensed. Skipping his study discussion group for Western civilization class, which was focusing that spring on the works of Lenin and Marx, Romney put on a blazer and attached a large sign to a pole: SPEAK OUT, DON'T SIT IN. Gene Tupper, a photographer who then worked for the *Palo Alto Times*, snapped an indelible image that showed Romney in his white shirt and dark jacket, thick hair sweeping over his forehead, appearing to lecture a protester as he brandished the sign. The picture ran the next day in the newspaper with a caption that read, "Governor's son pickets the pickets. Mitt Romney, son of Michigan Gov. George Romney, was one of the pickets who supported the Stanford University administration today in opposition to sit-in demonstrators." Mitt was at the forefront of a group of about 350 antiprotesters, who shouted at the antiwar group, "Down with mob rule!" "Out! Out!" and "Reason, not coercion!" When a university official announced that students participating in the sit-in would be disciplined, Mitt shouted, "Come out of the office and let school continue!"

Romney, asked about the confrontation years later, recalled Harris being on the opposite side and said, "We had animated discussions about political issues." To Harris, however, Romney was "a zero" who hadn't made a ripple on campus. But to those who knew him and admired his father, Mitt had earned a reputation as an organizer and was becoming a political figure in his own right; the image of him holding the sign at the antiprotest protest would linger in classmates' memories. After a childhood watching his father speak out and lead, Mitt plainly had a similar inclination. If he had left home with any intention of distancing himself from his parents and their politics, that time had passed. The sign he carried to the sit-in that day became a marker of sorts, pointing to the public path he would one day follow.

———————

One of the rallying cries for antiwar protesters did hit home for Romney. Mitt, like other Stanford students, had a deferment that meant he had little reason to worry about being drafted. But he intended to leave college after one year to serve the traditional thirty months as a missionary and then return to complete his studies. The missionary interlude created a potential problem. Most Mormons serving missions were declared "ministers of religion" by the church and, under an agreement with the Selective Service, granted an exemption from the draft. But the agreement was not absolute.

Such deferments for Mormon missionaries became increasingly controversial in the late 1960s, especially in Utah, leading the Mormon church and the government to limit the number of church missionaries who could put off their military service. That agreement called for each church ward, or district, to designate one male every six months to be exempted from potential duty for the duration of his missionary work. Thus, getting a deferment could be more difficult in a state such as Utah, where the huge Mormon population meant that there were sometimes more missionaries than available exemptions. Romney, however, benefitted from having lived in Michigan, where there were relatively few Mormons. Thus, the odds were high that he would receive a 4-D exemption as a missionary. Barry Mayo, who was counselor to the bishop of the Michigan ward where Romney attended church, said Mitt's deferment was never in much doubt. "There were some wards, mostly in the West, where the congregation was large and the number of youth was large," Mayo said. "The circumstances were very different here. Our congregation was small, and the number of youth were small. To the best of my knowledge we never had a situation where we had more than two young men wanting to go in any one year." Mayo said that no records are available from the period that would show how Romney's deferment was handled. But by serving as a missionary and being given the deferment, Romney ensured that he would not be drafted from July 1966 until February 1969. Romney's draft record from the time describes him as "minister of religion or divinity student."

As the war escalated and the demand for draftees grew, the Mormon exemption drew increasing fire. Some non-Mormons in Utah filed a lawsuit against the federal government in 1968, saying it was unfair for Mormons on a thirty-month mission to receive the same kind of deferment as those in other faiths who made a lifetime commitment to serve a religion. Richard Leedy, the lawyer who brought the suit, said he did so because "the substantial number of deferments to missionaries made the likelihood of us non-Mormons going to Vietnam a lot more likely."

Romney has denied that he sought to avoid the draft. Asked years later about his lack of military service, he said, "I was supportive of my country. I longed in many respects to actually be in Vietnam and be representing our country there, and in some ways it was frustrating not to feel like I was there as part of the troops that were fighting in Vietnam." But on another occasion he seemed to contradict himself, saying, "I was not planning on signing up for the military. It was not my desire to go off and serve in Vietnam, but nor did I take any actions to remove myself from the pool of young men who were eligible for the draft."

When Romney's student and religious deferments ended, his name was put into the lottery based on an individual's birthday. He drew the number 300 at a time when no one drawing higher than 195 was drafted. He would never serve, voluntarily or otherwise, in the military.

Years later, some of Romney's Stanford classmates would wonder what had happened to him. He had lost touch with most of them after freshman year ended, and some did not realize that he had intended to leave campus on a Mormon mission. But Mitt's path was preordained. Like nearly all nineteen-year-old men of his faith, Mitt would be called to serve somewhere around the world for two years. (Many women went on shorter missions at age twenty-one.) During his absence from Stanford, the campus would explode with protests that sometimes turned violent. He would never return there as a student.

His freshman year done, Mitt left campus and headed home to Michigan to spend time with his family—and Ann. While there, he did briefly consider breaking with family and religious tradition and not go on a mission. For one thing, he said, his Mormon beliefs at the time were "based on thin tissue." What he apparently meant, according to his friend Dane McBride, was that there were relatively few Mormons in Michigan and thus he didn't have the same kind of connection to the faith as someone growing up in a place like Mormon-heavy Utah. Though Romney's parents were deeply committed and conveyed their faith to their children, it was normal for a nineteen-year-old about to embark on a mission to have questions about his commitment, said McBride, who would later witness the growth of Romney's belief.

But Romney's biggest reason for not wanting to go may have been a fear that he would lose Ann. Countless missionaries before him had left behind girlfriends, only to learn in a letter that the relationship was over. He told her he might not go. But she was insistent. If he didn't, she told him, he would always regret it. Mitt, having sneaked home on many weekends from Stanford to see Ann, now faced the prospect of having to spend two and a half years apart from her. He would live in a location to be determined by the Mormon Church and try to convince strangers to convert to his faith. While his classmates rushed fraternities and prepared for sophomore year, and as a growing number of people his age were being shipped to Vietnam, Romney's life was heading in a very different direction.

The letter came as Romney completed his year at Stanford. "Your presiding officers have recommended you as one worthy to represent the Church of our Lord as a Minister of the Gospel," wrote Mormon Church president David O. McKay, whom members revered as a living prophet. From the very start, in the 1830s, the Latter-day Saints had sent out young envoys to preach the Gospel and try to win converts. It was a missionary who had convinced Mitt's great-great grandparents in England to convert and immigrate to America, and many Romneys had followed the tradition. George had done a mission in England. Now Mitt learned that he would be going to France. It sounded like one of the easier assignments. Some missionaries went to jungles and deserts and islands, while Mitt was off to one of the most cultured societies on the planet. But heavily Catholic France was a society mostly hostile to Mormons. Most French citizens, if they knew anything about Mormonism, were familiar with its history of polygamy and, in a country that takes its wine seriously, for its prohibition against alcohol.

The first Mormon missionary had arrived in France in 1849, but the missionaries had been evicted during the reign of Napoléon III and been evacuated during World War II. In the 1950s, a growing number of missionaries in France had questioned the tenets of Mormonism and embraced other faiths, a scandal that had resulted in nine members being excommunicated. The church had rebounded with a campaign to build chapels in France, and the first two, in Bordeaux and Paris, opened just before Romney arrived. Still, despite more than a century of missionary activity, Mormonism had barely taken root. There were 6,500 Mormons out of 49 million people in France by the time Romney prepared for his mission.

Less than two months after Romney left Stanford, he was on his way to the gritty seaport of Le Havre, best known to Americans for being occupied by the Germans during World War II. Horrific bombing had led to the deaths of thousands of residents and the destruction of much of the city. With the end of the war, the French government had undertaken one of the greatest rebuilding efforts in Europe. Over a twenty-year period, Le Havre had been remade into a modernist ideal. It was a long way from the sunny setting of Stanford and its Mission Revival–style architecture. In Le Havre, blocks of boxy concrete buildings surrounded the 351-foot-high spire of the Church of Saint Joseph. The spire served as a memorial to the war dead and a symbol of the traditional Catholic faith of the region. Feelings about Americans had veered from warmth to wariness as World War II receded into memory and the Vietnam conflict wore on. The mayor and other top city officials were Communists, adding to the anti-American sentiment but also fueling hope among the missionaries that some irreligious citizens might be curious about the Mormon message. Into this world of concrete, communism, and Catholicism came the gangly nineteen-year-old, selling something that very few people in Le Havre were interested in buying: a new religion.

Romney shared a one-bedroom apartment with three other missionaries. They put together makeshift beds by getting used mattresses from a ship in port and stacking plywood atop cinder blocks. There was no telephone, no television, and no radio. There were also no other Mormons in Le Havre, so the four American missionaries would hold worship in their apartment, taking turns preaching and singing and offering one another the sacrament of bread and water. Romney and his three fellow missionaries woke at 6 a.m., ate breakfast, and studied the Bible, the Book of Mormon, and French. They knocked on doors, with breaks for meals, and went to bed as required at 10 p.m. They traveled on Solex motorized bicycles, wearing their suits and carrying satchels containing pamphlets about Mormonism.

Romney's routine rarely varied. Joining with a fellow missionary, he went to a carefully mapped zone. The missionaries kept track of every door they knocked on to be sure they weren't duplicating efforts. It could be a mind-numbing and demoralizing day after day of rejection. Romney knocked on the door of an unsuspecting French family. *"Bonjour, Madame. Nous sommes deux jeunes Américains,"* Romney would say, a fellow missionary by his side. "Hello, ma'am. We are two young

Americans." He continued: "We're talking to people in your neighbor-hood about our faith and wonder if you'd like to—"

Decades later, Romney would recall what had often happened next: "Bang! The door shuts. And most people assumed we were salesmen and said, 'No, I don't want any,' and would shut the door. A lot of people would say, 'Americans? Get out of Vietnam!' Bang!"

Romney worked hard to memorize key French words and phrases that would help in his missionary work. A fellow missionary, Don-ald K. Miller, recalled that Romney would take a long, hot bath and emerge having memorized his lesson. Romney also stood out for his rarefied background. One of his fellow missionaries, Gerald Anderson, recalled how Romney, on a trip to Paris, stunned everyone with his familiarity with the fine French perfumes in a shop on the Champs-Élysées. Yet he could show a tougher side when the occasion war-ranted. Six months into his mission, Romney was in his apartment when a woman burst in to say some Frenchmen were beating up one of his fellow Mormons down the street. The barefoot Romney joined his roommates in rushing into the snowy night. They found a team of rugby players, drowning their sorrows after a lost match, hassling two female missionaries. The women had cried out *"Allez-y!"* which means "Go on," rather than *"Allez-vous-en,"* meaning "Go away." The male missionary who had leaped to their defense had been punched out. Romney ended up with a badly bruised jaw. "There were about 20 guys, very large and very muscular, and we were a group of very young and very small American guys," Romney would recall forty years later. "If you get into a fight with Muhammad Ali, you don't return the punch, you just put your arms up."

For two and a half years, Romney lived under the strict missionary regime. He was to call home only on Christmas and Mother's Day, although one of his fellow missionaries remembered him calling more often. There would be no drinking, no smoking, no sex, and no dating. He would be alone only in the bathroom—Mormon missionaries are always paired with a companion to reduce opportunities for mischief.

All of his time, all of his energy, would be devoted to trying to persuade the people of France to join the Mormon Church.

Mitt's fellow missionaries came from across the United States and Canada, including some from small towns where Mormons were as much a minority as they were in France. Some had viewed George Romney as their hero and were delighted to be paired with Mitt. He thus commanded a certain amount of respect from the start. And he looked the part: a tall, strikingly handsome young man, hair neatly trimmed, oozing confidence and speaking French better than most of his comrades. Marie-Blanche Caussé, a French Mormon who encountered Romney during his mission, thought the young man would go far. Romney "had a personality that was above average for the other missionaries. Usually the missionaries that are eighteen, nineteen years old stay in their corners, they don't speak good French, and you have to approach them, they're a little timid, but Mitt Romney . . . was very comfortable communicating with others."

At the urging of a church official from Utah, Romney encouraged his fellow missionaries to read *Think and Grow Rich*, a 1937 self-help book by Napoleon Hill that had been reissued in 1960. Hill, who interviewed hundreds of wealthy and famous Americans to learn the secret of success, concluded that wealth grew out of the rigorous application of personal beliefs and an ability to work with people of like-minded determination. Some of the chapter titles serve as guides to how Romney achieved success in his future careers in business and politics. Chapter 2 is titled "Desire: The Turning Point of All Achievement." It seems no coincidence that at a missionary conference Romney gave a talk with a similar theme, about "desire" and "how we can obtain anything we want in life—if we want it badly enough," as summarized in a missionary's journal.

One of the more unlikely lessons in the book was called "The Mystery of Sex: Transmutation." In this chapter, Hill advises that a successful person converts some of his sexual energy, "the most powerful of human desires," into other kinds of action. "Love, Romance, and Sex are all emotions capable of driving men to heights of super achievement," Hill wrote. "Love is the emotion which serves as a safety valve, and insures balance, poise, and constructive effort. When combined, these three emotions may lift one to an altitude of a genius." The message was welcomed by the missionaries, struggling as many were with their faith's prohibition of premarital sex. "We were red-blooded American boys. We were not eunuchs," said Dane McBride, one of Romney's fellow missionaries. "We joked about the fact that we didn't have much choice but to put sex drive into succeeding at something else. It fit our situation very well."

No one doubted Romney's determination. His fellow missionaries remember him as charming, charismatic, and passionate. In the "Conversion Diary," then a newsletter of the French mission, he is mentioned repeatedly for his standout numbers of hours spent knocking on doors, numbers of copies of the Book of Mormon distributed, and numbers of invitations for return visits. On the occasions when he was allowed to deliver his full pitch, it went something like this, according to McBride. Romney and his partner would explain that they were students from the United States who interrupted their studies to tell the French that they had a "great message" about Jesus Christ's ministry. They said that Christ's ministry "extended far beyond the small area in which he walked. . . . After his resurrection he visited a civilization living in the Americas at that time." As a result, the missionaries continued, "what we have in our Book of Mormon is another witness for Jesus Christ." The missionaries said that at a time when people questioned whether Jesus was the son of God, "we have very strong evidence he was who he said he was."

A number of those who heard them out, particularly those committed to their own faith, were offended. Some of those who had lost their faith were intrigued. The missionaries urged those people to return to a faith and to consider Mormonism and asked the others to think about converting. Relatively few accepted the message, however, and Romney grew frustrated. Convinced that door-to-door work was mostly unproductive, Romney came up with innovative ways to engage the French. In a letter to his parents, he talked about reaching out to people through "singing, basketball exhibitions, archeology [sic] lectures, street meetings. . . . Why even last Sat. night my comp [companion] and I went into bars, explaining that we had a message of great happiness and joy, and that we would like to talk to anyone who had a few minutes! Amazing how that builds one's courage!" Noticing some French people's interest in America, he staged "USA nights," complete with a slideshow. But the work took a toll on him. "I must admit that one gets really tired—I had never imagined that it's so hard to drive and drive."

Romney later said he converted ten to twenty people during his time as a missionary, but even that small-sounding number stood out among missionaries. Years later, Romney bluntly assessed the experience. "As you can imagine, it's quite an experience to go to Bordeaux and say, 'Give up your wine! I've got a great religion for you!' " he said. "It was good training for how life works. I mean, rejection of one kind or another is going to be an important part of everyone's life. Here I'd grown up as the son of a governor, from a wealthy home. No one had asked me about my religion, or cared, and now I was on the street, lower than a Fuller Brush salesman, in a place where Americans were not particularly liked, where I couldn't speak the language very well, and where selling religion, particularly Mormonism, was going to be very painful." The most successful conversion may have been of Romney himself. Having begun his mission with what he called thin ties to the faith, he became a stalwart believer. "On a mission your faith in Jesus Christ either evaporates or it becomes much deeper," Romney noted. "And for me it became much deeper." It sometimes seemed like "more of a teaching experience for me than it is for the people whose doors I knocked on."

While Mitt was changing and growing in France, many of his Stanford classmates were being transformed by the tumult of the late 1960s. His friend Mike Roake, a good Irish Catholic boy and navy ROTC student, would go on to question organized religion and seriously consider registering as a conscientious objector. Roake wondered what would have happened to Mitt had he never left. "Almost everybody I knew there changed," he said. "I know that as a thoughtful person Mitt would have been altered in some way." Paul Richardson, who lived across the dormitory hallway from Romney at Stanford, said that Mitt left for his mission just as the tumult of the mid-1960s was hitting campus. "History was changing," Richardson said. "It hadn't come full flood, and Mitt was taken on a different journey and different opportunity. . . . He didn't see the full unleashing of it because he was on assignment."

David Harris, the antiwar leader who had become president of the student body, went even further in the opposite direction from Mitt. He married his girlfriend, the singer Joan Baez, and continued to protest. Instead of seeking more deferments or fleeing to Canada, he resisted the draft. Given his prominence, he became a symbolic target and was arrested, convicted, and eventually jailed for twenty months, becoming one of the best-known symbols of resistance. Baez would later rally the rain-soaked masses at Woodstock with an account of her husband's imprisonment.

Years later, Harris wondered whether being so distant from the United States and the Stanford campus during those crucial years had affected Romney's political leanings in some way. Back on campus, Harris said, "There were plenty of people who started to the right of Mitt Romney who ended up as full-scale hippies." Instead, Romney— looking on from a safe remove—grew increasingly appalled at the growth of radicalism in the United States.

————————

While Mitt struggled to win converts in France, his father was having better luck at home. In between his jaunts across the country to test the waters for a presidential bid, George was guiding Ann Davies through her conversion into the Mormon faith. Ann was seventeen years old when Mitt left for France. She had matured into a beautiful young woman, athletic and confident, with shoulder-length hair. As she prepared to go to college the following year, she focused anew on what her relationship with Mitt meant for her religion. She told George she was interested in attending Mormon services. The governor headed straight for the Davies home. He asked Ann's parents for permission to send some U.S.-based missionaries to meet with her. Her mother was an easy sell. But getting clearance from Ann's father, whose rejection of organized religion ran deep, would be a much tougher challenge. Ultimately, Edward Davies and George Romney shook hands on an agreement: George could send the missionaries, provided Ann's mother sat in on the discussions. Ann's younger brother, Jim Davies, said their father had relented based on the trust he had in his daughter and the admiration he had for George Romney. Besides, the governor outranked him. In addition to being president of Jered Industries, which made maritime machinery, Edward Davies was the part-time mayor of Bloomfield Hills.

The missionaries came for six sessions, sitting with Ann in the family room on the lower level of the Davieses' split-level home, taking her through the Mormon conversion process. After learning the faith's story and embracing it, a convert is baptized and is said to be reborn and eligible to "inherit the kingdom of God." Besides her mother, Ann's friend Cindy Burton sat in on the lessons. Cindy was also the girlfriend of Ann's older brother, Rod, who was doing a study-abroad year in England. Little brother Jim wanted to sit in as well, but his parents decided he was too young. So Jim stood outside the family room window, listening in.

Before long, George Romney was picking Ann up and driving her to services at the Mormon chapel. With his magnetic presence, he must have added to the appeal. George, at about that time, was profiled in a *Life* magazine story that emphasized his ties to the faith. The article quoted a Michigan Mormon leader as saying George was "one of the greatest missionaries the church has ever had . . . miraculous things happen to him. We believe that we have the right to place our hands upon the heads of those who are sick, and anoint them with oil. One boy in particular, who had polio, made a very startling recovery after George participated in the blessing." George explained to the *Life* magazine writer that "the same pipeline is available to all. It's a procedure any human being can follow. I emphasize that. You simply seek such guidance as you can get from a source greater than yourself."

Ann, in the end, decided to be baptized and asked George to do the honors. Dressed in white, she followed George into the baptismal font, where she was immersed. By February 1967, Jim had also persuaded his parents to let him join the church, and again George performed the baptism. Even Rod's girlfriend, Cindy, decided to become a Mormon, though her father forbade it, warning her that she would become "a social outcast."

When Mitt lamented in letters home about the difficulty of gaining converts, his father tried to cheer him up. After he had been a missionary in Britain, George wrote to Mitt, he couldn't definitively say whether he had converted anyone, so "I can appreciate how discouraging your work is." George said Mitt should consider himself a success given the conversions he was doing on Mitt's behalf in Michigan. "I was thrilled to stand in for you in connection with Jim's baptism," George wrote back. "This makes two converts here that are certainly yours so don't worry about your difficulty in converting those Frenchmen! I am sure you can appreciate that Ann and Jim each are worth a dozen of them, at least to us." A few months later, even Rod, the family rebel who had been enjoying the pub-crawling life during his year abroad, returned home a baptized Mormon. Mitt had arranged for missionaries to contact him in England. Thanks largely to Mitt Romney, in less than one year the entire progeny of Edward Davies had joined the Mormon faith.

From afar, Mitt followed the progress of his father's soaring political career. George won reelection less than five months after Mitt arrived in France. By the beginning of 1967, George was considered a leading presidential candidate, and Mitt tracked his progress as closely as he could. A Gallup Poll showed George Romney leading former vice president Richard M. Nixon by 39 to 31 percent. Another poll said that if Romney were the nominee facing the Democratic incumbent, President Lyndon B. Johnson, Romney would win by eight points. Romney, as a perceived front-runner, came under fire for the Mormon Church's refusal to allow blacks to become ministers. He said that his church's view did not influence him on civil rights, casting himself as one of his party's strongest proponents of racial justice. Appearing in North Carolina, he took on segregationists who opposed civil rights measures on the grounds of states' rights, saying, "As far as I am concerned, states have no rights. Only people have rights . . . obstructionism masquerading as states' rights is the height of folly."

The major issue in the campaign was Vietnam. Romney, who had no foreign policy experience, had visited the country in 1965 and received a full briefing from U.S. officials, after which he had visited Mitt at Stanford. He had remained hawkish for the following year and a half, saying as late as April 7, 1967, "It is unthinkable that the United States withdraw from Vietnam." But the summer of 1967 deeply affected George as he prepared for a presidential bid. He began to reconsider his support of U.S. policy in Vietnam. At the same time, civil unrest rocked the nation, stoked by protests against racial injustice and the war. When racial disturbances broke out in Detroit in July 1967, the governor took a reconnaissance flight over the scene and reported that "it looks like the city has been bombed on the west side," with fires stretching for several miles. He asked President Johnson to send in federal troops and blamed the White House for delaying the response to his request.

Johnson, in turn, said his potential opponent in the 1968 election had "been unable to bring the situation under control." The troops did arrive, but by the time the riots were over, more than forty people were dead and thousands were injured across Michigan. Romney took the riots to heart, vowing to improve conditions for inner-city blacks and provide them with access to better housing. But his poll numbers had dropped sharply. He canceled a visit to Europe—on which he might have seen Mitt—and planned a tour of inner-city slums, expressing empathy for those far less fortunate than him. All of this further stoked his skepticism about an array of U.S. policies, and he seemed to have reaching a tipping point when he made a fateful decision to be interviewed by a local television personality.

Lou Gordon, a popular Detroit broadcaster, had recently begun a program on channel 50 called *Hot Seat*. Landing the governor and prospective presidential candidate was something of a coup. But when he took his seat on the set, Romney looked distracted. His family would explain later that he had just come from the state fair, where he had spent the afternoon with his grandchildren, and one had gone missing long enough to give the governor a good scare. Gordon's interview seemed, at first, to be uneventful. When Gordon got around to asking him about Vietnam, Romney swiveled in his chair, began speaking in a casual tone, and allowed a slight smile.

> *Gordon:* Isn't your position a bit inconsistent with what it was? And what do you propose we do now?
>
> *Romney:* Well, you know, when I came back from Vietnam, I had just had the greatest brainwashing that anybody can get. When you—
>
> *Gordon:* By the generals?
>
> *Romney:* Not only by the generals but also by the diplomatic corps over there. They do a very thorough job. Since returning from Vietnam, I've gone into the history of Vietnam all the way back into World War II and before. And, as a result, I have changed my mind in that particular. I no longer believe that it was necessary for us to get involved in South Vietnam to stop Communist aggression in Southeast Asia . . . and I've indicated that it was tragic that we became involved in the conflict.

Gordon didn't follow up. But Jeanne Findlater, Gordon's producer, knew news when she heard it. She was listening to the interview from the control room and remembered thinking, "Hot dog! That's good stuff; I'll use that." Chuck Harmon, Romney's press secretary, was at his desk the morning after the Gordon program aired. When a reporter called asking about the "brainwashed" line, Harmon, who hadn't seen the show, stalled long enough to get the transcript. Then his stomach sank. He and a few other aides went to Romney, advising him to backtrack and do damage control. But Romney refused. Coverage began slowly, with an AP story and then a small piece in *The New York Times*. Then it snowballed as rival campaigns—notably those of Richard Nixon and Lyndon Johnson—delighted in highlighting the comment. How can we trust this guy to sit across the table from the Russians, they asked, if he couldn't resist pressure from a few U.S. generals and diplomats?

Years later, George Romney downplayed the damage done by that one line. More to blame, he said, was that he got boxed out by Nixon from the right wing of his party and by Nelson Rockefeller, his one-time supporter, from the left. In reality, the remark was probably more of an accelerant than the cause, exposing how flimsy Romney's national support was. Before long, the former *Time* magazine cover boy became the punch line of a national political joke. The *Detroit News*, once a reliable supporter, blasted Romney's "blurt and retreat habits" and urged him to get out of the race.

Mitt watched it all closely. One day, he wrote to his father that he was preparing to deliver a lecture on American politics and asked for help in explaining the primaries. "I would be VERY happy if you would send me a brief brosure [*sic*] or explanation of the system as it stands in the states, perhaps mentioning some of the dis and advantages," Mitt wrote. "The rest of our system I know pretty well—only one thing I can't understand: how can the American public like such muttonheads?"

With dismay, Mitt also followed his father's political collapse. Every ten days or so, his family had sent him sheaves of newspaper clippings about George's race, and he often stayed up late at night reading them and culling them into files. The news got worse throughout the year: George was the favorite of 31 percent of Republicans in February 1967 but had dropped to 14 percent by November, in part due to the fallout from the television interview, the Gallup Poll found. The impact on the son could hardly be overstated. "Mitt was very passionate; he couldn't believe people were not portraying his dad the way they should be," said fellow missionary Byron Hansen. For months, while knocking on doors, Romney had defended the United States' involvement in Vietnam. But, he said years later, "When my dad said that he had been wrong about Vietnam and that it was a mistake and they had been brainwashed and so forth, I certainly trusted him and believed him." The younger Romney stopped giving blanket support to U.S. war policy. In other words, it seemed, the protesters at Stanford had gotten it right, even if Romney hadn't agreed with their tactics. Romney had initially believed the war was being fought for "for the right purposes," but his father made him realize that "I was wrong." A few years later, he put it even more starkly, using words that directly echoed his father. "I think we were brainwashed," he said. "If it wasn't a blunder to move into Vietnam, I don't know what is."

Mitt did not view the "brainwash" footage that caused his father such trouble until it was shown to him thirty-nine years later. But his sister Jane said the episode had a lasting impact on her brother. "The brainwash thing—has that affected us? You bet. Mitt is naturally a diplomat, but I think that made him more so. He's not going to put himself out on a limb. He's more cautious, more scripted."

Two and a half months after the devastating interview, George formally declared his candidacy for the presidency, and three weeks later he traveled to France with Lenore as part of an international fact-finding tour, including talks with leaders in Paris. Mitt met his parents at a Mormon church in Versailles on December 10, 1967. He helped translate for them, and George addressed the audience. George then went to Vietnam, insisting he would not be misled this time.

Shortly afterward, George headed for the first-primary state of New Hampshire, starting his day at 6:30 a.m. shaking hands outside a Nashua factory. He soldiered on for another seven weeks, but he couldn't shake the "brainwashing" episode. Years later, Romney's analysis would prove to have been prescient; President Nixon eventually adopted a policy of "Vietnamization," similar to what Romney had suggested. But his campaign was doomed. On February 28, with polls showing that former vice president Nixon was leading Romney in New Hampshire polls by a five-to-one margin, Romney flew to Washington and made his announcement. "It is clear to me that my candidacy has not won the wide acceptance with rank and file Republicans that I had hoped to achieve," he said. Privately, he told a friend, "It's a great relief." He would refocus on the governorship and his family.

———————————

Mitt's missionary years also kept him far from the great national tragedies of 1968. The civil rights leader Martin Luther King, Jr., was killed by a gunman in April 1968, and riots spread across the United States. Now Romney and other missionaries heard not only complaints about U.S. policy in Vietnam but also questions about why the United States was being rocked by such violence. King had been a popular figure in France, and the missionaries encountered a growing anti-American sentiment. "We were the only Americans they met," and the French people would ask, "Why are Americans such racists?" Dane McBride, the fellow missionary, said.

Romney, now two years into his mission and twenty-one years old, was still in Bordeaux in southern France when he learned he had won a promotion. It meant he would move to Paris and become assistant to the president of the Mormon mission to France, H. Duane Anderson. But as Mitt prepared to make the move into the mission's grandiose headquarters in the tony 16th arrondissement of Paris, the city was rocked by riots. Labor strikes had spread across France, and student unrest was also growing. An aircraft factory near the town of Nantes had been occupied by many of its 2,800 workers, followed by many other such sit-ins, accompanied by demands that President Charles de Gaulle leave office. Thousands of students occupied Paris landmarks such as the Odéon. Talk spread that revolution would lead to the ouster of the government. Communication was difficult. Mail and telephone service was suspended during much of May due to the strikes, and Mitt had not heard from his family for weeks. He finally found someone who had communications equipment that enabled him to get in touch with his father. The disorder appalled him and further solidified his respect for control in a civil society. He had seen what had happened at the single sit-in at Stanford and now at mass sit-ins across France. Turmoil seemed to be everywhere. "The feeling that we had and discussed was that the world is falling apart," McBride, his fellow missionary, said. "It was that there was disorder and anarchy, and we were very grateful for the order that was in our own lives because the life of the Mormon missionaries is pretty well ordered . . . and there's a security in that."

Amid the strife, the mission president, Duane Anderson, and his wife, Leola, tried to get in touch with the far-flung missionaries, including Romney, who they hoped would soon arrive in Paris in his role as an assistant. They were worried about their safety as well as the security of the missionaries. "Nobody has gas, many are out of money," Leola wrote in her diary. Electricity was shut off regularly, and Leola worried that the scattered missions were in danger. Romney, meanwhile, was apparently able to cross the border into Spain to find some funds. Day after day, the news worsened. On May 30, Leola wrote that the strikers continued to agitate for a new government. Amid talk of dissolution, crowds that Leola estimated at up to one million people filled the Avenue des Champs-Élysées. Leola saw no choice for the mission other than to "prepare for an evacuation."

Six days later, on her thirty-second wedding anniversary, Leola recorded the shock of the shooting of Robert F. Kennedy in California, where she and her husband had lived. "It's a horrid climax to a horrible experience here in France," she confided to her diary. The standstill in Paris had become excruciating. There were, she wrote, no "trains, metro, planes, buses." She felt as though she were sitting "on top of a smoking powder keg." In the following days, Romney finally made his way from Bordeaux to Paris, joining the Andersons at the mission. By mid-June, the strikes had subsided and gasoline became available. Duane Anderson seized the opportunity. He wanted to leave Paris and visit missions in the south that had been experiencing difficulties. He would take Leola along. Mitt, the newly arrived assistant to the president, would be their driver.

[FOUR]

A BRUSH WITH TRAGEDY

*This made me very painfully aware that
life is fragile . . . and that what we do with our time is
not for frivolity but for meaning.*

—MITT ROMNEY

The call from the new, untested leader of the local Mormon congregation had been urgent: "President, President! Two old women have become angry with each other, and the branch is choosing up sides! What do I do?" The spat, in the city of Pau in southern France, demanded the strong hand of Duane Anderson, who promised a swift departure from Paris. "Hang on," he said. "We'll be right there."

On their way south, Romney, Anderson, and the others stayed overnight in the city of Angoulême, arriving late on June 15, 1968. They found an empty parking lot next to the Grand Hôtel de France and left their Citroën DS for the night. Only the next morning did they realize why it had been vacant: this was where the outdoor market set up. Vendors' stalls surrounded their car. They were trapped. Anderson figured they'd have to wait hours to leave. But Romney had an idea. "Do you have a hundred francs on you?" Romney asked Anderson. He did. Romney changed the bill into five-franc coins. Then he waded into the market, announced *"Il nous faut partir!"*—"We must go!"—

and used the coins to grease their escape. Before long, the Citroën was back on the road.

After setting things straight in Pau, they headed back to Paris. Romney, Anderson, and Anderson's wife, Leola—everyone called her Sister Anderson—climbed into the front seat. In the back were David L. Wood, a twenty-one-year-old from Salt Lake City, and a Mormon couple from Bordeaux. Passing through the village of Beaulac, in a verdant landscape known for its fine vineyards, they came upon the aftermath of a serious car accident. A thirty-four-year-old man had smashed into a tree and been thrown from his vehicle. Police were still on the scene. Romney stopped to remove a roof rack lying in the middle of the two-lane highway, then continued on. The travelers, after pulling away, shared their fears of driving in France and what a white-knuckle experience it was. "We were all talking about how dangerous the highways were," Romney said.

At that moment, a Catholic priest in a Mercedes passed a truck at high speed, missed a curve near a post office, and smashed nearly head-on into their car. The collision collapsed the front of the Citroën, thrusting the engine into the front seat. Romney was pinned between the steering column and the driver's door. He was knocked out and trapped in the bloody wreckage. "It happened so quickly that, as I recall, there was no braking and no honking," Romney said. "I remember sort of being hood to hood. And then pretty much the next thing I recall was waking up in the hospital." They were driving slowly, but the other car had come at them in a flash, said Suzanne Farel, who was in the backseat. "My husband was the only one that got out of the car, and he went to get help," she said. Duane Anderson reached to cut the ignition. "Leola was slumped over beside me, and I tried in vain to arouse her," he later wrote in his wife's journal. Ambulances ferried the injured to a hospital in nearby Bazas, a small town famous for a grand medieval cathedral and a breed of choice cattle.

Romney's injuries appeared so severe that a police officer who responded to the scene made a grave notation in the young man's passport: *"Il est mort"*—"He is dead." In fact, Romney was unconscious but still breathing. Rescuers had to pry him out of the Citroën. When news of the crash reached Romney's parents, the details were thin. They were told he had survived but didn't know much more. George contacted Mitt's would-be bride, Ann Davies, inviting her to the Romney home as the family waited and prayed. "I remember the call coming in," said Jim Davies, Ann's brother. "I remember the shock of it." Duane Anderson was seriously injured, but he survived. His wife, Leola, who had been sitting between her husband and Romney, was not so lucky. Crushed by the impact, she lived for two and a half hours, dying in the hospital after unsuccessful attempts by the doctor and nurses to keep her alive.

———————

The days that followed, Duane Anderson would later say, were "a blur of pain." Word of the accident spread quickly through the Mormon world. Help began arriving within hours. Under instructions relayed to Paris from Salt Lake City, where church leaders were deeply worried, missionaries Joel H. McKinnon, who was Anderson's senior assistant, and Byron Hansen, the mission secretary, left Paris at midnight and drove through the rain, arriving in Bazas at 8:30 a.m. the day after the accident, according to Hansen's journal entry, which began, "Tragedy struck last night." Hansen recalled, "When we initially arrived, they thought Mitt had been killed." McKinnon and Hansen had to inform Anderson that his wife had died. The doctors had declined that somber task.

From the United States, George Romney took charge of his son's care. He asked a son-in-law, Bruce H. Robinson, a medical resident then married to Mitt's sister Jane, to fly to France and oversee Mitt's medical treatment. "I was making rounds that afternoon in Michigan, and George Romney called me and said, 'Mitt's been in a fatal car crash; he's survived so far, but we don't know the extent of his injuries,'" Robinson said. The family was also worried that French doctors would not know of Romney's allergy to penicillin. Robinson called his wife, who met him at the Detroit airport with a toothbrush and clothes. He flew through the night to Paris and then to Bordeaux, arriving on June 18. "Mitt was just coming out of his coma, but his face was all swollen, his eye was almost shut, and one arm was fractured," he said. As can happen after head injuries, Robinson said, Romney's breathing and heart rate had slowed, making them hard to detect and leading to premature declarations of his death. Lacking tests that exist today, the emergency medical personnel did what exams they could, finally establishing that he was going to pull through. He nearly did not. "He probably came within a hair of not surviving," Robinson said. In the days ahead, Romney recovered quickly and without surgery, benefiting from his youth and general good health. His emotions, though, were a muddle: he was in shock, profoundly grateful to be alive, and grieving Sister Anderson's death.

The driver of the Mercedes that hit them was a forty-six-year-old priest, Albert Marie, from the village of Sireuil. He had been traveling with his mother and another woman. Romney said the truck driver whom Marie passed estimated Marie's speed at about 120 kilometers per hour, or about 75 miles an hour. The Church of Jesus Christ of Latter-day Saints, having experienced a variety of run-ins with the French government over the previous century, was reluctant to inflame tensions with French officials or the Catholic Church by going after Marie. "Duane Anderson refused to press charges because he didn't want there to be difficulties between the two churches," said André Salarnier, a French Mormon who was living in Bordeaux and rushed to the hospital after the accident to help. David Wood said he remembered receiving some kind of settlement, but Romney has no such recollection. Romney has said he believes there was a criminal proceeding against Marie and that he recalls filling out an affidavit. But neither Romney nor the police in Bazas still have records from the case.

By all accounts, Romney himself was driving cautiously that day and deserved no blame. "We were conservative—he was below the speed limit," Wood said. Romney, asked how fast he was driving, said, "Oh, yeah, I was probably going less than the speed limit, so far as I know." Richard B. Anderson, a son of the Andersons who was twenty-seven and attending graduate school at Harvard at the time of the accident, said he does not hold Romney responsible at all. "Mitt was not in any way at fault," he said.

Duane Anderson, the church's top man in France, faced a long emotional and physical recovery. His chest had been crushed, his ribs and wrist were shattered, and his liver and spleen were damaged. After a few days in the local hospital, the church rented a private train car to transport him back to Paris. When they arrived at the mansion that served as the mission home, Romney and Robinson helped Anderson, who was confined to a wheelchair, up to his bedroom by an elevator. When he entered the room, the full weight of what had happened finally hit him. "He became hysterical with grief," Robinson said. "I had to put my arms around him, and Mitt did, too, as he sobbed uncontrollably." Anderson wrote a few months later, "I felt that the world had come to an end and there was nothing left to live for. Every cell in my body screamed with anguish."

For Mitt, the fatal accident was a turning point. He was still a young man, twenty-one years old, with a young man's sense of his own invincibility. Now Sister Anderson, who'd been giving him advice on his love life shortly before the crash, was gone, a brutal lesson in man's impermanence. Romney, as he sorted through his grief, would also see his responsibilities in the mission grow. "It was a very difficult and heart-wrenching experience to lose someone that I respected and admired, and to see someone who I loved—the mission president— lose the love of his life," Romney recalled. "It is still a tender spot in my heart."

Church leaders called J. Fielding Nelson, the president of the Mormon mission in Geneva, Switzerland, with a pressing request. France was on its own, they told him, a disparate network of two hundred or so young missionaries with no guiding voice. Duane Anderson was returning to the United States to heal and to bury his wife, and it wasn't immediately clear when he would return. So Nelson, as the closest church elder, was asked to pack his bags and go to Paris. He did not know what he would find.

A mission president is like a father figure, responsible for scores of vulnerable, often immature young adults in a foreign country. Any number of problems could arise. The missionaries could crash their scooters or bicycles and need medical care. Their pairings with other Mormons for door-to-door proselytizing could turn volatile. A loved one might die while they were away. And then there were the frequent forceful rejections to their evangelizing. In France, the social upheaval of 1968 added yet another complication. Through it all, it fell to the mission president to hold things together and keep the missionaries safe. "The Lord, and those parents, are trusting you with their kids," Nelson said. "You're responsible for them." But when Nelson arrived in Paris, he found no such disarray. The two young leaders, Mitt Romney and Joel McKinnon, still nursing their injuries and their grief, had stepped into the vacuum to manage the enterprise. That meant giving missionaries assignments, overseeing the financial operations and other administrative aspects of the mission, and helping people in the field deal with whatever problems arose. "You had this shock experience that affected all of us emotionally, but the work had to go on," said fellow missionary Dane McBride. "You saw this exceptional leadership in Mitt to inspire, uplift, bring people to focus, remember what they're about."

Indeed, it was in this period, in the latter part of 1968, that glimpses of Romney's capacity to lead first began drawing notice. Those who worked with him remember a young man who took on the demeanor of someone much older. He resisted suggestions that he return home to recuperate, believing he owed it to the mission to stay. He was back at full speed in a matter of weeks. "His resilience was truly astounding," McKinnon said. "He didn't seem to be particularly pensive or particularly concerned about the accident, as to what had happened to him and how close he'd come to death." He was even, perhaps, too eager to work. McKinnon's sole focus was keeping the mission functioning, but Romney saw the crisis as an opportunity to make some changes in how things were done. At times they clashed; McKinnon felt overwhelmed by Romney's tide of ideas. "I just got to the point where I kept saying, 'No, no, no,'" McKinnon said. "His mind did not rest." But he admired Romney's drive. "Mitt had the ability to just sort of see all the things that needed to be done and begin to figure out how to do them," he added.

Inside, however, things were different. Emotions swirled in Romney's head. He shared his feelings only with those closest to him, exchanging calls and letters with his parents. But he wanted to keep all of that from view, obscured behind a mask of resolve. It was an early glimmer of a trait that would shine brighter in his adult life: a fierce instinct to protect his privacy and keep others at bay. "I didn't want to, if you will, carry to all of the people in France a posture that would suggest that I didn't have the emotional strength and the spiritual confidence to carry on," he said.

In the final months of his mission, Romney worked to transform everyone's mourning into something more constructive. He called on all the missionaries to make an extra effort in their proselytizing, to recognize the sacrifice the Andersons had made. "We put our shoulder to the wheel," he said. Duane Anderson returned in August 1968 with his son and his son's family, describing the mission home as "painfully lonely." He commended Romney and the others for "carrying on beautifully in our absence through some very trying times"—citing one missionary who had suffered an emotional breakdown and had to return to the United States and another who had committed an unspecified transgression and been excommunicated from the church. Despite Romney's new stature within the mission, he did lose one small battle after Anderson settled back in. He objected to the amount of garlic the mission home's Spanish chef used for daily meals and lobbied for less, but the Andersons pulled rank and backed the chef. The garlic lovers won.

In his final months in France, Romney helped accelerate the number of conversions credited to the mission. He personally brought few new members into the church over his two and a half years. But his leadership helped the church achieve its goal of two hundred new recruits in what had been a challenging year. He found inspiration in the story, a parable really, of a Utah chemist, Henry Eyring, who, hobbled by cancer, had nonetheless once struggled to help his church weed an onion patch, only to learn that the row he had worked on didn't need weeding. The mistake didn't bother him, though. "I wasn't there for the weeds," Eyring said. "He came to respond to the call of service," Romney said, "and I think that's what happens to young men or young women who go on a mission." Indeed, the closing months of his mission brought Romney closer to the faith his family had helped build. He had matured, both as a man and as a Mormon, and, following the accident, was able to see more clearly that he desired to live a life of purpose. "This made me very painfully aware that life is fragile, that we're only here for a short time and that life is important," he said, "and that what we do with our time is not for frivolity but for meaning."

Something else had changed in him, too. The boy who had grown up immersed in American car culture now feared automobiles. "I was frightened of driving a car or being in a car and had a sense of vulnerability that I had not experienced before," he said. Romney apparently had good reason for concern. On an icy day in December 1968, he was driving a Peugeot through the city of Le Mans when it was hit from behind by a garbage truck. Romney saw the truck coming in the rearview mirror and braced himself. The truck "slammed into the back of my vehicle," he said, "which caused it to slam into the car in front of us, and they kept going—bang, bang, bang, bang!" No one was seriously injured, but Romney, who would return to the United States a few weeks later, had had it with French roads.

On his way home from France, right before Christmas 1968, Romney stopped first in England, where Ann Davies's older brother, Rod, who had recently converted to Mormonism, had been called as a missionary. Romney gave Rod his old shirts, shoes, and suits, and they spent a day together knocking on doors before Romney flew home. But Romney wasn't especially concerned about converting the English. All he cared about was Rod's sister. Mitt and Ann had agreed to get married once upon a time, but that was a childhood promise, made more than two and a half years earlier. Romney hoped the deal was still on, but he couldn't say for sure.

Throughout his mission, Romney never lost sight of his primary goal: holding on to Ann Davies. Just as George Romney had doggedly pursued Lenore, Mitt was determined not to let Ann slip from his grasp. He had grown over the past two and a half years—he'd come face-to-face with death, drawn closer to his faith, and seen his father wage and lose a presidential campaign. All of that, combined with the distance from Ann, had brought clarity: he wanted a life like his father's, with Ann at his side and a family in his future. But he had ample reason to fear that his grand plan would crumble into pieces. When Romney moved into his Paris apartment with fellow Mormon missionaries, his eyes were immediately drawn to a wall covered with letters. They were "Dear John" breakup notes that other missionaries had received from their girlfriends back home. Staring at the wall, Romney worried, "Is this what's in store for me?" Ann Davies had said yes to his informal marriage proposal, but she had been just sixteen years old. Romney's many months away had tested their youthful romance.

Under the rigid rules for missionaries, Romney was forbidden to telephone Ann more than a couple of times a year. His two visits with her were brief and supervised. Ann, meanwhile, was living the life of a coed at Brigham Young University. The Provo, Utah, campus was crawling with men who had just returned from their own missions, with sharpened skills of persuasion and a determination to find a wife made more urgent by the Mormon ban on premarital sex. Not for nothing was the place nicknamed B-Y-Woo.

The letter Romney dreaded arrived in the fall of 1968, just months after the crash. It wasn't a classic breakup letter, but it was close. Ann wrote to say that she hadn't experienced feelings for any of the BYU men pursuing her—except one. His name was Kim Cameron, and he was a basketball player and a student government leader. Cameron reminded her, she wrote, of him. "That was terrifying," Romney said. "I went, 'Oh, my goodness, this is it!'" The letter threw him into despair. "He became really, really distraught that she had indicated she had gone out with this guy," McBride said. The saga dragged on for weeks. Mitt wrestled with whether to call her. Instead he poured his heart into letters, auditioning sweet nothings with fellow missionaries before putting them to paper. "It was the only time I've ever seen him where he just couldn't focus on anything else," McBride said. "He was just kind of worthless."

Ann's roommate at BYU was Cindy Burton, a friend from Michigan who would go on to marry Ann's older brother. Now Cindy Davies, she said that for a time she had thought Ann might end up marrying Cameron. "I think that's probably right," Cameron recalled. "Emotionally, I felt very close to her." Romney feared the same. He implored her to wait for him. Ann had written back assuring Mitt that it was him whom she loved, easing his mind. But he couldn't be certain. As he flew home, he worried about what awaited him. "I didn't know how we would feel," he said. Ann joined the Romney clan in meeting him at the airport. Enveloped in hugs from his family, Mitt kept his focus squarely on Ann. Sitting with her in the third-row seat of his sister's Oldsmobile Vista Cruiser, he wasted little time. "Gosh, this feels like I've never been gone," he recalled telling her. "I can't believe it."

"I feel exactly the same way," she said.

"You want to get married?" he asked.

"Yeah."

When they made it home, he told his parents about their plans for an immediate wedding. His father was delighted. His mother was not. A pillar of Detroit society, Lenore Romney knew marriage was not something to be rushed. But that was only part of her hesitation. "I think Lenore had a hard time letting go of her youngest son," Cindy Davies said. This was, after all, the baby her doctors had said she could never have. Though George had quickly forged a loving bond with Ann, it took Lenore longer. "Her relationship with Ann wasn't as warm," Davies said. "She held back more." They agreed to wait three months to walk down the aisle. In the meantime, Romney ditched Stanford for BYU to be with Ann. Besides, that was where his new friends from the church would be enrolling. He would be comfortable there. Romney joined the honors program at BYU and began to dive into his studies, his time in France driving him to want to "accomplish things of significance," he said. "I said, 'Boy, I want to do something with my life if I can.' So when I came home, I was a much better student."

In the spring of 1969, he finally got his longtime wish, marrying Ann in a wedding ceremony that stretched over two days. On March 21, exactly four years after their first date, Mitt, then twenty-two, and Ann, who was nineteen, exchanged rings in a small civil ceremony before an improvised altar in her parents' home. About sixty people attended the ceremony, which was officiated by a church elder, Edwin Jones, the man after whom teenage Mitt had patterned his hairstyle. Ann—happy, tearful, and carrying a handful of orchids—was escorted by her father. Afterward the newlyweds paused for pictures and punch and then headed to the Bloomfield Hills Country Club for a reception dotted with boldfaced names from the auto industry and government. Three hundred guests came. Mitt cut the cake, posed for countless photos, and helped his new bride fix her veil. But there was one thing he wouldn't do. When a photographer wanted to capture a kiss for posterity, he refused. "Not for cameras," he said.

The next morning, the wedding party and guests flew to Salt Lake City, most of them on a plane the Romneys had chartered for the occasion. In the spired Mormon temple, Mitt and Ann were "sealed" for eternity. Because they were not Mormons, Ann's parents were not allowed inside to witness the ceremony. Afterward, the family hosted a reception at a hotel across the street from Temple Square, attended by a number of church leaders and Utah political figures. Once the ceremonies and celebrations were complete, Mitt and Ann Romney returned to BYU to begin building a life together in a modest $62-a-month basement apartment in a complex within walking distance from campus. They fit in easily among the many young Mormon couples who had started their families while at school in Provo, sharing classes and potluck meals. George and Lenore had bought them a car as a wedding gift. A year to the day after their first marriage ceremony, their first son, Taggart, named for a friend at BYU, was born. The new parents were thrilled.

Romney, despite having soured on the Vietnam War, felt at home within the conservative culture at BYU, which prohibited many rock-

and-roll bands, liberal speakers and student organizations, and even long hair on male students. During Romney's time at the school, the president of the university enlisted students to spy on professors deemed to be liberals. Students who displayed peace signs were told to take them down.

He was invited to join the Cougar Club, an all-male service club on campus—BYU's version of a fraternity—that sought out students who had shown leadership potential. A few dozen students participated in the club, which, until Romney became president around 1970, had raised modest amounts of money for the university through bake sales, luaus, and auction sales. But Romney, put off by the protests, vandalism, and violence that had engulfed other college campuses around the country, wanted to transform the club into something greater, an entity that could provide more robust support to a school that he and the other members loved. "We felt very differently about our university . . . and we wanted it to succeed," said McBride, who also joined the club. "We were proud of it." So Romney set an ambitious goal: instead of just selling cookies and sponsoring parties, the Cougar Club, in collaboration with the university administration, would endeavor to raise $100,000 a year by directly soliciting alumni and their families for contributions. Romney secured names of potential donors from the school, signed up volunteers, and established phone banks. The plan worked, and they achieved their goal. The Cougar Club has since become a major booster for university athletics. It was Romney's vision, and he made it happen.

In 1971, after two and a half years in Provo, Romney earned a degree in English literature, graduating with "highest honors" and delivering an address to students at commencement that year. In his speech, Romney invoked scripture and said that for the blessed like them, the expectations were high. "I pray that this graduating class will choose a different kind of life, that we may develop an attitude of restlessness and discomfort, not self-satisfaction," he said. "Our education should spark us to challenge ignorance and prepare to receive new truths from God." Though they would follow many of the same paths, Romney's degree from BYU set him apart from his father, who had reached the heights of business without ever graduating from college. Four decades earlier, George Romney had dreamed of going to Harvard University to obtain a business degree, but he had given up the dream in order to pursue and marry Lenore. Things had worked out well for George, but now, he felt, times were different. He sat down with Mitt and laid out his view: not only should Mitt get a business degree; he should also try simultaneously to get a law degree from Harvard. It was a competitive world, and one needed the best education to thrive, he believed. Mitt wasn't sure at first but agreed to consider the idea. He talked it over with Ann and soon set the plan into motion. They would move to Massachusetts.

Howard C. Serkin worked his way down the long rows of terraced desks, searching the alphabetized name cards for his seat in Aldrich Hall. He was no stranger to pressure after four years as a nuclear submarine officer in the navy, but that morning his stomach was in knots. It was the first day of classes at Harvard Business School in 1972. He found his seat and introduced himself to the fellow with the glossy dark hair in the next chair. "He looks up. He smiles. I say, 'Hi, I'm Howard Serkin.' He says, 'Hi, I'm Mitt Romney,'" Serkin recalled. "Stupid me, I say, 'Where are you from?' He says, 'I'm from Michigan.' At that point, I thought, 'Oh, my God,' and then I knew: He was . . . George's son."

Romney's privileged pedigree was common knowledge to many of his classmates at Harvard Business School and Harvard Law School, where he was simultaneously enrolled through a joint degree program. But he was only one of many children of the wealthy and politically influential. His business school class included the son of Kurt Waldheim, the UN secretary-general; and Michael Darling, whose family had given Darling Harbour in Sydney its name. The class behind his had included George W. Bush, whose father was then the chairman of the Republican National Committee. At the law school, Romney counted among his classmates Susan Roosevelt, the great-granddaughter of Theodore Roosevelt; and Edward F. Cox, who was frequently trailed by Secret Service agents and news photographers when he appeared on campus with his wife, Tricia, the daughter of President Nixon. "When we all got there, for the first week or so, everyone—even the rich and famous—walked around saying, 'What the hell am I doing here? Why did they pick me?'" said Janice Stewart, a member of Romney's business school class. "After several weeks, I figured it out: Everyone I talked to were all internally driven human beings. They had fire in the belly." She added, "It was expressed in any number of ways, but it was always there, always present. And Mitt's got it big."

Harvard's joint MBA/JD program was relatively new at the time—it had been launched two years earlier—and intensely rigorous. Typically, business school was completed in two years and law school in three; dual-degree students earned both degrees in four years, spending their first year at one of the schools, their second at the other, and their final two shuttling between the two. Out of Romney's 800 business school classmates and 550 law school classmates, only 15 earned degrees through the program. "We viewed ourselves as kind of an elite guerrilla band," said Howard B. Brownstein, who graduated from the program with Romney in 1975 and then worked with him at the same firm. "We were small and a little different."

Academically, the law school was more theoretical, the business school more practical. Harvard Law, where Romney's professors included Stephen Breyer, now an associate justice on the Supreme Court, relied largely on textbooks and instruction. The business school revolved around the case study method, in which students dissected real-life business decisions to learn to think like managers and executives. Romney excelled at both, and together the two tracks of Romney's graduate school experience provided excellent preparation for his future career in consulting and private equity. His legal training honed skills he had possessed since childhood: asking challenging questions, playing the role of devil's advocate, and using an adversarial process to get answers. His new ability to analyze and reconcile conflicting points of view and data would become an important asset in his future high-stakes investing.

The Harvard campus of his day was an exciting place, crackling with talent and the collision of ideas, though it had quieted from the tumultuous days of the late 1960s. Antiwar sentiment persisted, and many veterans of the strikes and sit-ins were still on campus. But as he had at Stanford, Romney ran with a different crowd. In an age when many of his peers were challenging authority, his enthusiasm and optimism stood out. "There was nothing jaded about him, nothing skeptical, nothing ironic," said Garret G. Rasmussen, who, by virtue of alphabetical seating, sat near Romney their first year at law school. "He was all positive, and it was a very refreshing style." Even in the casual environment of graduate school, Romney presented a more buttoned-down image, dressing more formally than his fellow students. "Most of us dressed like borderline slobs," recalled William L. Neff, a member of Romney's law school study group. "He was a little neater than that."

By the time Romney arrived at Harvard, his father had run a major corporation, been elected three times as Michigan's governor, sought the presidency, and been appointed to President Nixon's cabinet. But despite strongly resembling the elder Romney—the full head of strikingly dark hair, square jaw, dazzling smile—Mitt did little to draw attention to his parentage. Classmates said that the only hint was George's faded gold initials on a beat-up old briefcase that Mitt carried around. Mitt's father, meanwhile, had been drawing attention in Washington, making waves in a White House that from the beginning had viewed him more as an adversary than a collaborator.

After George Romney abandoned his bid for president, Nixon surprised many by appointing him secretary of housing and urban development. But the two former rivals never really made up. Romney had refused to release his delegates to Nixon at the 1968 GOP convention. The snub "was an incident that Nixon could never forget," Nixon aide John Ehrlichman later wrote. Ehrlichman believed the decision to put Romney in the cabinet was purely strategic. "Nixon," he wrote, "needed a few moderate Republicans to balance the Cabinet. What better revenge than to put Romney into a meaningless department, never to be noticed again." But Romney did not toil quietly in obscurity. He fought hard to fulfill his vow to improve race relations, push-ing for integration of suburban housing. "We've got to put an end to the idea of moving to suburban areas and living only among people of the same economic and social class," said Romney, who still owned the family home in exclusive Bloomfield Hills, Michigan. It was a volatile issue, and his advocacy was unpopular in the Republican Party. Mitt's mother, Lenore, tried to follow her husband into politics, mounting an unsuccessful Republican campaign for a U.S. Senate seat in Michigan in 1970. Some analysts attributed her defeat to her husband's push for integrated housing. George Romney initially believed he had Nixon's support for his housing policies, only to learn that the president, at the urging of his aides, was keeping his distance from Romney's plans for political reasons. Romney was torn between speaking out about his disagreements and being a team player.

Eventually, the anger within the party at George Romney led Nixon to want to push him out. Romney declined Nixon's suggestion that he become ambassador to Mexico, his birthplace. When Nixon met with Romney in late 1970, the president, concerned about los-ing Michigan as well as urban voters across the country in the 1972 election, couldn't bring himself to ax his old foe. Instead he tried to bully Romney into capitulating on a variety of issues. Romney be-came increasingly infuriated at his lack of authority. Then, in early August 1972, flooding following a hurricane devastated Wilkes-Barre, Pennsylvania. Nixon announced that he was sending George Rom-ney to assess the situation—but didn't bother to inform Romney, who learned about it from the media. As Romney angrily stewed about the slight, his wife, Lenore, decided to secretly contact the White House. Unaware that Ehrlichman had been assigned to keep Romney in line, Lenore wrote the Nixon aide on August 8, 1972, saying, "It was a stun-ning blow to have the president send a communication through the press that he was 'ordering' George to Wilkes-Barre and demanding a report. . . . It is demoralizing to know . . . that your President has such low regard for your own dignity and service."

Arriving in Wilkes-Barre, Romney was confronted by flood victims who believed the Nixon administration had abandoned them. Rom-

Check Out Receipt

BPL- Honan-Allston Branch Library
617-787-6313
http://www.bpl.org/branches/allston.htm

Monday, Jan 6 2014 5:43PM

Item: 39999076901543
Title: The real Romney
Material: Book
Due: 01/27/2014

Item: 39999069418596
Title: The war for late night : when Leno
went early and television went crazy
Material: Book
Due: 01/27/2014

Total items: 2

Thank You!

ney's blunt contrarian style welled up, and he brusquely dismissed a suggestion from the state's Democratic governor that the federal government pay off the mortgages of hurricane victims, calling the idea "unrealistic and demagogic." That prompted a sixty-three-year-old grandmother, Min Matheson, to challenge Romney during a press conference. "You don't give a damn whether we live or die," she told him, thrusting a photo of the devastated area in his face.

The confrontation received wide media coverage. Though Romney's "brainwashing" comment about his early views on Vietnam had become his best-known sound bite, the clash in Wilkes-Barre, at the time, almost surpassed it. Nixon feared that the fallout from Romney's trip could hurt his reelection chances in such a large, crucial state. Two days after his visit to Pennsylvania, Romney arrived for a rare personal meeting with Nixon in the Oval Office. In the course of an emotional hour, Romney let loose with his many frustrations about serving under Nixon and repeatedly tried to quit, according to a conversation captured on Nixon's secret tape-recording system. "I have no effective voice in the policy areas or the operational areas relating to my own department!" he thundered. Nixon, however, didn't want Romney leaving in the midst of a reelection campaign. In a soothing voice, the president told him that his leaving immediately would hurt both of them. Romney relented, agreeing to put off his resignation. He stayed silent about his disagreements with Nixon, but only until Nixon won a second term. Then, in a tart resignation letter to Nixon on November 9, 1972, he said that politicians had become too fixated on simply winning elections to lead effectively. "Their basic function is to compete for the responsibility to govern," too afraid of alienating voters to tackle "real issues," he wrote. He envisioned forming a "coalition of concerned citizens" with the goal of creating "an enlightened electorate." Romney's critique was in some ways out of step with an American political culture that was moving toward scripted, media-driven campaigns. He briefly considered running for a U.S. Senate seat in Utah in 1974 but elected not to and never again sought national office. His accumulated political wisdom, though, would prove useful yet.

Mitt Romney was already, at twenty-four, married and the father of two young sons—their second boy, Matthew, was born in October 1971—as he threw himself into graduate work at Harvard. His social circle was generally made up of other men and women who, like himself, lived off campus with their families. Mitt and Ann had, with his parents' help, bought a house in Belmont, a leafy Boston suburb. His responsibilities at school and at home consumed most of his time. But he was still a presence on the graduate school social scene. He was an occasional visitor to the Lincoln's Inn Society, a Harvard Law School social club where students could eat, relax, and meet other students. He sometimes attended weekend parties and group dinners at Cambridge restaurants. The restrictions of his Mormon faith—church members are instructed to avoid alcohol, caffeine, cigarettes, and drugs—never interfered. "He didn't mind if we were drinking coffee or having a beer, but that wasn't what he did," Serkin said. "We respected him for being true to what he believed in, and I found him to be completely open and tolerant to everybody else." Romney also involved himself in the Harvard Law School Forum, a student group that brought prominent speakers to campus. One guest Romney recruited was his father. When George Romney arrived to speak, orange juice—prominently labeled as such—was added to the usual mix of soda, coffee, tea, and other caffeinated refreshments.

Romney's classmates were widely aware that he was Mormon but said he never proselytized. Mark E. Mazo, one of Romney's law school study group partners, recalled that Romney offered to discuss his faith with any classmates interested in learning more about it. "He mentioned it once and only once, and it never came up again," he said. On occasion, Mitt and Ann invited classmates to what's called family home evening, a Mormon tradition in which families set aside time each week to spend together. "You got the feeling you were dealing with a guy with a very strong moral fiber who is very devoted to church and family," Brownstein said. "You're not going to hear from Mitt a joke at anyone's expense, and you're not going to hear any swearwords. You know when you meet him and when you're with him that you're dealing with a very serious-minded guy."

When Romney left Harvard in 1975, he had graduated with honors from the law school and was a Baker Scholar at the business school, a distinction reserved for the top 5 percent of the class. But he had long been a hot commodity to prospective employers, even before he entered the job market. Consulting firms and investment banks were always on the hunt for future employees among Harvard's best and brightest, and the select group enrolled in the university's competitive dual-degree program seemed an obvious place to start. They were the elite of the elite.

Not long after he arrived on the Cambridge campus, Romney appeared on the radar of the Boston Consulting Group, then one of the hottest companies in the emerging field of business consulting. Charles Faris was assigned the task of wooing Romney to BCG. Over the course of Romney's four years at Harvard, Faris kept in frequent contact with him, treating him to occasional lunches and dinners and inviting him to company events. As Romney neared graduation, Faris found plenty of competition when he tried to hire him. "He was an outstanding recruit with exceptional grades, and he was the very charming, smooth, attractive son of a former presidential candidate," Faris said. "So everybody was bending over backward to get their hands on him." Faris's flattery and persistence paid off. Shortly after Romney left Harvard, he began working at BCG, a fitting first job for a freshly minted Ivy League graduate.

Romney was hedging his bets, though, not wholly confident that he would make it in the business world. He passed the Michigan bar exam in July 1975 and was admitted to practice law there the next year. He figured it would provide him a landing place if he didn't cut it in business. Romney recalled thinking, "That's where my friends are, and the industry that I know. I love cars." But the safety valve wouldn't be necessary. Companies nationwide were clamoring to hire BCG's consultants, who analyzed mountains of financial data with an eye to lowering costs, improving production, and gaining market share. Romney rapidly established a reputation as a rising star.

BCG consultants marketed themselves as objective outsiders, an excellent fit for Romney's rational turn of mind. "At BCG, analysis was king, clients were paying a lot of money, and you were expected to come in with really significant insights," said Lonnie M. Smith, who attended Harvard Business School with Romney and later worked with him at the firm. For Romney, whose young family was expanding quickly, that meant often working nights and weekends and traveling frequently. Faris, who became Romney's mentor at the firm, spent two summers flying regularly with him to Europe, where they worked for a U.S. client with operations overseas. "He worked his butt off," Faris said.

Romney was part of what several of his colleagues affectionately called the "Mormon mafia," a coterie of smart, talented, hardworking Mormon men at the firm who eventually rose to leadership positions. "For me and everybody there, including Mitt, it was a very formative time and probably more powerful than business school or law school," Smith said. But as the 1970s wore on, one rival firm began to eclipse BCG. It was called Bain & Company, the namesake of a former BCG executive, Bill Bain. Four years after starting his own firm, Bain had positioned Bain & Company as one of the nation's premier consulting outfits. And Mitt Romney wanted in.

FIVE

FAMILY MAN, CHURCH MAN

We've tried to civilize the boys.
Unfortunately, it's been very difficult with Mitt.

—ANN ROMNEY, JOKING ABOUT HER HUSBAND'S RAMBUNCTIOUS SIDE

It was shaping up to be a hard Christmas for Mark and Sheryl Nixon. They had recently moved their family to the Boston area for his job and didn't know many people. And then, on the night of April 4, 1995, they got the kind of phone call every parent dreads. Four of their six children, including two sons in high school, Rob and Reed, had been driving back from a youth gathering at the Mormon meetinghouse in Marlborough, a city about forty-five minutes west of Boston. Shortly after leaving the parking lot, Reed lost control of the red Oldsmobile minivan. The car sideswiped a utility pole, struck two trees and a sign for a condominium complex, and flipped over. Six others in the minivan escaped with bumps and bruises, but Rob and Reed, both in the front seat, were pinned upside down, their necks shattered. In a flash, the two Nixon boys, standouts on the high school cross-country team, became quadriplegics. "I could see my legs, and they were kind of crooked, off to the side," Reed would say later. "And I couldn't feel them."

After a number of major surgeries, hundreds of thousands of dollars' worth of treatment, and six months in rehab, Rob and Reed returned home in October 1995. Rob's injuries had been less severe, allowing him eventually to move his arms and breathe on his own. Reed, though, was completely paralyzed and put on a ventilator. The family suddenly needed a major addition on their house. They needed a special van to transport their sons. Their financial and emotional burdens were vast. Shortly before the holidays that year, Mark Nixon, a professor of accounting at Bentley University outside Boston, got a call at his office. It was Mitt Romney. He said he wanted to help. Would they be home on Christmas Eve?

That morning, a Sunday, the Nixons opened their door to find not just Mitt but Ann Romney and their sons. They held large boxes. Inside were a massive stereo system for Rob—"beyond anything he would ever hope to have," Mark said—and a VCR for Reed. They'd also brought Reed a check, not knowing what else to get him. The Romneys stayed a while. Their sons helped set up Rob's new stereo. "What a Christmas surprise for the boys," Sheryl wrote in her journal at the time.

The Nixons were floored. They shared a faith with Romney but didn't really know him—they weren't strangers, but neither were they friends. At that point, Romney held no formal leadership position in the Mormon church. He bore no direct ecclesiastical obligation to help. Many people within and outside the church assisted the Nixons during this difficult chapter in their lives, but the Romneys' generosity still stands out. What impressed the Nixons more than anything was that Mitt and Ann, despite their own packed holiday calendars, made a point of delivering the gifts themselves, spending time with the family, and, by bringing their children with them, leading by example. "I knew his schedule. I knew how busy he was. And their whole family came," Mark said. "He was actually teaching his boys, saying, 'This is what we do. We do this as a family.'" Sheryl added, "We've never forgotten it. It stood out so much in our minds and helped us to want to be better parents, too."

That wasn't all. Romney had also told Mark not to worry about Rob's or Reed's college education; he would pay for it. The Nixons, in the end, didn't need the help. But Romney continued to quietly lend his hand. He participated in a 5K road race and fund-raiser for Rob and Reed at Bentley the next spring. He contributed substantial financial gifts toward golf tournament fund-raisers in subsequent years. Then, in 2007, when Reed graduated from Bentley with a degree in finance after ten years in school, the Romneys sent him a Bentley desk clock engraved with a special message of congratulations. "It wasn't," Mark said, "a onetime thing."

The Romneys' Mormon faith, as they began building a life together, formed a deep foundation. It lay under nearly everything—not just their acts of charity but their marriage, their parenting, their social lives, even their weekly schedules. The Romneys' family-centric lifestyle was a choice; Mitt and Ann plainly cherished time at home with the boys more than anything. But it was also a duty. Belonging to the Mormon church meant accepting a code of conduct that placed supreme value on strong families—strong heterosexual families, in which men and women often filled defined and traditional roles. The Romneys have long cited a well-known Mormon credo popularized by the late church leader David O. McKay: "No other success can compensate for failure in the home." That was how Mitt had grown up in Michigan. But for Ann, who had been reared in a family in which organized religion was viewed with skepticism, raising a devout Mormon brood would be a new experience, one she would learn and master along the way.

When the Romneys arrived in the Boston area in 1971, they established a home in Belmont, a well-to-do suburb that was fast becoming a magnet for Mormon families. Over the next decade, they would have three more boys in addition to Tagg and Matt. Joshua was born in 1975, Benjamin in 1978, and then Craig in 1981, when Mitt was thirty-four years old and had begun making his mark at Bain & Company.

Like many Mormons, the Romneys established a routine for their new family. Sundays were for church, reflection, volunteer work, family dinners, and, in the fall, watching the New England Patriots on

TV. Monday nights were for the Mormon ritual of family home evening, in which the Romneys would gather for Gospel lessons, stories, and activities. Ann once said that Mitt would sometimes tell stories about animals, and the children would act them out. "For us, family night was less about lessons and more about having fun together," Mitt Romney said. Tuesday evenings brought church families together for basketball games and cookouts. Friday nights were reserved for date nights for Mitt and Ann, often consisting of dinner and a movie, and Saturdays the family performed chores at home. Before high school every day, the boys joined other children at a neighbor's house for "seminary," where they discussed scripture for forty-five minutes.

The parental roles were clear: Mitt would have the career, and Ann would run the house. In an era when many women had professional aspirations, homemaking became Ann's calling. She had left Brigham Young University before graduating to go east with Mitt, later finishing her bachelor's degree with a concentration in French. She would become active in charities such as the United Way, work with inner-city youth, compete in equestrian events, and take on various responsibilities at church. But the home was her workplace, and she was the chief executive. (Her husband's preferred term for her was CFO, or Chief Family Officer.) "So far as the family, she has a leadership point of view, and she's not afraid to express it," said Douglas Anderson, a longtime friend of the Romneys.

With five boys, the domestic tasks piled up like laundry—the meals, the cleaning, the heaps of whites and colors. There were countless school, sports, and church activities, including the Eagle Scout badges that she helped their three youngest sons achieve. And with epic battles raging in kickball, basketball, and football, there were numerous cuts, breaks, and bruises to mend. Mitt once said that motherhood was its own profession. "It's one which is challenging, it's demanding," he said. "It requires being a psychologist, a psychoanalyst, an engineer, a teacher." Once Ann forgot to close the sunroof on a BMW coupe that was one of Mitt's favorite cars. It poured, and the inside was soaked. But Mitt didn't blow up. "I know who does the cooking here," John Wright, a close friend, neighbor, and fellow church member, recalled

him saying. "I know who prepares my meals." Wright said the response captured Mitt's genuine appreciation of Ann's importance in
the life of their family. In other words, it wasn't a patronizing line;
to the Romneys, as to many Mormon families, maintaining a strong,
functional home was always the first priority.

Besides, Ann's cooking, a skill she had absorbed from her mother
and grandmother, was legendary. She loved to provide cooking dem-
onstrations and once even ran a small cooking school. Within the family, everyone had favorite dishes. One of the most popular was Ann's
"monkey bread," a treat during the Thanksgiving holiday. Mitt, meanwhile, had his own ideas about what he would—and wouldn't—do
as a father, evidently counting on Ann's maternal generosity. "I was
willing to change the urine-soaked diapers, but the messier types gave
me dry heaves," he told *GQ* magazine in 2007. "So my wife allowed
me to escape that."

If Ann Romney had to learn how to run a Mormon household, one
thing she already knew was how to contend with boys. She had grown
up with only brothers. Still, presiding over five sons as they got older
and more physical was a challenge. Tagg once told an interviewer that
his mother, on account of his many childhood scrapes, joked that the
hospital was going to name a wing after him. "I've broken almost every bone in my body," he said. "I've had my head stitched up five or six
times. I've broken my shoulder, my elbow, my ankle, my femur, most
of my toes, most of my fingers."

One winter day, Wright's son David was helping clear snow at the
Romneys' home. Tagg accidentally gashed David with a shovel above
his right eye, which required four stitches. Wright's wife, Laraine, was
a nurse, and predisposed to concern. But Ann, having witnessed mis-
haps like that many times before, was sanguine, just as she had been
a few months earlier when Tagg had broken David's nose on the basketball court with an errant elbow. "Ann just laughed," Wright said of
the shoveling accident, "because this was something that happened all
the time to her boys." Indeed, Ann was not an overprotective mother

who worried over every little thing, friends said. That would have been impossible anyway, with her sons, like their father, always out experimenting, building things, boating, and skiing. "They were not kids she could hold back if she wanted to," Wright said.

In time, each of the boys would develop his own niche within the family. As Tagg would later describe his brothers, Matt, the second oldest, was "the jokester, always pushing people's buttons." Josh was "the typical middle child, wanting lots of attention and getting a lot of it." Ben, the fourth child, remained "very reserved and quiet, a little aloof from the situation," while Craig relished his role as "the ultimate baby, everyone's favorite brother." Tagg said he fit the mold as the oldest: "Type A and too tightly wound."

Though distinct personalities, the five Romney boys, by many accounts, came to represent the wholesome Mormon ideal: they were disciplined, well mannered, clean cut, giving, and the embodiment of G-rated fun. "They were very impressive young men," said Philip Barlow, who worked closely with Romney in Romney's early years as a local church leader. Wright said the Romneys, unlike some other church families he knew, were not overly strict or prone to threatening grave consequences for disobedience. They set high standards and sought to demonstrate the long-term payoffs of adhering to Mormonism's moral compact. "Mitt tried to teach his boys to be leaders and develop a sense of self-confidence," he noted. "They've grown in their faith by their father's example." Indeed, all five sons would, in time, follow their father's path and serve on missions, leaving as boys and returning as caring, compassionate men, their mother would later say.

Mitt Romney said he came into his own as a parent as the boys got old enough to tease, roughhouse with, and play pranks on one another. Unlike diaper duty, all that was very much in his wheelhouse. "Growing up in that household was so much fun, because of the jokes, the laughter, bathroom humor, the physical, you know, fisticuffs, wrestling, games," Romney recalled. "It was just an enormously great experience." Tagg said his father, when he was home, was always on the floor with everyone else. "He was right there in the mix with us," he recalled. In a campaign ad Romney would air years later, Ann would describe her husband as just another teenager: "We've tried to civilize the boys. Unfortunately, it's been very difficult with Mitt."

His antics were not confined to their Belmont home or even to their family. For fifteen years, the Romneys owned a modest weekend house on the waterfront on Cape Cod, where Mitt would cook his kids pancakes and they would sometimes entertain friends. Grant Bennett, a friend from church, remembered being invited down with his family one weekend to go waterskiing. Bennett was up on the skis and Mitt was behind the wheel of the boat, swinging him around in a series of figure eights. The pattern intensified until Bennett grew tired of getting whipped around the water. He flashed a thumbs-down, the sign for "slow down." "He turns around and smiles at me and speeds up and starts to make the figure eights tighter and tighter," Bennett recalled. Eventually, Bennett dropped the rope out of sheer exhaustion. "I went up and said, 'Mitt, you only have one speed. It's full speed or nothing.'"

Mitt had learned to water-ski on the Great Lakes and had loved the water and loved his boat ever since. The family took frequent day or weekend trips to lakes, to the sea, and up to the White Mountains of New Hampshire, sometimes with other families. Romney helped teach Wright's children to water-ski on those trips. Even in the mountains, the water was a draw. On one excursion with their families up to New Hampshire's Loon Mountain, Wright said, they became obsessed with trying to dam up a stream to create their own waterfall. That, Wright said, was the kind of hands-on project Romney loved.

On one trip to waters closer to home, Romney got himself into trouble. In June 1981, Romney and Wright went to Lake Cochituate, about a half hour west of Boston, intending to do some boating with their families. A park ranger told him he couldn't put his boat into the lake because the license number was too difficult to read, Romney would later say. Romney then asked what the fine was, and the ranger told him: fifty bucks. To Romney, it was a no-brainer—he'd easily pay that in exchange for a day of fun. But when he began to lower his boat into the water, the ranger became incensed. "The ranger took it as a personal attack," Wright said. The ranger pulled out a pair of handcuffs and took Romney, dripping wet in his bathing suit, into custody for disorderly conduct. The case was soon dismissed after Romney and his lawyer pushed back hard. But that day, the lake outing was over before it had begun.

For all the hijinks in the Romney household, there were serious moments, too, and at times moments of friction. Mormonism may have muted the boys' teenage rebellion, but it wasn't an antidote entirely. Still, friends and family describe their home life as remarkably harmonious.

Tagg said some of his best childhood memories are of the nights the boys would gather with their parents in the dark and just talk, often on a couch at the foot of Mitt and Ann's bed. The tradition grew out of the boys' habit of wandering into their parents' room in the middle of the night. Over time, the discussion drifted to the evening hours before bedtime, with the darkened room giving everyone license to talk freely. "It was just a time to totally be yourself and completely open up," Tagg said. Ann and Mitt would offer their advice on whatever someone brought up, and so would the brothers. The tradition would continue as the boys aged, with the points of debate shifting from school to where they would go to college and then, later, to raising children of their own and their careers. Matt, Josh, and Craig would later go into real estate development or management, Tagg would work in private equity, and Ben would become a radiologist.

As Mitt had, Tagg spent his early years idolizing his father. But then he spent part of his adolescence wanting nothing to do with him. His rebellion began when he was young, around age eleven, Tagg said, as his father, in his eyes, went from being "superman" to "supernerd." "Overnight," Tagg said, "everything about him bugged me." The way he wore his jeans so short. The way his hair was never mussed. The way he insisted the boys wake up early on Saturdays for chores. Even the way he said good morning. "It bothered me that he would be so nice about it," Tagg said.

Family members say Mitt had a tough time dealing with rejection by his oldest son, the only Romney boy to experience this degree of teenage angst. After all, Mitt's relationship with his own father had not suffered such strain. The tensions lifted after a few years, as Mitt learned to give Tagg more space and Tagg began to regret how he'd been behaving. By the time he was fifteen, the arguments subsided, and Tagg came around to a new appreciation of his father. One night not long after, Tagg was struggling with tremendous peer pressure from friends at school, who were starting to do things he didn't want to do. Mitt came to his room to ask what was wrong, but Tagg didn't feel like talking. So Mitt sat there for about two hours, chatting about the Boston Red Sox and waiting for his son to open up. "Finally," Tagg said, "he asked enough questions and stayed long enough that I felt comfortable in saying what I was feeling." As for his rebellion, Tagg said, "I matured and understood that he had faults, like anybody. But I did recognize that he was special."

To Mitt, the special one in the house was Ann, with her wide smile, piercing eyes, and steadying domestic presence. And woe to the boy who forgot it. Tagg said there was one rule that was simply not breakable: "We were not allowed to say anything negative about my mother, talk back to her, do anything that would not be respectful of her." On Mother's Day, their home would be fragrant with lilacs, Ann's favorite flowers. Tagg didn't get it back then, but he came to understand. From the beginning, Mitt had put Ann on a pedestal and kept her there. "When they were dating," Tagg said, "he felt like she was way better than him and he was really lucky to have this catch. He really genuinely still feels that way." What makes his parents' relationship work, he said, is their distinct characters: Mitt is driven first by reason, while Ann operates more on emotion. "She helps him see there's stuff beyond the logic, he helps her see that there's more than just instinct and feeling," Tagg said. "We're all a little bit in between the two."

Mitt and Ann's relationship would grow and change as their family entered the public eye. But she has remained his chief counselor and confidante, the one person who can lead Mitt to a final decision.

Though she did not necessarily offer input on every business deal, friends said, she weighed in on just about everything else. "Mitt's not going to do something that they don't feel good about together," said Mitt's sister Jane. Tagg said they called their mom "the great Mitt stabilizer." Ann would later be mocked for her claim that she and Mitt had never had an argument during their marriage, which sounded preposterous to the ears of many married mortals. Tagg said it's not that his parents never disagree. "I know there are things that she says that he doesn't agree with sometimes, and I see him kind of bite his tongue. But I know that they go and discuss it in private. He doesn't ever contradict my mother in public." Friends of the Romneys back up that account, saying they cannot recall Mitt ever raising his voice toward Ann. In that way, the relationship between Mitt and Ann differed from that of Mitt's parents. Despite their lifetime of devotion, George and Lenore had no problem airing their disagreements, especially in later years, according to Tagg. "Listen, they fought like cats and dogs," he said.

Nowhere was Ann's special status more evident than on long family car trips. Mitt imposed strict rules: they would stop only for gas, and that was the only chance to get food or use the restroom. With one exception, Tagg explained. "As soon as my mom says, 'I think I need to go to the bathroom,' he pulls over instantly, and doesn't complain. 'Anything for you, Ann.' " On one infamous road trip, though, it wasn't Ann who forced Mitt off the highway.

The destination of this journey, in the summer of 1983, was his parents' cottage on the Canadian shores of Lake Huron. Mitt would be returning to the place of his most cherished childhood memories. The white Chevy station wagon with the wood paneling was overstuffed with suitcases, supplies, and sons when Mitt climbed behind the wheel to begin the twelve-hour family trek from Boston to Ontario. As with most ventures in his life, he had left little to chance, mapping out the route and planning each stop. Before beginning the drive, Mitt put Seamus, the family's hulking Irish setter, in a dog carrier and attached it to the station wagon's roof rack. He had improvised a windshield for the carrier to make the ride more comfortable for the dog.

Then Romney put his sons on notice: there would be predetermined stops for gas, and that was it. The ride was largely what one would expect with five brothers, ages thirteen and under, packed into a wagon they called the "white whale." Tagg was commandeering the way-back of the wagon, keeping his eyes fixed out the rear window, when he glimpsed the first sign of trouble. "Dad!" he yelled. "Gross!" A brown liquid was dripping down the back window, payback from an Irish setter who'd been riding on the roof in the wind for hours. As the rest of the boys joined in the howls of disgust, Mitt coolly pulled off the highway and into a service station. There he borrowed a hose, washed down Seamus and the car, then hopped back onto the highway with the dog still on the roof. It was a preview of a trait he would grow famous for in business: emotion-free crisis management. But the story would trail him years later on the national political stage, where the name Seamus would become shorthand for Romney's coldly clinical approach to problem solving.

———————

If Romney is exceedingly comfortable around family and close friends, he's much less so around those he doesn't know well, drawing a boundary that's difficult to traverse. It's a strict social order—us and them—that has put coworkers, political aides, casual acquaintances, and others in his professional circles, even people who have worked with or known him for years, outside the bubble. As a result, he has numerous admirers but, by several accounts, not a long list of close pals. "He's very engaging and charming in a small group of friends he's comfortable with," said one former aide. "When he's with people he doesn't know, he gets more formal. And if it's a political thing where he doesn't know anybody, he has a mask." For those outside the inner circle, Romney comes across as all business. Colleagues at work or political staffers are there to do a job, not to bond. He has little patience for idle chatter or small talk, little interest in mingling at cocktail parties, at social functions, or even in a crowded hallway. He is not fed by, and does not crave, casual social interaction, often displaying little desire to know who people are and what makes them tick. "He wasn't overly interested in people's personal details or their kids or spouses or team building or their career path," said another former aide. "It was all very friendly but not very deep." Or, as one fellow Republican put it, "He has that invisible wall between 'me' and 'you.'"

This sense of detachment is a function partly of his faith, which has its own tight social community that most outsiders don't see. Indeed, the stories of Romney's humanity and warmth come mostly from people who know him as a fellow Mormon. His abstention from drinking also makes parties and other alcohol-fueled functions distinctly less appealing. He is the antithesis of the gregarious pol with a highball in one hand and cigar in his mouth, offering a colorful dose of political lore under a dim bar light. When he does have to show his face after dark at nonchurch social events, his visits can last a half hour or less. For the Romneys, who became known as generous dinner hosts, especially for newcomers to the area, their home has long been a preferred social setting.

Romney's discomfort around strangers would later become more than just a curiosity; it would be an impediment on the campaign trail. Lacking an easy rapport with voters, he would come across as aloof, even off-putting. "A lot of it is, he is patrician. He just is. He has lived a charmed life," said one former aide. "It is a big challenge that he has, connecting to folks who haven't swum in the same rarefied waters that he has." His growing wealth, the deeper he got into his career, only widened the disconnect. At the time of the now-legendary road trip with Seamus, Romney was already on his way to a new, and spectacularly lucrative, phase in his career. He was about to take over a new enterprise in private equity called Bain Capital. The idea was to buy or take a controlling stake in companies, retool them with Bain's analytical expertise, and then sell them at a profit. The venture would, in the latter part of the 1980s and into the 1990s, bring in millions of dollars for Romney and his partners. As a result, his family, already financially comfortable, was entering an entirely new social class: they were rich.

After seven years on the East Coast, they had moved from a modest three-bedroom home in Belmont to a handsome natural-shingle house with white trim on a big corner lot near the private Belmont

Hill School, which all five boys would attend. Then, in 1989, Mitt and Ann allowed their first bout of conspicuous spending, plunking down $1.25 million for a stately Colonial on nearly two and a half acres up the road, enlarging and renovating it, and installing a pool and tennis court. At the time, Tagg was in France on his Mormon mission. After his parents sent him a photo, he asked his father, "How can you afford that house?" The family's modest getaway on Cape Cod gave way, in 1997, to a stunning waterfront retreat on the shores of Lake Winnipesaukee in New Hampshire. They would also purchase a sprawling ski retreat in the mountains of Park City, Utah, and an oceanfront home on the Pacific coast north of downtown San Diego, California. Mitt was wealthy to a degree well beyond what his parents had ever been.

Romney acknowledged in 1994 that his good fortune afforded them a lifestyle few could reach, one that allowed Ann to stay home without a second thought. "I tell my kids, 'We won the lottery. Don't think this is normal. Don't think that your life will have the kind of plenty that ours has had,'" he said. Still, the Romney boys had no idea how much their dad was truly worth. Perhaps that's because Mitt eschewed many of the trappings of wealth. The family had no cook or full-time maid. His sons urged him to buy a luxury car, but he refused, continuing to drive a dented Chevy Caprice Classic nicknamed the Gray Grunt. And he was frugal to the core, wearing winter gloves patched with duct tape and cracking down on anyone in the house who left the water running or the lights on.

They did join the Belmont Hill Club, a small private tennis and swimming club near their home. Ann got to be a good tennis player; Mitt less so. "I sometimes thought God put him on Earth so I could beat him six-love," said Joseph J. O'Donnell, a longtime friend and neighbor from Belmont. Mitt compensated on the tennis court with skills he did possess: strategic thinking and gamesmanship. "His strategy is simply to hit the ball back one more time than you do," Wright said. "He would encourage me—'John, hit it harder, hit it harder'—trying to get me to hit it out."

Even as he began shouldering more responsibility at Bain, Romney would assume several leadership positions in the Mormon church. But he could handle it. "Mitt," said Kem Gardner, a fellow church official from this period, "just had the capacity to keep all the balls up in the air." Or, as Tagg put it, "Compared to my dad, everyone's lazy." Helen Claire Sievers, who served in a church leadership position under Romney, got a glimpse of his work habits during weekend bus trips to the Mormon temple near Washington, D.C. Church groups would leave late on a Friday, drive all night, and arrive early on Saturday morning. Then they'd spend all day Saturday in temple sessions before turning around and driving home, to be back by Sunday morning. It was a grueling itinerary, Sievers said, so everyone used the time on the bus to sleep or read quietly. Everyone but Romney. "Mitt was always working. His light was on," she said. "He was taking advantage of every moment." Similarly telling was the time the Romneys were renovating their Belmont house before moving in. Romney had asked Grant Bennett to meet him at the new home to review church business. Bennett thought it was strange, knowing the family wasn't in the house yet. When Bennett showed up, Romney pried open a large storage container on the property. Inside, in front of a heap of their belongings, Romney had created a makeshift office with his desk, a chair, and his papers. "He opened it and sat down and worked," Bennett recalled. If Romney's drive was legendary, so was his conspicuous caution. Years before he launched his first political campaign, Romney sensed that he would, like his father, enter public life. John Wright remembered an instance early on in Mitt's professional career when he and Mitt had been approached by a businessman who wanted them to invest in some real estate deals. There was nothing illegal about them, Wright said, but they seemed unsavory enough that Mitt balked. "He said, 'If I ever ran for office, that's not something that I would want people to know about.'"

For all he took on, Romney did set some limits. His growing responsibilities at Bain and at church were demanding more and more of him, keeping him away from home. There were stretches where he would miss the Monday-night family sessions for weeks at a time. But one thing Romney would not do was take work home. "Every night when I was home, I put my work at the door," he recalled. "When I was home, I was home. For me, life is about my wife and my kids, and everything else I was doing was earning enough money to support them and was a necessary part of living." The boys made as much as possible of the time their father was around. "To us, he was just Dad," Tagg said. "He wasn't a business guy. He wasn't a politician. He was just Dad."

From the moment they first settled out east, the Romneys wove themselves into the local Mormon tapestry, which had been expanding as church members—doctors, university professors, scientists, and entrepreneurs—came to the Boston area for school and work. Romney's religious pedigree perhaps made the family's integration smoother, but other area church members had their own esteemed Mormon lineages. Mormon congregations, typically groups of four hundred to five hundred people, are known as wards, and their boundaries are determined by geography. That is, unlike Protestants or Catholics, Mormons do not choose the congregations to which they belong. It depends entirely on where they live. Wards, along with smaller congregations known as branches, are organized into stakes. Thus a stake, akin to a Catholic diocese, is a collection of wards and branches in a city or region. Because of the Mormon church's rapid growth, wards and stakes in the Boston area have often changed and split in recent decades to account for all the new members.

In another departure from many other faiths, Mormons do not have paid full-time clergy. Members of stakes and wards in good standing take turns serving in leadership roles. They are expected to perform their ecclesiastical duties on top of career and family responsibilities. But despite the all-volunteer nature of the Mormon priesthood, its lay leaders are very much part of the church's rigid hierarchy. Those called to serve as stake presidents and bishops or leaders of local wards are fully empowered as agents of the church, and they carry great authority over their domains. Their selection is carefully vetted by church headquarters in Salt Lake City. "It really is quite a tremendous amount of trust that's placed in the leadership," said Tony Kimball, who worked closely with Romney as a local church leader.

Mitt Romney first took on a major church role around 1977, when he was called to be a counselor to Gordon Williams, then the president of the Boston stake. Romney was essentially an adviser and deputy to Williams, helping oversee area congregations. His appointment was somewhat unusual in that counselors at that level have typically been bishops of their local wards first. But Romney, who was only about thirty years old and just at the dawn of his Bain career, was deemed to possess leadership qualities beyond his years. "He was obviously younger than most people who had had that calling," Grant Bennett said. Romney's responsibilities only grew from there; he would go on to serve as bishop and then as stake president, overseeing about a dozen congregations with close to four thousand members all together. Those positions in the church amounted to his biggest leadership test yet, exposing him to personal and institutional crises, human tragedies, immigrant cultures, social forces, and organizational challenges that he had never before encountered.

His leadership in the church coincided with a period of profound change and growth within Mormonism, which was contending—at times uneasily—with shifting social currents in America. Abortion had been legalized. Feminists were pushing for gender equality. The gay rights movement was gaining steam. And African Americans were still facing barriers to equality, despite the civil rights gains of prior decades. In 1978, after what it called a revelation from God, the church reversed decades of discrimination and allowed black men to hold the priesthood. Up until then, church practice reflected the text and interpretations of Mormon scripture, which deemed dark skin to be a curse upon those descended from sinners. Romney later described the reversal as "one of the most emotional and happy days of my life." He said he had been driving near his home when he learned the news. "I heard it on the radio and I pulled over and literally wept," Romney once recalled. But though the church had liberalized its views on race, women were another story. In a high-profile crackdown fifteen years later, in 1993, the church disciplined six Mormon activists and scholars for challenging Mormonism's official teachings and history, including feminists who questioned the church's treatment of women.

The Church of Jesus Christ of Latter-day Saints is far more than a form of Sunday worship. It is a code of ethics that frowns on homosexuality, out-of-wedlock births, and abortion and forbids premarital sex. It offers a robust, effective social safety net, capable of incredible feats of charity, support, and service, particularly when its own members are in trouble. And it works hard to create community, a built-in network of friends who often share values and a worldview. For many Mormons, the all-encompassing nature of their faith, as an extension of their spiritual lives, is what makes belonging to the church so wonderful, so warm, even as its insularity can set members apart from society.

But a dichotomy exists within the Mormon church, which holds that one is either in or out; there is little or no tolerance for those who, like so-called cafeteria Catholics, pick and choose what doctrines to follow. And in Mormonism, if one is in, a lot is expected, including tithing 10 percent of one's income, participating regularly in church activities, meeting high moral expectations, and accepting Mormon doctrine—including many concepts, such as the belief that Jesus will rule from Missouri in his second coming, that run counter to those of other Christian faiths. That rigidity can be difficult to abide for those who love the faith but chafe at its strictures or question its teachings and cultural habits. For one, Mormonism is male-dominated—women can serve only in certain leadership roles and never as bishops or stake presidents. The church also makes a number of firm value judgments, typically prohibiting single or divorced men from leading wards and stakes, for example, and not looking kindly upon single parenthood.

The portrait of Romney that emerges from those he led and served with in the church is of a leader who was pulled between Mormonism's conservative core views and practices and the demands from some quarters within the Boston stake for a more elastic, more open-minded application of church doctrine. The Boston area had been known as a comparatively liberal redoubt of Mormon thought, where church members would sometimes openly question church tenets. Romney was forced to strike a balance between those local expectations and the dictates out of Salt Lake City. Some believe that Romney artfully reconciled the two, praising him as an innovative and generous leader who was willing to make accommodations, such as giving women expanded responsibility, and who was always there for church members in times of need. To others, he was the product of a hidebound, patriarchal Mormon culture, inflexible and insensitive in delicate situations and dismissive of those who didn't share his perspective. One thing is clear: Romney was heavily involved in all aspects of local church life.

In the early-morning darkness of August 1, 1984, flames shot some fifty or sixty feet into the sky. By daybreak, as word spread through the local Mormon community, members' hearts sank. Their gleaming new Belmont chapel, nearly complete, had been gutted by fire. And the blaze, they feared, was no coincidence. For a faith that had known its share of persecution, the suspicious nature of the fire was deeply unsettling. "I don't know when I've felt lower," Kent Bowen, a local church leader, said later.

Tensions had been growing among Belmont's Mormons and others in town. Some locals objected to the $1.6 million chapel going up on the church-owned wooded plot on Belmont Hill—opposition that church members felt had anti-Mormon overtones. Neighbors said their property values would decline. The local zoning board had initially refused to allow parking on the site, before a compromise was reached. The church urgently needed the new building. Several wards were now crammed into the Cambridge meetinghouse. Space had run out, and parking was a nightmare. The new Belmont chapel not only would host new wards in Belmont and neighboring Arlington, it would be a jewel befitting a flourishing church. Members had given additional money for the building fund, including widows on fixed incomes who insisted on helping. Some church members had worked as a group doing inventory at area department stores to raise funds. Others had started a small consulting firm and donated the proceeds to the cause.

It fell to Mitt Romney to heal his hurting congregation. A few years earlier, in 1981, Romney had been called to lead a Cambridge ward, which had led to his becoming bishop of the Belmont ward when it was created in 1984. As bishop, Romney was intimately involved in families' lives, counseling and guiding them through marital problems, illness, unemployment, and other struggles. He orchestrated church efforts to help the needy within the congregation. He led lessons on scripture and delivered sermons on Sundays. And he interviewed members to determine their fitness to enter the sacred Mormon temple outside Washington, D.C. But the fire presented a new challenge for Romney, who cut short his vacation on Cape Cod to return to the scene. His ward was barely a few months old, and already it faced a serious crisis. Not only did the congregation lack a gathering place, members felt demoralized. Romney said later that, in addition to the parking dispute with the town, there had been other hints that the Mormons weren't welcome. "Some people in Belmont thought of Latter-day Saints as bizarre, and we were not part of the church community," he recalled in an interview with a Mormon magazine.

Then something unexpected happened. As the church embers cooled, so did the tensions. The day of the fire, offers of help began pouring in from other churches in Belmont and from town officials. As Grant Bennett recalled, "Many of the religious communities in town approached Mitt and said, 'We want you to know that we welcome the Mormons to Belmont, and we think what's happened is terrible, and you are welcome to meet in our building.'" Romney, knowing the fire had set the chapel project back months, wanted to accept all the offers, but some of the other church buildings simply wouldn't work. So he and his fellow leaders accepted help from three of them—a local Catholic parish, an Armenian church, and a Congregational church. They also accepted the town's offer to use the town hall.

It was a complex situation, both logistically and emotionally, and church members said that Romney handled it deftly. "One of the things that impressed me was how fast he thought on his feet," recalled Connie Eddington, a church member from Belmont. Romney, mindful that they were worshipping in borrowed space, established ground rules for his congregation. He outlawed food, which church members had routinely used to keep their children quiet on Sundays. "It was hard, but we did it," Eddington said. "I had little ones, and they did not have Cheerios." And Romney organized a cleaning plan, assigning Mormon families to return to the churches early Monday morning to mop the floors and leave the facilities cleaner than when they'd found them. The experience of sharing buildings with other denominations resulted in the Mormons feeling more at home in Belmont, and Belmont becoming more accepting in return. "It turned out to be a huge blessing to all of us to worship in their churches that year," Eddington said. "Everyone's feelings softened toward one another." It also led to new friendships and new traditions, some of which persist almost thirty years later. One of the churches whose space they used, First Armenian Church, invited members of the Mormon congregation to sing and play instruments in its annual Christmas Eve concert. "There are still members of our congregation who go to that service," Bennett said. When the rebuilt Belmont chapel held an open house the year after the fire, nearly three thousand people came.

Within the family, Romney's zany side was well known, causing both laughter and eye rolls. He loved to goof around with his sons, loved making jokes, even if they sometimes fell flat. As a missionary, he had sometimes assumed the voices of cartoon characters in letters home. Now, as he took on larger roles within the church as an adult, his fellow Mormons got glimpses of this mirthful instinct.

On one Saturday morning at Romney's home in Belmont, Philip Barlow and another counselor were meeting with Romney, their bishop, to review the state of their congregation. For some reason that Barlow can't remember, Romney brought up the singer Michael Jackson. "I was a little surprised at his pop-icon consciousness," Barlow recalled. But he was even more surprised at what came next. "He just said, 'Oh, yeah!' and he stood up theatrically and started to ooze out a pretty credible rendition of 'Billie Jean' and moonwalked gracefully backwards," Barlow said. He couldn't believe what he had just seen.

Those who worked closely with Romney said he was serious about his faith but frequently made wisecracks or injected levity into their work. Ken Hutchins, who held leadership positions under Romney during Romney's tenure as stake president, said that church meetings could sometimes feel like a duty. But not when Romney was leading them. "He had an engaging personality, and that didn't stop at the door just because he was ministering over spiritual things," Hutchins said. "You'd go away, and you'd say, 'I gotta tell my wife about that.'" Romney's church colleagues said he was warm, accessible, and a good listener, if not terribly good at remembering names. "He was reasonable, accommodating, and imaginative," Barlow said. In fact, Barlow was so taken with Romney's analytic mind and executive ability that he wrote to his mother at the time and told her that his bishop could be president of the United States. Douglas Anderson, the family friend, is a Democrat who doesn't share Romney's politics, but he said, "His leadership has been obvious to the people who know him best and who've known him longest."

At times Romney was a willing delegator. "He let the people around him kind of fulfill their responsibilities and drew out of them what their strengths were," Hutchins said. "And that made them feel better about who they were." But there were many times when Romney just wanted to take care of things himself. This seemed to be his nature: headstrong and self-assured, at times stubbornly so. He put it to a church friend once that Romneys were built to swim upstream. In other words, leave it to him when things got sticky. When Tony Kimball became Romney's executive secretary in the Boston stake, one of the first things Romney told him was that he need not keep Romney's schedule. He would keep his own. Later, while Romney was still stake president, a group of nine or ten Laotian youths from the Lowell area, north of Boston, needed a ride to a church gathering in Cambridge. "Next thing I know, he's driving them in a van," said David Gillette, who led the church's Boston mission program from 1991 to 1994. "He could have asked a hundred guys to do that for him, but he did it himself."

This hands-on mentality extended far beyond the confines of his official church duties. One Saturday, Grant Bennett got up on a ladder

outside his two-story Belmont Colonial intent on dislodging a hornets' nest, which had formed between an air-conditioning unit and a
second-floor window. Things didn't go so well. The hornets went right
at him, and he fell off the ladder, breaking his foot. The next day,
Bennett was forced to skip a leadership meeting with Romney at the
church. Romney noticed his absence, learned what had happened, and
went over that afternoon to see if there was anything he could do. He
and Bennett chatted for a few minutes, and then Romney left. Around
nine thirty that Sunday night, Romney reappeared. Only this time, it
was dark out, Romney was in jeans and a polo shirt instead of his suit,
and he was carrying a bucket, a piece of hose, and a couple of screw-
drivers. "He said, 'I noticed you hadn't gotten rid of the hornets,'"
Bennett recalled. "I said, 'Mitt, you don't need to do that.' He said,
'I'm here, and I'm going to do it. . . . You demonstrated that doing it
on a ladder is not a good idea.'" Romney went at it from inside the
house, opening the window enough to dislodge it. Soon the hornets'
nest was gone.

Everyone who has known Romney in the church community seems
to have a story like this, about him and his family pitching in to
help in ways big and small. They took chicken and asparagus soup to
sick parishioners. They invited unsettled Mormon transplants to their
home for lasagna. Helen Claire Sievers and her husband once loaned
a friend from church a six-figure sum and weren't getting paid back,
putting a serious financial strain on the family. Suddenly they couldn't
pay their daughter's Harvard College tuition. Romney, who was stake
president at the time, not only worked closely with Sievers's family and
the loan recipient to try to resolve the problem, he offered to give Sievers and her husband money and tried to help her find a job. "He spent
an infinite amount of time with us, all the time we needed," Sievers
said. "It was way above and beyond what he had to do." Romney has
also upheld his obligation to tithe, which means he has personally
given millions of dollars to the Mormon church over the years.

On Super Bowl Sunday 1989, Douglas Anderson was at home in
Belmont with his four children when a fire broke out. The blaze spread
quickly, and all Anderson could think of was racing his family to

safety. "There was no thought in my mind other than 'Get my kids out,'" he said. "I was not thinking about saving anything." He doesn't remember exactly when Romney, who lived nearby, showed up. But he got there quickly. Immediately, Romney organized the gathered neighbors, and they began dashing into the house to rescue what they could: a desk, couches, books. "Whatever they could lift off the main floor," Anderson said. "They saved some important things for us, and Mitt was the general in charge of that." This went on until firefighters ordered them to stop. "Literally," Anderson said, "they were finally kicked out by the firemen as they were bringing hoses and stuff in." After the fire was finally out, Anderson, Romney, and other church members shared a spiritual moment on the front steps of the charred house. A few weeks earlier, Romney had given a talk on what he called his father's favorite Mormon scripture, which reads in part, "Search diligently, pray always, and be believing, and all things shall work together for your good." By coincidence, Anderson and his teenage daughter had been in their study discussing that very passage when the fire began. Outside on the steps, Anderson recalled, "we talked about how even in a case like this, if we tried to be true to the faith, it could turn out to be a positive thing." Over the many years since, Anderson said, his family has seen that come true.

Romney's acts of charity extended beyond just the church community. After his friend and neighbor Joseph O'Donnell lost a son, Joey, to cystic fibrosis—he died in 1986 at age twelve—Romney helped lead a community effort to build Joey's Park, a playground at the Winn Brook School in Belmont. "There he was, with a hammer in his belt, the Mitt nobody sees," O'Donnell said. Romney didn't stop there. About a year later, it became apparent that the park would need regular maintenance and repairs. "The next thing I know, my wife calls me up and says, 'You're not going to believe this, but Mitt Romney is down with a bunch of Boy Scouts and kids and they're working on the park,'" said O'Donnell, who coached some of Romney's sons in youth sports. "He did it for like the next five years, without ever calling to say, 'We're doing this,' without a reporter in tow, not looking for any credit."

Though the Romneys established themselves as the go-to family when people required help, there were times they were the ones who needed support. On one occasion, Romney, feeling soreness in one of his legs, believed he had pulled a tendon. A couple of days went by, and his leg began changing color. "By the time he got to the hospital, he had some kind of very serious, massive infection," Grant Bennett recalled. When the severity of it became clear, Romney asked if members of the church would come to the hospital and administer a priesthood blessing, which is given to the sick. In his own moment of vulnerability, Romney looked to the same source of strength that he so often drew on for others.

In the spring of 1993, Helen Claire Sievers performed a bit of shuttle diplomacy to resolve a thorny problem confronting church leaders in Boston: resentment among progressive Mormon women at their subservient status within the church. Sievers was active in an organization of liberal women called Exponent II, which published a periodical. The group had been chewing over the challenges of being a woman in the male-led faith. So Sievers went to Romney, who was stake president, with a proposal. "I said, 'Why don't you have a meeting and have an open forum and let women talk to you?'" she recalled. The idea was that although there were many church rules stake presidents and bishops could not change, they did have some leeway to do things their own way.

Romney wasn't sure about holding such a meeting, but he ultimately agreed to it. Sievers went back to the Exponent II group and said they should be realistic and not demand things Romney could never deliver, such as allowing women to hold the priesthood. On the day of the meeting, about 250 women filled the pews of the Belmont chapel. After an opening song, prayer, and some housekeeping items, the floor was open. Women began proposing changes that would include them more in the life of the church. In the end, the group came up with some seventy suggestions—from letting women speak after men in church to putting changing tables in men's bathrooms—as Romney and one of his counselors listened and took careful notes.

Romney was essentially willing to grant any request he couldn't see a reason to reject, Sievers said. "Pretty much, he said yes to everything that I would have said yes to, and I'm kind of a liberal Mormon," she said. "I was pretty impressed." Tony Kimball said that when they reviewed the list a year later, right before Romney left the stake presidency, he was amazed at how many of the women's suggestions had been implemented. Many were small, procedural matters, but they added up to a significant concession. One shift Romney allowed was to let women who led auxiliaries within the stake speak to congregations in a monthly address on behalf of the stake presidency. The role had historically been afforded only to men who served on the twelve-member High Council. Sievers was afterward assigned to address a Boston congregation. She felt it was about time women had a chance to stand up as the men always did; say to the congregation, "I bring you greetings from the stake presidency"; and deliver the sermon. Ann Romney was not considered to be sympathetic to the agitation of liberal women within the stake. She was invited to social events sponsored by Exponent II but did not attend. She was, in the words of one member, understood to be "not that kind of woman."

Mitt Romney showed flexibility, too, in choosing his leadership teams. While he was stake president, one of his counselors went through a divorce. The counselor asked to be released from his duties, knowing the church wanted only married men in the role. But Romney refused. His executive secretary at the time was Tony Kimball, who was single and therefore, in Kimball's own words, also an "iffy" choice for someone serving in a position of authority. "Mitt was kind of proud of the fact that he had a counselor who was now divorced and an executive secretary who was single," Kimball said. Romney had once said that he felt strongly about tapping single Mormons for leadership posts within the stake. "They feel needed and wanted," he said. "And that's part of our church experience." That church experience under Romney's leadership was not so rosy for everyone, though.

As both bishop and stake president, he at times clashed with women he felt strayed too far from church beliefs and practice. To them, he lacked the empathy and courage that they had known in other leaders, putting the church first even at times of great personal vulnerability.

Peggie Hayes had joined the church as a teenager along with her mother and siblings. They'd had a difficult life. Mormonism offered the serenity and stability her mother craved. "It was," Hayes said, "the answer to everything." Her family, though poorer than many of the well-off members, felt accepted within the faith. Everyone was so nice. The church provided emotional and, at times, financial support. As a teenager, Hayes babysat for Mitt and Ann Romney and other couples in the ward. Then Hayes's mother abruptly moved the family to Salt Lake City for Hayes's senior year of high school. Restless and unhappy, Hayes moved to Los Angeles once she turned eighteen. She got married, had a daughter, and then got divorced shortly after. But she remained part of the church.

By 1983, Hayes was twenty-three and back in the Boston area, raising a three-year-old daughter on her own and working as a nurse's aide. Then she got pregnant again. Single motherhood was no picnic, but Hayes said she had wanted a second child and wasn't upset at the news. "I kind of felt like I could do it," she said. "And I wanted to." By that point Mitt Romney, the man whose kids Hayes used to watch, was, as bishop of her ward, her church leader. But it didn't feel so formal at first. While she was pregnant she earned some money organizing the Romneys' basement. The Romneys also arranged for her to do odd jobs for other church members, who knew she needed the cash. "Mitt was really good to us. He did a lot for us," Hayes said. Then Romney called Hayes one winter day and said he wanted to come over and talk. He arrived at her apartment in Somerville, a dense, largely working-class city just north of Boston. They chitchatted for a few minutes. Then Romney said something about the church's adoption agency. Hayes initially thought she must have misunderstood. But Romney's intent became apparent: he was urging her to give up her soon-to-be-born son for adoption, saying that this was what the church wanted. Indeed, the church encourages adoption in cases where "a successful marriage is unlikely."

Hayes was deeply insulted. She told him she would never surrender her child. Sure, her life wasn't exactly the picture of Rockwellian harmony, but she felt she was on a path to stability. In that moment, she also felt intimidated. Here was Romney, who held great power as her church leader and was the head of a wealthy, prominent Belmont family, sitting in her gritty apartment making grave demands. "And then he says, 'Well, this is what the church wants you to do, and if you don't then you could be excommunicated for failing to follow the leadership of the church,'" Hayes recalled. It was a serious threat. At that point Hayes still valued her place within the Mormon church. "This is not playing around," she said. "This is not like 'You don't get to take Communion.' This is like 'You will not be saved. You will never see the face of God.'" Romney would later deny that he had threatened Hayes with excommunication, but Hayes said his message was crystal clear: "Give up your son, or give up your God."

Romney left her apartment. Not long after, Hayes gave birth to a son. She named him Dane. At nine months old, Dane needed serious, and risky, surgery. The bones in his head were fused together, restricting the growth of his brain, and would need to be separated. Hayes was scared. She sought emotional and spiritual support from the church once again. Looking past their uncomfortable conversation before Dane's birth, she called Romney and asked him to come to the hospital to confer a blessing on her baby. Hayes was expecting him. Instead, two people she didn't know showed up. She was crushed. "I needed him," she said. "It was very significant that he didn't come." Sitting there in the hospital, Hayes decided she was finished with the Mormon church. The decision was easy, yet she made it with a heavy heart. To this day, she remains grateful to Romney and others in the church for all they did for her family. But she shudders at what they were asking her to do in return, especially when she pulls out pictures of Dane, now a twenty-seven-year-old electrician in Salt Lake City. "There's my baby," she says.

In the fall of 1990, Exponent II published in its journal an unsigned essay by a married woman who, having already borne five children, had found herself some years earlier facing an unplanned sixth pregnancy. She couldn't bear the thought of another child and was contemplating abortion. But the Mormon church makes few exceptions in forbidding women to end a pregnancy. Church leaders have said that abortion can be justified in cases of rape or incest, when the health of the mother is seriously threatened, or when the fetus will surely not survive beyond birth. However, even those circumstances "do not automatically justify an abortion," according to church policy.

The woman feared excommunication. She had come to love the church and couldn't bear the thought of losing it, she said. It was her whole life. Then her doctors discovered she had a serious blood clot in her pelvis. She thought initially that this would be her way out—of course she would have to get an abortion. But the doctors, she said, ultimately told her that, with some risk to her life, she might be able to deliver a full-term baby, whose chance of survival they put at 50 percent. One day in the hospital, her bishop—later identified as Romney, though she did not name him in the piece—paid her a visit. He told her about his nephew who had Down syndrome and what a blessing it had turned out to be for their family. "As your bishop," she said he told her, "my concern is with the child." The woman wrote, "Here I—a baptized, endowed, dedicated worker, and tithe-payer in the church—lay helpless, hurt, and frightened, trying to maintain my psychological equilibrium, and his concern was for the eight-week possibility in my uterus—not for me!"

Romney would later contend that he couldn't recall the incident, saying, "I don't have any memory of what she is referring to, although I certainly can't say it could not have been me." Romney acknowledged having counseled Mormon women not to have abortions except in exceptional cases, in accordance with church rules. The woman told Romney, she wrote, that her stake president, a doctor, had already told her, "Of course, you should have this abortion and then recover from the blood clot and take care of the healthy children you already have."

Romney, she said, fired back, "I don't believe you. He wouldn't say that. I'm going to call him." And then he left. The woman said that she went on to have the abortion and never regretted it. "What I do feel bad about," she wrote, "is that at a time when I would have appreciated nurturing and support from spiritual leaders and friends, I got judgment, criticism, prejudicial advice, and rejection."

One woman who had been active in the Exponent II organization was Judy Dushku, a longtime scholar of global politics at Suffolk University in Boston. At one point while Romney was stake president, Dushku wanted to visit the temple outside Washington to take out endowments, a sacred rite that commits Mormons to a lifetime of faithfulness to the church. She had never entered a temple before and was thrilled at the chance to affirm her dedication to a faith she'd grown up with and grown to love. Earlier in her life, temples had been off-limits to Mormons who, like Dushku, were married to non-Mormons. Now that rule had changed, and she was eager to go. But first she needed permission from her bishop and stake president.

After what she described as a "lovely interview" with her bishop and after speaking with one of Romney's counselors, she went to see Romney. She wasn't sure what to expect. Despite Romney's willingness to allow some changes in 1993, he and Dushku had clashed over the church's treatment of women. "He says something like 'I suspect if you've gotten through both of the interviews, there's nothing I can do to keep you from going to the temple,'" Dushku recalled. "I said, 'Well, why would you want to keep me from going to the temple?'" Romney's answer, Dushku said, was biting. "He said, 'Well, Judy, I just don't understand why you stay in the church.'" She asked him whether he wanted her to really answer that question. "And he said, 'No, actually, I don't understand it, but I also don't care. I don't care why you do. But I can tell you one thing: You're not my kind of Mormon.'" With that, Dushku said, he dismissively signed her recommendation to visit the temple and let her go. Dushku was deeply hurt. Though she and Romney had had their differences, he was still her spiritual leader. She had hoped he would be excited at her yearning to visit the temple. "I'm coming to you as a member of the church, essentially expecting you to say, 'I'm happy for you,'" Dushku said. Instead, "I just felt kicked in the stomach."

The world map hung at the church's Boston branch neatly captured the growing diversity within the faith. Bright stars marked the countries from which branch members hailed: Haiti, Nigeria, Mexico, Lebanon, and many others. These immigrants were injecting fresh energy into Mormonism and remaking its face. They were also posing new tests for local church leaders. In his roughly eight years as Boston stake president, Romney oversaw a dizzying expansion. In the past, most area Mormons had been white and living in the suburbs. But the church had begun a major push to recruit new members from immigrant communities in urban enclaves in and around Boston. This had exposed Romney and other church leaders to poverty, lifestyles, and cultural traditions with which they'd had little experience. "It has been a great challenge to local members to provide leadership and support to begin six new branches in this stake in the past seven years," Romney said in 1991. "But it has also been a great source of joy to see so many people join the church and see them progress in the Gospel."

Romney and other Mormon officials made a conscious effort to take the church to the people, instead of letting the people come to the church. In the cities they created so-called storefront branches dedicated to certain nationalities and languages. Missionaries worked with Laotians, Cambodians, Portuguese- and Spanish-speakers, Chinese, Haitians, and others. Area wards and branches became a potpourri of world cultures. "Love those people," Romney told Keith Knighton in calling him to be president of the Boston branch in the late 1980s. "Just simply love them." David Gillette, who ran the Boston mission program in the early 1990s, worked closely with Romney and other stake leaders to bring new faces into the church. From 1991 to 1994, Gillette said, they baptized some 1,600 new members. "We taught in seven different languages." A whole new stake had to be created to account for all the newcomers. Though Gillette's missionaries were the ones baptizing Mormon converts, Romney played a major supporting role in establishing new congregations, figuring out where they would meet and how to pair them with more established local wards and branches. Because he had learned French on his own mission, he was also able to personally counsel many Haitians, both in Boston and at suburban meetinghouses. He would sprinkle in some French, too, when regaling missionaries and area congregations with stories from his two and a half years in France.

It was fitting that the Boston area, which had helped spread Mormonism in its infancy in the mid–nineteenth century, was again becoming a source of growth and change for the faith. "You feel humbled by the character of great men and women who walked here 150 years ago as the church was starting," Romney said during his stake presidency. "We see it starting over and over again as each brother and sister finds the truth and is baptized."

If Romney's Mormon mission in France had deepened his connection to the faith his family had helped establish, his leadership in the Boston church, which lasted until 1994, gave him the chance to apply it, to put its teachings into practice, and to bring its spiritual message into new quarters. In time, more secular leadership demands, in business, sport, and politics, would command ever greater shares of his attention. But Mitt Romney's faith and family would remain two unshakable pillars in his life. "To understand my faith, people should look at me and my home and how we live," he would say years later. "What my church teaches is evidenced by what I have become, and what my family has become."

SIX

THE MONEYMAKER

I never actually ran one of our investments.
That was left to management.

—MITT ROMNEY ON HIS WORK AT BAIN CAPITAL

By the time Mitt Romney walked into the Faneuil Hall offices of his mentor and boss, Bill Bain, in the spring of 1983, the thirty-six-year-old was already a business consulting star, coveted by clients for his analytical cool. He was, as people had said of him since childhood, mature beyond his years and organized to a fault. Everything he took on was thought through in advance, down to the smallest detail; he was rarely taken by surprise. This day, however, would be an exception. Bain, the founder of Bain & Company and a legendary figure in the consulting trade, had a stunning proposition: he was prepared to entrust an entirely new venture to the striking young man seated before him.

From the moment they'd first met, Bill Bain had seen something special, something he knew in Mitt Romney. Indeed, he had seen *someone* he knew when he interviewed Romney for a job in 1977: George Romney. "I remember him [George] as president of American Motors when he was fighting the gas guzzlers and making funny ads. . . . So when I saw Mitt, I instantly saw George Romney. He doesn't look exactly like his dad did, but he very strongly resembles his father."

Beyond appearances, Mitt had an air of great promise about him. He seemed brilliant but not cocky. All of the partners were impressed, and some were jealous. More than one partner told Bain, "This guy is going to be president of the United States someday."

Bain had hired Romney away from the Boston Consulting Group, after a former colleague of Bain's at BCG had told him, "I don't believe you have anybody better than he is." During his interview with Bain, Romney had posed a series of smart questions, including this one: with all of his success at BCG, Romney asked Bain, how had he found the nerve to leave? "It was a flattering question," Bain said. "Mitt had a presence so that you took him seriously, but he acted as if he took me very seriously."

They had come to BCG by very different paths: Romney, a child of privilege, had been heavily recruited as a graduate of Harvard's law and business schools; Bain had scraped his way up from humbler roots in Tennessee, and although he would retain a trace of his down-home accent, his bearing was patrician and intensely competitive. Before Bain had bolted to form his own firm in 1973, he had been seen as the heir apparent at BCG. He had taken with him several colleagues and some of his prime clients. He also carried with him an idea, the seed from which his namesake firm would grow.

The Bain Way, as it became known, was intensely analytical and data-driven, a quality it shared with some other firms' methods. But Bill Bain had come up with the idea of working for just one client per industry and devoting Bain & Company entirely to that company, with a strict vow of confidentiality. As a result, the companies shared more information, and consultants could plow to greater depth. Bain thought the value of his firm's advice could be maximized by sticking with clients, guiding restructuring plans from the page of a consultant's report to the executive suite and the factory floor. It meant actively seeking out relationships with CEOs and mastering the projects that mattered to them. Success would be measured by the improved performance of the business. It doesn't sound like a revolutionary notion, but in many ways it was.

From the start Romney was perfectly adapted to the Bain Way and became a devoted disciple. Patient analysis and attention to nuance were what drove him. It took a healthy ego to go into a business and tell an owner how to run his own firm better, and most clients lauded Romney's efforts. For six years, he delved into numerous unfamiliar companies, learned what made them work, scoped out the competition, and then presented his findings. He was successful enough, swiftly enough, that he could have taken his talents to any consulting firm or perhaps risen to the top of Bain & Company. An increasing number of clients preferred Romney over more senior partners. He was plainly a star, and Bain treated him as a kind of prince regent at the firm, a favored son. Just the man for the big move he now had in mind.

And so Bain made his pitch. As the two men talked in his office, just beyond the nearby marketplace of shops and restaurants teeming with tourists, Bain laid out a surprising vision for Romney's future— one that could make Romney and his partners a lot of money. Up to that point, Bain & Company could only watch its clients prosper from a distance, taking handsome fees but not directly sharing in profits. Bain's epiphany was that he would create a new enterprise that would invest in companies and share in their growth, rather than just advise them.

Starting almost immediately, Bain proposed, Romney would become the head of a new company to be called Bain Capital. With seed money from Bill Bain and other partners at the consulting firm, Bain Capital would raise tens of millions of dollars, invest in start-ups and troubled businesses, apply Bain's brand of management advice, and then resell the revitalized companies or sell their shares to the public at a profit. It was a risky leap, but Bain was convinced that it would work—and that Romney, who was as prudent as he was ambitious, would be the ideal leader. It sounded exciting, daring, new. It would be Romney's first chance to run his own firm and, potentially, to make a killing. It was an offer few young men in a hurry could refuse.

Yet Romney stunned his boss by doing just that. He saw the opportunity, of course, but he also saw risks. First, he felt comfortable in his life. He already had a great job and had five young sons at home. Second, he and the partners in the new firm would be expected to contribute significantly to the investment fund, and thus, if deals went south, they could lose their own money. Romney explained to Bain that he didn't want to risk his position, earnings, and reputation on an experiment. He found the offer appealing but didn't want to make the decision in a "light or flippant manner." So Bain sweetened the pot. He guaranteed that if the experiment failed, Romney would get his old job and salary back, plus any raises he would have earned during his absence. Still, Romney worried about the impact on his reputation if he proved unable to do the job. Again the pot was sweetened. Bain promised that, if necessary, he would craft a cover story saying that Romney's return to Bain & Company was needed because of his value as a consultant. "So," Bain explained, "there was no professional or financial risk." This time Romney said yes.

Years later, when Romney launched his 2012 campaign, he revealed none of the calculation involved in his decision. His description of how he began the work that would earn him hundreds of millions of dollars was almost quaint. "I left a steady job to join with some friends to start a business," Romney said. "It had been a dream of mine to try and build a business from the ground up. We started in a small office." But this start was nothing like that of the typical small business. It was a nearly risk-free opportunity with substantial financial backing—and a lucrative fallback plan in case of failure. And he had the security of knowing that he would still report to Bain and have offices just down the hall from his mentor.

Just as Romney took the reins, the entire Bain operation moved from its funky surroundings at Faneuil Hall, where the younger partners had enjoyed the hustle and bustle of the marketplace and the buildings with exposed brick and pipes, to a newly opened office tower at Copley Place in Boston's Back Bay. Crammed into a sterile suite on the seventh floor, with metal desks and outdated chairs, the partners shared a small room and Romney had his own office. The consulting business of Bain & Company was at one end of the hall, and the spin-off, Bain Capital, was at the other. The setting was businesslike and spare, and that suited Romney. He struck some of his partners as being uncomfortable amid crowds, ill at ease when riding an elevator with everyday shoppers. He was at home in the conference room, free of distractions. An early riser, he was usually the first in the office and often forgot to eat lunch. He was ruthlessly efficient with his time, both to get the work done at the office and to ensure that he could make it to church activities and his kids' athletic events. If he took a briefcase home with him, he said, he left it in the car.

Romney, fully in charge, searched for partners who fit his comfort zone. He promised his new hires that they would all be key players, with every major decision hashed out openly in meetings. Romney once described his style as a two-step process. One step was to "wallow in the data," following the "Bain Way" of deep analysis. Then he encouraged vigorous debate: "Get people of different background and experience who disagree with each other and are willing to debate and argue." He was, in a sense, replicating the way he had once dissected cases with his fellow students at Harvard Business School. In business, the method often worked well. Later, in politics, Romney would find that the model didn't fit so neatly.

Most of Romney's hires were of a certain type: they were spreadsheet geeks ranking at the top of their class, often alumni of Stanford, where Romney had spent an undergraduate year, or Harvard. Though Bain & Company was a much-sought-after workplace in the 1980s, Bain Capital was little known and risky. Some of Romney's recruits were attracted by the pace of life in a relatively provincial city like Boston and passed over glamorous jobs in New York City or on the West Coast. One former partner described the group as a cast of brilliant, socially awkward young men—and they were all men—anxious to prove that they were just as worthy as peers who had gone to big Wall Street firms. Bob White and Josh Bekenstein were hired by Bain & Company out of Harvard Business School, and Geoffrey Rehnert came from Stanford Law School. Romney was only thirty-seven when he took the reins of Bain Capital, and many of his new employees were a decade younger.

Robert Gay came by a different route. Gay, one of the few Bain employees who shared Romney's Mormon religion, worked for a Wall Street firm but had become disgusted by deals being done largely for big fees, whether merited or not, and by the job losses or factory closures that often followed mergers and acquisitions. Afraid that he would become the kind of person he "despised," Gay leapt at a chance to become one of Romney's partners. Financially, it made "no sense," he said later. Bain was a fraction of the size of Wall Street firms, with fifteen employees and only a handful of deals completed. But Gay was convinced that Romney would enable him to exercise much more "influence for good" at Bain than he could elsewhere. So he took a pay cut and signed on. Gay and the other partners had one thing in common: they were utterly loyal to Romney, and that loyalty would only grow as they all grew rich together.

Thus began Romney's fifteen-year odyssey at Bain Capital. Boasting about those years when running for senator, governor, or president, Romney would usually talk about how he had helped create jobs at new or underperforming companies and say that he learned how jobs and businesses come and go. He'd typically mention a few well-known companies in which he and his partners had invested, such as Staples. But the full story of his years at Bain Capital is far more complicated and has rarely been closely scrutinized. Romney was involved in about a hundred deals, many of which have received little notice because the companies involved were privately held and not household names. The most thorough analysis of Romney's performance comes from a private solicitation for investment in Bain Capital's funds written by the Wall Street firm Deutsche Bank. The company examined sixty-eight major deals that had taken place on Romney's watch. Of those, Bain had lost money or broken even on thirty-three. Overall, though, the numbers were stunning: Bain was nearly doubling its investors' money annually, achieving one of the best track records in the business. Most of that success came from a handful of little-known but incredibly successful investments. But the venture had begun with plenty of failures—and lessons.

Romney was on yet another road show to attract investors, firing up an overhead projector and explaining how prudently he would handle their money if they bought into the fund that would make up Bain Capital's first pool of capital. It could be a hard sell; his firm had no track record to speak of, just that magic name Bain. Romney and a number of Bain & Company partners had put $14 million of their own money into the fund and then worked to drum up support from outsiders. He and his road-show partner, Coleman Andrews, were well short of their funding goal. Then, one day, Romney got a tip from a Bain & Company executive, Harry Strachan, that some wealthy families in war-torn regions of Central America were looking for a place to invest, a safe harbor for their cash.

Romney was intrigued but worried. Already thinking of a career in politics, he wanted to be sure that the funds would not later be regarded as tainted. El Salvador, for example, was a scene of regular massacres and assassinations of political figures. Strachan said Romney "expressed to me that I had to put my hand in the fire for him, that none of the people we were introducing to him were involved in illegal drug money, right-wing death squads, or left-wing terrorism." Strachan assured Romney that he would carefully vet the investors, including one whose nephew reputedly had a questionable background; he concluded that the investors' money was clean. With that assurance, Romney agreed to meet with the Central American investors at a Miami bank, where he approved the investment. Years later, when he was asked about reports that some of the family members of investors might have had ties to paramilitary groups, he said he was satisfied that the individuals who had put money into Bain Capital had gotten the funds from legitimate sources. "We investigated the individuals' integrity and looked for any obvious signs of illegal activity and . . . found none," Romney said. For Bain, the Central American money was crucial, providing $6.5 million of $37 million in the fund.

Now the question was where to invest. Romney dispatched his new partners to study potential deals of the two types prevalent in the market: buyouts, which involved purchasing existing companies, and venture capital investments in younger businesses that had yet to take off. In the early 1980s, venture capital was a niche field, dominated by Wall Street financiers backing high-tech companies. It could take years for such infant concerns to become profitable; payoffs were unpredictable, but winners could be bonanzas. Juggernauts such as Apple Computer had started with venture funding in the 1970s, and companies such as Cisco Systems were taking off in the 1980s. But Romney and his partners were loath to do battle in high tech, where they didn't feel they had an edge. "We thought we'd lose if we tried to invest in technology-based start-ups," Bekenstein said.

Romney was, by nature, deeply risk averse in a business based on risk. He worried about losing the money of his partners and his outside investors—not to mention his own savings. "He was troubled when we didn't invest fast enough, he was troubled when we made an investment," Andrews said. "He never wanted to fall short on commitments or representations made to investors." So rather than jump boldly into new fields, he focused his attention on more mundane corners of the economy: makers of wheel rims, photo albums, and handbags, to cite three examples. Sorting through possible investments, Romney met weekly with his young partners, pushing them for deeper analysis and more data and giving himself the final vote on whether to go forward. They operated more like a group of bankers carefully guarding their cash than an aggressive firm eager to embrace giant deals. Tellingly, Romney called the group that reviewed deals the credit committee, instead of the investment committee (the usual term), and his banker's jargon stuck. Perhaps the caution came from his family background. Generations of boom and bust had lifted the fortunes of his ancestors and then impoverished them until his father, George, had made it big. And even George had staked his reputation on being liberal on social issues but conservative with other people's money. Wherever the impulse came from, Romney was so relentless in playing the role of devil's advocate that his partner Bob White would joke about wanting to "punch him in the nose."

Some partners suspected that Romney always had one eye on his political future. "I always wondered about Mitt, whether he was concerned about the blemishes from a business perspective or from a personal and political perspective," one partner said years later. The partner concluded that it was the latter. Whereas most entrepreneurs accepted failure as an inherent part of the game, the partner said, Romney worried that a single flop would bring disgrace. Every calculation had to be made with care.

One winter evening in 1985, Romney sat in a drab ten-by-ten-foot conference room in Bain Capital's office, flapping his tie to mimic a rapidly beating heart. His colleagues knew that when Romney flapped his tie, he was feeling pressure. At the time, he was so worried about Bain Capital's future that, according to one colleague, he raised the possibility of returning the millions they had received from investors and going back to their old jobs. Dressed in a crisp blue shirt with a white collar and gold collar pin, Romney appeared to be the model of a successful 1980s financier. But his shirt, according to his former colleague Geoffrey Rehnert, was drenched dark with sweat under his arms. "Mitt was struggling," Rehnert said. "And he wasn't used to struggling."

Romney and his partners, in avoiding high tech, had taken on some challenging, if obscure, investments. One of Bain Capital's first deals was the $2 million it put into Key Airlines in 1984. Key ran shuttle routes from Las Vegas deep into the Nevada desert, used mainly by government personnel. Bain wanted to expand the operation and added contract flights for tourism to places like Mexico and the Caribbean. Two years later, Bain merged Key with a start-up airline, Presidential Airways, which went public, and Bain ultimately more than doubled its investment, to $5.4 million. (The company's fortunes did not last; Presidential went bankrupt in 1989.) Another early deal was MediVision, which ran surgical centers for outpatient eye surgery. An effort to expand by building centers around the country was slow, but buying facilities worked out better, and the company was ultimately packaged with a medical supply start-up and sold to a larger firm at a profit. Holson Burnes, a company that made photo accessories, also struggled. Bain had bought the photo album maker in 1986 and after a few tough years had merged it with a frame maker, but the combination didn't immediately pay off. There were cost cuts, including layoffs, and product problems. An analog version of a digital photo frame—a contraption that, with a press of a button, would flip from one photo in a stack to the next—didn't work well at first. It wasn't until 1992 that Bain would take the company public—well beyond Bain's goal of a three-to-five-year investment—and ultimately double its money on the $10 million investment.

Despite the struggles, 1986 would prove to be a pivotal year for Romney. It started with a most unlikely deal. A former supermarket executive, Thomas Stemberg, was trying to sell venture capitalists on what seemed like a modest idea: a cheaper way to sell paper clips, pens, and other office supplies. The enterprise that would become the superstore Staples at first met with skepticism. Small and midsize businesses at the time bought most of their supplies from local stationers, often at significant markups. Few people saw the profit margin potential in selling such homely goods at discount and in massive volume. But Stemberg was convinced and hired an investment banker to help raise money. Romney eventually heard Stemberg's pitch, and he and his partners dug into Stemberg's projections. They called lawyers, accountants, and scores of business owners in the Boston area to query them on how much they spent on supplies and whether they'd be willing to shop at a large new store. The partners initially concluded that Stemberg was overestimating the market. "Look," Stemberg told Romney, "your mistake is that the guys you called think they know what they spend, but they don't." Romney and Bain Capital went back to the businesses and tallied up invoices. Stemberg's assessment that this was a hidden giant of a market seemed right after all.

Romney hadn't stumbled on Staples on his own. A partner at another Boston firm, Bessemer Venture Partners, had invited him to the first meeting with Stemberg. But after that, Romney took the lead; he finally had his hands on what looked like a promising start-up. Bain Capital invested $650,000 to help Staples open its first store in Brighton, Massachusetts, in May 1986. In all, it invested about $2.5 million in the company. Three years later, in 1989, Staples sold shares to the public, when it was just barely turning a profit, and Bain reaped more than $13 million. It was a big success at the time. Yet it was very modest compared with later Bain deals that reached into the hundreds of millions of dollars.

For years Romney would cite the Staples investment as proof that he had helped create thousands of jobs. And it is true that his foresight in investing in Staples helped a major enterprise lift off. But neither Romney nor Bain directly ran the business, though Romney was active on its board. At the initial public offering, Staples was a firm of 24 stores and 1,100 full- and part-time jobs. Its boom years were still to come. Romney resigned his seat on the board of directors in 2001 in preparation for his run for governor. A decade later, the company had more than 2,200 stores and 89,000 employees.

Assessing claims about job creation is hard. Staples grew hugely, of course, but the gains were offset, at least partially, by losses elsewhere: smaller, mom-and-pop stationery stores and suppliers were being squeezed, and some went out of business entirely. Ultimately, Romney would approvingly call Staples "a classic 'category killer,' like Toys 'R' Us." Staples steamrollered the competition, undercutting prices and selling in large quantities. When asked during the 1994 Senate campaign about his job creation claim—that he had helped create ten thousand jobs at various companies (a claim he expanded during his 2012 presidential campaign to having "helped to create tens of thousands" of jobs)—Romney responded with a careful hedge. He emphasized that he always used the word "helped" and didn't take full credit for the jobs. "That's why I'm always very careful to use the words 'help create,'" he acknowledged. "Bain Capital, or Mitt Romney, 'helped create' over 10,000 jobs. I don't take credit for the jobs at Staples. I helped create the jobs at Staples."

Howard Anderson, a professor at MIT's Sloan School of Management and a former entrepreneur who has invested with Bain, put it more plainly: "What you really cannot do is claim every job was because of your good judgment," he said. "You're not really running those organizations. You're financing it, you're offering your judgment and your advice. I think you can only really claim credit for the jobs of the company that you ran." Stemberg, however, begrudges Romney nothing. If Romney gets the blame for jobs lost on Bain Capital's watch, he said, "Why not give him credit for every job ever created at Staples? One could argue that he was instrumental in Staples both getting started and, more importantly, being successful. It goes both ways."

The same year Romney invested in Staples—digging into a true start-up—he also inked the biggest transaction, by far, that Bain Capital had put together up till then. And with this $200 million deal, he waded fully into the high-stakes financial arena of the time: lever-aged buyouts, or LBOs. Whereas a venture capital deal bet on a new business, pursuing an LBO meant borrowing huge sums of money to buy an established company, typically saddling the target with big debts. The goal was to mine value that others had missed, to quickly improve profitability by cutting costs and often jobs, and then to sell. Romney thought he and his team could add to that equation by mak-ing management and operational changes to grow the company's busi-ness. Romney would later acknowledge the shift in his thinking about whether to stress venture capital investments or LBOs. Initially, he thought that putting money into young firms "would be just as good as acquiring an existing company and trying to make it better." But he found that "there's a lot greater risk in a start-up than there is in acquiring an existing company." He was unnerved by the prospect of investing money when "the success of the enterprise depended upon something that was out of our control, such as 'Could Dr. X make the technology work?' or would the market develop in the future that had not yet come to fruition?"

He was much more comfortable in an environment where the is-sue wasn't whether an idea would pan out but whether the numbers worked. He knew himself, knew that his powers ran less to the creative than to the analytical; he was not at heart an entrepreneur. Perhaps that was what led him to push the pause button at the outset with Bill Bain. But he now felt ready to take on much bigger financial risks, mostly by making leveraged bets on existing companies, whose market was known and which had business plans he could parse and master.

Romney found an ideal target in the auto industry, a field he knew well from his Michigan upbringing. It was a wheel-rim maker for trucks called Accuride, part of the empire of tire giant Firestone. There was no "Dr. X" factor to fear here. The wheel-rim business was as unsexy, and as solid, as it got. Firestone wanted to sell the Kentucky-based business because it wasn't part of its major product line. Bain told management and unions that current employees would not be laid off, a pledge the firm did not repeat in many other deals. Bain vowed to grow the business by making substantial changes, from revamping production and giving executives greater pay incentives to offering discounts to customers who agreed to give Accuride all their business. That was part of what one Romney colleague at Bain & Company called the "loyalty effect," giving employees, customers, and investors incentives to make a company successful.

Ultimately, Bain won the bidding for Accuride with a $200 million offer, an offer that was highly leveraged. Romney put just $5 million of Bain's money at risk; the rest was borrowed from banks, and, as was typical in a buyout, Accuride would be responsible for repayment. It was like buying a house with 1 percent down and someone else on the hook for the mortgage. No wonder the LBO business was alluring. The only catch was that the value of the business—the house—had to go up to make it all work. Accuride did, and the bet paid off handsomely for both Bain and the wheel-rim maker. The company's earnings rose by 20 percent in the first year on Bain's watch, and the number of plant jobs increased by 16 percent, to 1,785. Eighteen months later, Bain sold Accuride to the mining conglomerate Phelps Dodge Corp., turning its $5 million into $121 million. This was its first big LBO hit.

Romney couldn't have predicted how successful the Accuride deal would be. But his decision to try to pull it off put Bain Capital on the map. Bain's partners believed they had joined the big time, even before the profits were in, and felt ready to celebrate. Romney, famous for being tightfisted (and not a drinker), was not the sort to host the kind of champagne-soaked party typical of buyout firms in the 1980s; he ran a spartan operation. Rehnert recalled being among the first at Bain to have a cell phone in his car. Romney was aghast, asking why he was wasting money on the device, which at the time was unwieldy and unreliable. Why not wait to use a landline or stop at a pay phone?

he asked. Rehnert also recalled going with Romney to Au Bon Pain, a fast-food restaurant. Romney got his lunch and began carefully counting his change. Rehnert said he asked him why he bothered. "I throw mine in the fountain over there," he told Romney. Romney looked stunned, even "viscerally pained," not realizing his colleague was joking. But Romney did have a fondness for fine meals at fancy restaurants. So on the night the Accuride deal was sealed, Romney took his partners to L'Espalier, one of Boston's finest French restaurants. The crew left behind their mishmash of metal desks and dined on haute cuisine, celebrating what they considered a great success after the ups and downs of the firm's first two years.

Billions of dollars were being made in the field of leveraged buyouts in the roaring eighties, and Romney was fully in the game, continuing to ratchet up his favored strategy. On the campaign trail in 2011, Romney said his work had "led me to become very deeply involved in helping other businesses, from start-ups to large companies that were going through tough times. Sometimes I was successful and we were able to help create jobs, other times I wasn't. I learned how America competes with other companies in other countries, what works in the real world and what doesn't." It was a vague summary of what was a very controversial type of business. In his 2004 autobiography, *Turnaround*, Romney put it more bluntly: "I never actually ran one of our investments; that was left to management." He explained that his strategy was to "invest in these underperforming companies, using the equivalent of a mortgage to leverage up our investment. Then we would go to work to help management make their business more successful."

Romney's phrase "leverage up" provides the key to understanding this most profitable stage of his business career. While putting relatively little money on the table, Bain could strike a deal using largely debt. That generally meant that the company being acquired had to borrow huge sums. But although the strategy had worked with the Accuride deal, there was no guarantee that target companies would be able to repay their debts. At Bain, the goal was to buy businesses that were stagnating as subsidiaries of large corporations and grow them or shake them up to burnish their performance. Because many of the companies were troubled, or at least were going to be heavily indebted after Bain bought them, their bonds would be considered lower grade, or "junk." That meant they would have to pay higher interest on the bonds, like a strapped credit card holder facing a higher rate than a person who pays off purchases more quickly. High-yielding junk bonds were appealing to investors willing to take on risk in exchange for big payouts. But they also represented a big bet: if the companies didn't generate large profits or could not sell their stock to the public, some would be crippled by the debt layered on them by the buyout firms. This was the world that Romney now embraced wholeheartedly.

The arcane domain of corporate buyouts and junk-bond financing had entered the public consciousness at the time, and not always in a positive way. Ivan Boesky, a Wall Street arbitrageur who often bought the stock of takeover targets, was charged with insider trading and featured on the cover of *Time* magazine as "Ivan the Terrible." Shortly after Romney began working on leveraged deals, a movie called *Wall Street* opened. It featured the fictional corporate raider Gordon Gekko, who justified his behavior by declaring "I am not a destroyer of companies. I am a liberator of them! . . . Greed, for lack of a better word, is good. Greed is right. Greed works. Greed clarifies, cuts through, and captures the essence of the evolutionary spirit."

Romney, of course, never said that greed is good, and there was nothing of Gekko in his mores or style. But he bought into the broader ethic of the LBO kings, who believed that through the aggressive use of leverage and skilled management they could quickly remake underperforming enterprises. Romney described himself as driven by a core economic credo, that capitalism is a form of "creative destruction." This theory, espoused in the 1940s by the economist Joseph Schumpeter and later touted by former Federal Reserve Board chairman Alan Greenspan, holds that business must exist in a state of ceaseless revolution. A thriving economy changes from within, Schumpeter wrote in his landmark book, *Capitalism, Socialism and Democracy*, "incessantly destroying the old one, incessantly creating a new one. This process of Creative Destruction is the essential fact about capitalism. It is what capitalism consists in and what every capitalist concern has got to live in." But as even the theory's proponents acknowledged, such destruction could bankrupt companies, upending lives and communities, and raise questions about society's role in softening some of the harsher consequences. As Greenspan once put it, "the problem with creative destruction is that it is destruction and there is a very considerable amount of turmoil that goes on in the process."

Romney, for his part, contrasted the capitalistic benefits of creative destruction with what happened in controlled economies, in which jobs might be protected but productivity and competitiveness falter. Far better, Romney wrote in his book *No Apology*, "for governments to stand aside and allow the creative destruction inherent in a free economy." He acknowledged that it is "unquestionably stressful—on workers, managers, owners, bankers, suppliers, customers, and the communities that surround the affected businesses." But it was necessary to rebuild a moribund company and economy. It was a point of view he would stick with in years ahead. Indeed, he wrote a 2008 op-ed piece for *The New York Times* opposing a federal bailout for automakers that the newspaper headlined, "Let Detroit Go Bankrupt." His advice went unheeded, and his prediction that "you can kiss the American automotive industry goodbye" if it got a bailout has not come true.

Of course, economic theories are one thing, and the impact of leveraged buyouts another. Some can have bitter consequences, yielding profits to investors but nothing good for the purchased company. Romney preferred negotiated buyouts over hostile takeovers, in part because he wanted to avoid the risk of negative headlines if a deal wasn't welcomed by a targeted company. Romney and his partners also fervently believed that with their Bain background in tough-minded business analysis, they could make companies more valuable by working with them, something that would be difficult after a hostile deal. Romney's low-profile approach was not the kind to capture headlines or the imagination of Hollywood, but his success was undeniable.

As a result of the payoff from the Accuride deal, Romney's first investment fund would return to his investors all the money they had bet and more, even if some of the other companies they bought went bad. Romney and his partners were on their way. "We were pretty happy," said Josh Bekenstein. "It was pretty exciting to send more than a hundred percent of the fund back to the limited partners." Under Romney's arrangement with Bill Bain, about half of the profit of the first fund went back to Bain & Company partners who had invested in the new firm. Tellingly, the first pool of money for Bain partners in the fund had been called Meadowbrook, after a Bill Bain street address. But from then on, Romney and his Bain Capital partners controlled most of the profits, and there was no doubt about who was in charge. The second Bain investment fund was named Tyler, after the road Romney lived on in Belmont at the time.

Bain Capital had become a hot property. So much money poured into Romney's second investment fund that the firm had to turn away investors. Romney set out to raise $80 million and received offers totaling $150 million. The partners settled on $105 million, half of it from wealthy customers of a New York bank. It was a stunning turnabout from the scramble to raise the first fund a few years earlier; now they were competing with the big investment houses. During a break at a photo shoot for a brochure to attract investors, the Bain partners playfully posed for a photo that showed them flush with cash. They clutched $10 and $20 bills, stuffed them into their pockets, and even clenched them in their grinning teeth. Romney tucked a bill between his striped tie and his buttoned suit jacket. Everything was different now.

———————

It was time for another road show, but the days of soliciting prospects for scarce cash in obscure locales was mostly over. This time Romney and his partners headed to Beverly Hills, California. Arriving at the intersection of Rodeo Drive and Wilshire Boulevard, they headed to the office of Michael Milken, the canny and controversial junk-bond king, at his company, Drexel Burnham Lambert. Romney knew

Milken was able to find buyers for the high-yield, high-risk bonds that were crucial to the success of many leveraged buyout deals. At the time of Romney's visit, it was widely known that Drexel and Milken were under investigation by the Securities and Exchange Commission. But Drexel was still the big player in the junk-bond business, and Romney needed the financing. A former Drexel executive recalled that Romney met Milken a couple of times but the details of the deal making were left to Milken's deputies.

Romney had come to Drexel to obtain financing for the $300 million purchase of two Texas department store chains, Bealls and Palais Royal, to form Specialty Retailers Inc. On September 7, 1988, two months after Bain hired Drexel to issue junk bonds to finance the deal, the Securities and Exchange Commission filed a complaint against Drexel and Milken for insider trading. Romney had to decide whether to close a deal with a company ensnared in a growing clash with regulators. The old Romney might well have backed off; the newly assertive, emboldened Mitt decided to press ahead. At the time, Drexel was losing customers who didn't want to be associated with the company, but Romney decided he needed the company's access to money and stuck with Milken. "There were other people who were definitely deciding not to do business with Drexel if they had an alternative available," said Marc Wolpow, a former Drexel executive involved in the Romney deal who later worked at Bain. "To Mitt's credit, he didn't back away when the Drexel mess occurred." Prosecutors who worked on the case said the effort by Romney and his Bain partners to get financing from Drexel had never been part of their case against Drexel and determined that Bain could have gotten similar financing elsewhere. Romney, asked years later why he had risked being associated with Drexel, said, "We did not say, 'Oh my goodness, Drexel has been accused of something, not been found guilty. Should we basically stop the transaction and blow the whole thing up?'"

But after deciding to stick by Milken, Romney soon feared that the deal could blow up for another reason. He learned that the Drexel case was going before U.S. District Judge Milton Pollack. The judge's wife, Moselle Pollack, was the chairman of and major stockholder in

Palais Royal. Concerned that questions would be raised about a judicial conflict of interest, Romney called Drexel's chief executive officer, Fred Joseph, "to make sure there would be no problem with the deal." Drexel's lawyers, seeing the potential conflict as an opportunity to stall the government's prosecution, then sought to have Judge Pollack taken off the case.

Romney's determination to go ahead with the deal dismayed federal securities officials. They feared that Drexel was trying to use the appearance of conflict to get the case thrown out of court, according to James T. Coffman, who played a key role in the civil case against Drexel in his role as assistant director of enforcement at the Securities and Exchange Commission. "By doing the deal he enabled Drexel to use the claim of conflict of interest on the part of the judge—which I think at a minimum reflects a lack of concern about the impact of his financing activities on the administration of justice," Coffman said about Romney. But the judge said there was no conflict of interest and, after a series of hearings, was allowed to remain on the case. The matter became moot when Drexel subsequently pleaded guilty. In the end, Romney's deal with Drexel went through. Around the time it closed, Drexel pleaded guilty to six criminal counts of securities and mail fraud and paid $650 million in fines. Three months later, Milken was indicted on racketeering charges. Prosecutors alleged that Milken had manipulated stock prices and defrauded clients in order to complete his junk-bond deals. Milken eventually pleaded guilty to securities and reporting violations and served twenty-two months in prison.

Romney's deal with Drexel, meanwhile, turned out well for both him and Bain Capital, which put $10 million into the retailer and financed most of the rest of the $300 million deal with junk bonds. The newly constituted company, later known as Stage Stores, refocused in 1989 on its small-town, small-department-store roots. Sales climbed over the next few years, but a plan to sell stock to the public in 1992 failed to get off the ground. Four years later, in October 1996, the company successfully sold shares to the public at $16 a share. By the following year, the stock had climbed to a high of nearly $53, and Bain Capital and a number of its officers and directors sold a large part of their holdings. Bain made a $175 million gain by 1997. It was one of the most profitable leveraged buyouts of the era.

Romney sold at just the right time. Shares plunged in value the next year amid declining sales at the stores. The department store company filed for Chapter 11 bankruptcy protection in 2000, struggling with $600 million in debt, and a reorganized company emerged the following year. So ended the story of a deal that Romney would not be likely to cite on the campaign trail: the highly leveraged purchase, financed with junk bonds from a firm that became infamous for its financial practices, of a department store company that had subsequently gone into bankruptcy. But on the Bain balance sheet, and on Romney's, it was a huge win.

Not every deal worked out so well for Romney and his investors. Bain invested $4 million in a company called Handbag Holdings, which sold pocketbooks and other accessories. When a major customer stopped buying, the company failed and two hundred jobs were lost. "It was a terrible situation," Robert White, a Bain Capital partner at the time, said. Bain invested $2.1 million in a bathroom fixtures company called PPM and lost nearly all of it. An investment in a company called Mothercare Stores also didn't pan out; the firm had eliminated a hundred jobs by the time Bain dumped it. White said Bain lost its $1 million and blamed "a difficult retail environment."

In some cases, Bain Capital's alternative strategy of buying into companies also ended in trouble. In 1993, Bain bought GST Steel, a maker of steel wire rods, and later more than doubled its $24 million investment. The company borrowed heavily to modernize plants in Kansas City and North Carolina—and to pay out dividends to Bain. But foreign competition increased and steel prices fell. GST Steel filed for bankruptcy and shut down its money-losing Kansas City plant, throwing some 750 employees out of work. Union workers there blamed Bain, then and now, for ruining the company, upending their lives, and devastating the community.

Then, in 1994, Bain invested $27 million as part of a deal with other firms to acquire Dade International, a medical diagnostics equipment firm, from its parent company, Baxter International. Bain ultimately made nearly ten times its money, getting back $230 million. But Dade wound up laying off more than 1,600 people and filed for bankruptcy protection in 2002, amid crushing debt and rising interest rates. The company, with Bain in charge, had borrowed heavily to do acquisitions, accumulating $1.6 billion in debt by 2000. The company cut benefits for some workers at the acquired firms and laid off others. When it merged with Behring Diagnostics, a German company, Dade shut down three U.S. plants. At the same time, Dade paid out $421 million to Bain Capital's investors and investing partners.

This was one of the basic tenets of LBO investing: even with large debts to pay off and operating challenges ahead, acquired firms were often obligated to make big payouts to the investment firms that bought them. Romney said later that he regretted that dividend payments to Bain had hurt some of the companies he and his partners had bought. "It is one thing that if I had a chance to go back I would be more sensitive to," Romney told *The New York Times* in 2007. "It is always a balance. Great care has got to be taken not to take a dividend or a distribution from a company that puts that company at risk," he said, adding that taking a big payment from a company that later failed "would make me sick, sick at heart."

In that time of immense success and growing personal renown, disaster loomed. Everything Mitt Romney had worked for was at risk as the calendar turned to 1991. His reputation, his business, his political future—all of it was on the line.

Romney's mentor, Bill Bain, and his former firm were in deep trouble. Bain and the seven other founders had come up with a plan to cash out part of their ownership stakes. To do so, they had Bain & Company take out more than $200 million in loans. The idea was to give the founders a large payout and provide partners with a substantial stake in the company. But the result was that a mountain of debt was piled on the firm at exactly the wrong time. The economy was slowing, the company's revenues were declining, and banks were failing. It was the ultimate embarrassment: here was a company that prided itself on advising others how to prudently improve their businesses, and it was facing near-disaster by failing to follow some of the very principles it preached. Within the firm, partners saw their own earning potential slashed—all in the name of enriching the founders. The company's existence was at risk.

The senior partners saw Romney as the best person to deal with the crisis: he had the trust of Bill Bain but had not been part of the founder group whose plan had caused all the trouble. They needed someone who could wring concessions from creditors and keep the firm together. A group of the partners went to Romney with a stark message. The consulting company was "going over a cliff," they told him. "We need you to come in and run the organization." Though it had been seven years since Romney had left Bain & Company, he realized that his own fate was at stake. If Bain & Company went bankrupt, it could cast such a shadow that his own Bain Capital might also collapse. Bill Bain agreed to let Romney take charge.

Once he saw the books, Romney realized that the problems were even greater than he had imagined. On the day in 1991 he took control, layoffs began. Eventually, 260 people around the world, 18 percent of the Bain & Company workforce, would lose their jobs. The salaries and benefits of many remaining employees were slashed. But that wasn't nearly enough. Romney learned that the company had just sent a $1 million check to one of Bain & Company's landlords. "We have bad news," Romney's message to the landlord said. "The check has been sent but it's not going to clear because we've cancelled payment." Romney canceled many other checks for rent payments, real estate, and to "all sorts of suppliers," doing whatever he could to meet the payroll. Romney's worry was not just about the financial ledger at Bain & Company. He worried that some company executives would be subject to prosecution. The Massachusetts legal code calls for up to a year in prison for a company official who willfully violates wage law. "It's a crime in this state to employ someone knowing or having reason to know that you won't be able to pay them at the end of the pay period," Romney later told the conservative radio talk-show host and writer Hugh Hewitt. "It's a crime and we were perilously close to not being able to meet payroll. So, we watched this like a hawk." There were, he said, "some really frightening months."

Romney pitted creditors against one another and beseeched banks to go easy on loan repayments, or everyone would lose. At one point, on a Saturday in Bain Capital's office, Romney and a couple of his partners met with a banker from Goldman Sachs to try to persuade the Wall Street giant to help restructure the consulting firm's debt. The banker, known for his sharp-elbowed style in his dealings with troubled companies, sat across the table from Romney. The discussion grew increasingly heated. As Romney made his pitch, the Goldman banker said, "Shut up," and told him that Bain & Company's best hope was to file for bankruptcy protection, according to a Bain partner. Romney rose to his feet, and the other partners in the room thought he might launch across the table and hit the banker; they had rarely seen him so angry.

As far as Romney was concerned, bankruptcy was not an option. He would embrace that idea when it came to reorganizing other failing companies, given his belief in the creative destruction of capitalism. But if Bain & Company went bankrupt, hundreds of jobs would be lost and Romney's effort would be seen as a failure. That left him to deal with a cascade of troubles: checks were at risk of bouncing, loans couldn't be paid, and bankers were balking at Romney's plan. He began demanding concessions from almost everyone owed something by Bain & Company. He convinced the Federal Deposit Insurance Corporation, which insures bank deposits, to forgive roughly $10 million of $38 million in loans owed to the failed Bank of New England. Several major lenders agreed to take 80 cents on the dollar rather than risk default. If Romney hadn't come up with the rescue plan, his aides later said, the losses to the FDIC and other institutions would have been much greater.

He was toughest when it came to negotiating with the partners at Bain & Company. He told the founding partners they had to give up about $100 million, or half the money they'd been planning to take out of the firm. "He was willing to make very tough decisions that had to be made and force them on people," said former Bain & Company partner Harry Strachan. Then, one autumn Saturday, Romney summoned about forty partners from Bain & Company to an urgent meeting. He had a take-it-or-leave-it offer: agree to pay cuts and promise to stay with the company for a year, and he would do everything he could to fix the firm. If everyone stayed at a lower salary but joined together, he felt confident that revenues would rebound and they would soon be earning more than ever. Romney said he would leave the room for thirty minutes to let the partners think about it. He knew that many could easily go to other firms for higher pay. He told the partners that anyone who rejected his plan should leave before he returned to the room. Only one person left.

Much of what Romney did to save the company was never made public. But in the end the breaks on loans, the layoffs, financial concessions from partners, and other measures helped Bain & Company save enough money to stay afloat and eventually thrive again. It was the very definition of the kind of "turnaround" for which Romney would claim expertise. Given the embarrassing circumstances of how the company nearly failed, the story of the rescue of Bain & Company has not fit neatly into Romney's campaign narrative. But it was one of his most impressive displays of executive talent and toughness; in some ways, it was his finest hour at Bain. "If Bain & Company went bankrupt, nobody would ever have taken us seriously," former Bain Capital partner Geoffrey Rehnert said. "They would've been known as the clowns who charged a lot in consulting fees and went bankrupt."

Shortly after Romney returned to full-time work at Bain Capital, he faced a new crisis that tested his toughened management skills. He was serving on the board of Damon Corp., a medical testing firm based in Needham, Massachusetts, as a result of a $4 million Bain Capital investment he had approved back in 1989. Bain Capital held an 8 percent stake in the company. Initially, the deal had seemed to go as Romney had envisioned. He had helped take Damon public in

1991, and the company had paid down some of its debt from the buy-out and sold off unwanted businesses. In August 1993, Bain helped sell the company to Corning and nearly tripled its money. Romney personally reaped $473,000. The day after the merger was completed, Damon's Needham plant was shut down, and 115 people were laid off.

It was later revealed that during Romney's tenure on the board, federal investigators had been looking into whether the company had defrauded Medicare by overbilling for blood tests. Indeed, the same month that Damon was acquired, it received subpoenas from the federal government regarding an investigation of the matter. Romney said he had first learned about allegations of overbilling by a rival firm in December 1992 and been moved to insist that the board hire an outside attorney to investigate Damon's billing practices. Romney said that the board had taken "corrective action" and investors had "received a good return on their investment because we were able to blow the whistle."

As the details of the federal case were made public, however, Romney's version appeared questionable. He later faced criticism from his political opponents about whether he had really played any role in uncovering the fraud—and whether he could have reported the overbilling practices to authorities earlier. In 1996, Damon pleaded guilty to criminally defrauding Medicare and Medicaid of $25 million and paid $119 million in fines, the most at that time for a health care fraud case. United States Attorney Donald Stern of Massachusetts, who handled the case, called it "corporate greed run amok." Four company executives would be charged with conspiring to defraud Medicare; one received a three-month jail sentence. Prosecutors said a former Damon employee had blown the whistle on the billing practices. And the government credited Corning, not Romney or his fellow Damon directors, with cleaning up the situation.

The matter received little notice until a decade later, when Romney was in pursuit of the Massachusetts governorship. His Democratic opponent, Shannon O'Brien, accused him of lax oversight at Damon and failing to report the fraud. "There is a mess in corporate America, and its name is Mitt Romney," she said during a debate with him. But although some of the fraud occurred during Romney's time on Damon's board, he and other board members were never implicated in the case. The Damon case raised a recurring question about corporate governance: how much responsibility does a board member have for what happens at a company he helps oversee? In his comments about Damon, Romney seemed at times to hold two views of the matter, both of them to his benefit. On the one hand, he said he hadn't known what was going on at Damon; on the other, he said he'd helped to put a stop to practices later found to be fraudulent. One thing that looked good at Damon was the bottom line for Bain. In the end, it was a profitable deal for both the firm and Romney, however tainted by legal troubles and layoffs it may have been.

Romney excelled at courting investors to put millions of dollars into his funds. He was a shrewd analyst of proposed ventures. But he was never an expert at finding new deals. Indeed, he brought few investment proposals to the table, and when he did, they often flopped. One day, for example, Romney arrived at work with what he thought was a great idea. It came from a friend, Reed Wilcox, whose background was almost uncannily like Romney's: Wilcox, a Mormon, had both law and business degrees from Harvard, was a Brigham Young University alumnus, had worked at Boston Consulting Group, and would later do missionary work in France. Now Wilcox was with a company that had developed a technology that enabled photographs of children to be used to create dolls that looked like them. The two-foot-high, $150 dolls became popular and were featured in magazines. Romney authorized a $2.1 million investment from Bain Capital in 1996, and he personally loaned money for the operation. The company, called Lifelike, had manufacturing operations in Colorado and Hong Kong. But sales dropped when the economy sputtered in 2001; and by 2003 there were production and quality problems, and hundreds of consumers complained to Colorado's attorney general that they hadn't received their dolls in time for Christmas 2003. By early 2004, Lifelike was bankrupt. It owed nearly $2 million to its Hong Kong manufacturer and hundreds of thousands of dollars more to advertising agencies and other creditors. In May of that year, a judge approved an auction of the company's assets, which went for $1.1 million. Bain lost its money. Lifelike became a deal Bain partners wanted to forget; they shook their heads when asked about it and erased mention of it from the company's web site.

Romney also came up with the idea of investing $5 million in a car parts company called Auto Palace/ADAP. Romney served on the board, and Bain lost about half of the investment. The boss's poor track record on finding his own deals reinforced the view that Romney's strength was analyzing other people's proposals. He brought his characteristic grilling to hundreds of proposals over the years. During one of Bain Capital's weekly business review meetings, one partner, Stephen Pagliuca, pitched the acquisition of the high-tech research firm Gartner Group. Romney went quickly to the heart of the matter: what did Gartner want to be? It was losing millions of dollars as a unit of the advertising firm Saatchi & Saatchi. Could it be a sustainable business? "Mitt was great at finding what the key issue on the deal was and then pressure testing it immediately," Pagliuca said. "I'd lose a lot of sleep on those nights." Gartner would go on to be one of Bain's early big winners. After taking the company public, Bain eventually saw a 1,500 percent return on its investment, turning $3.5 million into $55 million. Over the years Gartner increased its workforce from about 700 to 4,400 employees.

Romney made up for his weaknesses by hiring—and paying well—partners who excelled in finding and closing deals. "I don't think Mitt's favorite thing in the world is the backroom negotiation of deals," Stemberg, the Staples founder, said. "But one thing you've got to remember . . . Mitt Romney has always surrounded himself with great people who know how to execute his vision." That would be a theme Romney would reprise as a politician. Faced with a problem, he would often suggest forming a committee, just as he had formed a partnership group, and let the experts hash it out while he listened and posed questions.

Occasionally, however, Romney showed he could make a deal happen. The most unlikely case occurred when he tried to buy Domino's Pizza. Mark Nunnelly, the Bain Capital partner leading the deal, thought it might be helpful to bring Romney along to meet Domino's Pizza CEO Tom Monaghan. After all, Domino's was based in Ann Arbor, Michigan, and Romney was a son of a Michigan governor and had grown up in the state. Nunnelly figured Romney would be his secret weapon in wooing Monaghan. They flew to Michigan to meet with him. Monaghan, a staunch Catholic raised by nuns and a blunt character who had once owned the Detroit Tigers, promptly made clear his displeasure that Romney's brother, Scott, was running against his favorite candidate for Michigan's attorney general. The atmosphere turned cold. Monaghan asked Nunnelly if he had read his book, *Pizza Tiger*. Nunnelly had not. About forty-five minutes into what Nunnelly called a "dead meeting," he figured it was a waste of an August day.

At that point Romney suddenly stood up. Eyeing a model car on Monaghan's desk, he said, "I love '57 Chevys," switching to his car-guy persona. Almost immediately, the atmosphere warmed. For an hour and fifteen minutes, they talked about silver crankshafts, Detroit, and the car business. Nunnelly had to urge Romney to wrap it up or they'd miss their plane home. In the fall of 1998, Bain led a $1.1 billion buyout of the pizza chain, putting down about $385 million in cash and borrowing the rest. It outbid everyone else. As Bain took ownership, Romney later recalled, he thought about what he had just done. "We're the biggest schmoes who said, 'We'll pay more than anybody else,'" he said. The "schmoes" apparently knew what they were doing. They took the company public in 2004, and Bain reaped more than $100 million in that first sale of stock, plus a $10 million fee for ending its management contract with Domino's. Two Bain partners served on the board. Over time Bain sold all of its shares, earning a 500 percent return.

Romney wasn't one to socialize much with work colleagues. His ruthlessness with his personal time was meant to show how to balance work and family, but some partners who felt pressure from him to work eighty-hour weeks believed there was no way to follow the boss's example. One partner left after his wife said he didn't have enough time for his family. Another partner marveled at how different Romney was from most people in the stratosphere of the investment world. He didn't go out for a beer, of course, but he also rarely went out with the guys in any social venue. He was all business or all family. One partner chalked it up to Romney's introverted personality. Another called Romney "the Tin Man" for his inability to bond. He tried to compensate for his habit of social detachment by showing up at key moments, whether at the funeral of a young partner's father or an important basketball game for another partner's son.

One day in July 1996, Romney's partner Bob Gay sent an urgent message. Gay's fourteen-year-old daughter, Melissa, was missing. Romney went into high gear. "I don't care how long it takes. We're going to find her," Romney said. He shut down the Boston office and sent fifty-six employees to New York City to help find Melissa. Another 250 people from Wall Street firms joined in. The quest became big news. "Investment Firm Shuts to Help Find Girl," said a headline in *The Boston Globe.* The story reported that Romney and his partners had "decided that finding a missing daughter was more important than operating a $1 billion investment firm." The Bain crew set up a distribution system for 200,000 brochures with Melissa's picture, established a toll-free tip line, and hired private investigators.

The search took Romney onto the seediest streets of New York City. Soon Melissa's image was distributed everywhere, but there was still no sign of her. Romney then arranged for Bob Gay to tell his story on a local news program. "Shortly thereafter, through a traced telephone call asking if there was a reward, my daughter was safely secured," Gay said later. Melissa had taken a train from Connecticut without telling her parents she was going to a concert, and she was found later in the week at a house in New Jersey.

Romney later said that the search had changed his perspective on life. When Bain Capital ranked its annual accomplishments, the search for Melissa was number one. He said he would never forget talking with runaways in an effort to learn about Melissa's whereabouts. "It was a shocker," said Romney, who had rarely walked into the urban underbelly of America. "The number of lost souls was astounding." The search had put Bain into the public eye in a way that was unusual for the private firm. Romney would heighten the profile even further when he first ran for president. Eleven years after Melissa was found, he authorized a campaign ad called "Searched," which featured Bob Gay saying of his friend, "Mitt's done a lot of things that people say are nearly impossible. But for me, the most important thing he's ever done is to help save my daughter." Romney, meanwhile, would say that the time spent searching for Melissa had been "more valuable than some financial home runs that made the front page of *The Wall Street Journal*. I mean, money is just money."

The amount of money now being earned at Bain Capital was skyrocketing, and much of it came from a handful of giant deals. During Romney's fifteen years there, the firm invested about $260 million in its ten top deals and reaped a nearly $3 billion return. That was about three-quarters of its overall profit on roughly one hundred transactions during Romney's tenure. In one of his most specific explanations of how he made his fortune, Romney wrote in his autobiography, *Turnaround*, that he had not taken "the standard investment banker approach of snatching up companies with high, sustained growth, hanging on for six months, and then flipping them for a profit." Instead, he wrote, "We were looking for troubled companies, businesses that were not performing as well as we think they could." Romney said he or his partners would then "go to work to help management make their business more successful." He wrote that most of the companies were ones that "no one has heard of—TRW's credit services, the Yellow Pages of Italy." Those weren't just any two deals. They were two of the most lucrative of Romney's career, and luck played a big part in both. One was a quick hit with a huge payout, and the other benefited unexpectedly from the Internet frenzy.

The first deal was for a credit-reporting service owned by the defense and aerospace company TRW. At first it seemed a mundane transaction, even though it was the biggest investment of Romney's career, with Bain and its partners putting in about $100 million. Romney had little to do with the deal other than helping approve the investment. His partners included a savvy deal maker at another Boston private equity firm, Scott Sperling, of the Thomas H. Lee Company. The Bain and Lee groups had agreed to invest fifty-fifty and had taken preliminary control of the credit company spin-off, pending assurances that a fast new database system would be successful. They were thinking that they could make three times their money in about five years—a conservative return in their business.

With the newly named credit company Experian in hand, the Bain-Lee group was looking for other credit services to acquire, hoping to build a credit-reporting empire. One such inquiry went to a British company, Great Universal Stores, which was known for its Burberry chain but also had a credit unit. But it turned out that Great Universal didn't want to sell its credit service; it wanted to build it into something bigger. So Great Universal came back to the Bain-Lee group with a surprising response: it wanted to buy Experian. The Bain and Lee partners were stunned. A mere seven weeks after buying it, Romney and his partners flipped the company. Bain's $100 million investment returned at least $300 million. *The Wall Street Journal*, without mentioning Romney's name, said the windfall would "go down as one of the quickest big hits in Wall Street history." Romney personally pocketed millions of dollars. The profit margin and quick turnaround on the deal stunned investment houses around the world. Two of Romney's associates used an identical phrase in describing how it turned out: it was like being "hit with the lucky stick."

The second deal cited by Romney took longer but involved even more good timing and luck. It began with a renowned Italian investor named Phil Cuneo, who had the idea of buying the Italian version of the yellow pages. The companies that produced the bulky phone books were being bought and sold by investment houses around the world at the time. The deal was managed by Bain Capital partner Mark Nunnelly. It was unusual for Bain to invest in an Italian company, and Romney pressed Nunnelly. "Gee, Mark, are you sure?" he asked. "Gee, Mark, have you looked under every rock?" He was flapping his Hermès tie, mimicking a fast-beating heart; this one made him nervous.

Romney eventually agreed to the deal, and Bain Capital contributed $51.3 million. It seemed a solid investment in a firm with a staid and stable business model. But months after closing the deal, Cuneo and his Bain associates realized that they had acquired a company that might benefit from the surging interest in dot-com businesses. That had not been part of the original justification for the investment. "We stumbled on the Internet bubble," Cuneo said. "We started seeing what was happening in the U.S." Indeed, before long, they found they had bought a pot of gold. The yellow pages company owned a web-based directory that had the potential to be the Italian version of America Online or Yahoo! Romney was a well-known skeptic of Internet plays—correctly fearing that some would burst in an inevitable bubble—but this one seemed to have fallen into his lap at just the right time. The company, known by its acronym, SEAT, began to soar on the stock market, and Cuneo and Bain Capital began positioning it as "an Internet star."

In just under three years, in September 2000, the partners sold the investment, earning a windfall that far exceeded anyone's initial expectations. Bain's $51.3 million investment in the Italian yellow pages returned at least $1.17 billion, according to a Romney associate familiar with the deal. There is no public documentation of the how the profits were distributed, but at that time at least 20 percent of the return would have gone to Bain Capital. Of that, Romney's typical payout at the time was 5 to 10 percent. That means this one obscure deal would have given him a profit of $11 million to $22 million. If Romney made a side investment in the deal, as was standard among Bain partners, he would have made even larger gains. One Romney associate said Romney's total profit could have been as much as $40 million. (A Romney spokesman did not respond to questions about the deal.) A Romney associate marveled that the deal "wasn't like being hit by the lucky stick; it was being thrashed by it."

It was those kinds of deals that enabled Bain Capital to report the highest returns in the business in the 1990s. That is why the firm, citing Romney's track record, was able to raise its share of profits on deals to a stunning 30 percent—up from the industry standard of 20 percent—on top of its 2 percent up-front fee. Investors were willing to pay the higher share in the belief that Romney and his Bain team were worth it. Their history—averaging an 88 percent annual return over the fifteen years—said they were. "The returns were just eye-popping, and Bain Capital was able to command a premium," former Bain Capital partner Geoffrey Rehnert said. Romney himself would sometimes boast as well, comparing the Bain results with the 3 to 4 percent payout of "passbook" savings accounts. Not many of Bain's customers carried passbooks, of course; the minimum stake for investors was generally $1 million.

Romney's own wealth had increased exponentially. His net worth would grow to at least $250 million, and maybe much more, a trove that would enable him to foot a large part of the bill for his 2008 presidential campaign. Asked about a report that his wealth at one point reached as high as $1 billion, Romney said, "I'm not going to get into my net worth. No estimates whatsoever." The extent of Romney's personal take was helped by a favorable tax rate. Most of Romney's income came from capital gains at the Bain funds, not from salary. Under federal tax law, capital gains are taxed at a much lower rate than regular income. In 1999, when Romney stepped down as the head of Bain Capital, the top tax rate for income was 39.6 percent, while the top rate for capital gains was about half that, at 20 percent. This differential meant that Romney, like most high earners, was paying a lower tax rate on most of his earnings than some of the lowest-level workers at his firm and many working-class Americans.

For fifteen years, Romney had been in the business of creative destruction and wealth creation. But what about his claims of job creation? Though Bain Capital surely helped expand some companies that had created jobs, the layoffs and closures at other firms would lead Rom-

ney's political opponents to say that he had amassed a fortune in part
by putting people out of work. The lucrative deals that made Romney
wealthy could exact a cost. Maximizing financial return to investors
could mean slashing jobs, closing plants, and moving production overseas. It could also mean clashing with union workers, serving on a
board of a company that ran afoul of federal laws (as in the Damon
case), and loading up already struggling companies with debt.

There is a difference between companies run by buyout firms and
those rooted in their communities, according to Ross Gittell, a professor at the University of New Hampshire's Whittemore School of Business and Economics. When it comes to buyout firms, he said, "The
objective is: make money for investors. It's not to maximize jobs." Romney, in fact, had a fiduciary duty to investors to make as much money
as possible. Sometimes everything worked out perfectly; a change in
strategy might lead to cost savings and higher profits, and Bain cashed
in. Sometimes jobs were lost, and Bain cashed in or lost part or all of
its investment. In the end, Romney's winners outweighed his losers
on the Bain balance sheet. Marc Wolpow, a former Bain partner who
worked with Romney on many deals, said the discussion at buyout
companies typically does not focus on whether jobs will be created.
"It's the opposite, what jobs we can cut," Wolpow said, "because you
had to document how you were going to create value. Eliminating re-
dundancy, or the elimination of people, is a very valid way. Businesses
will die if you don't do that. I think the way Mitt should explain it is,
if we didn't buy these businesses and impose efficiencies on them, the
market would have done it with disastrous consequences."

Romney has stood by his assertion that he helped create a "net, net"
of tens of thousands of jobs, and Bain Capital officials said in 2011 that
his claim is accurate. However, neither Romney nor Bain provided
documentation of that claim. Nor is the claim something that can
be verified independently with anything approaching certainty. Many
companies that Romney held briefly were in private hands and changed
owners numerous times. They were saddled with debt, restructured,
and split up. Some companies under Romney's control prospered, and

some failed; some produced new jobs, and others shut down and left people out of work. It is possible that many of the companies in which Romney invested might otherwise have gone out of business entirely.

The best example of Romney's turnaround skills came when he saved his old firm, Bain & Company. In that case, 260 people were laid off, salaries and benefits were cut, and a painful restructuring plan was put into place. But without Romney's work, the entire enterprise might have sunk, taking Bain Capital with it. In the long run, Romney enabled Bain & Company and Bain Capital to grow—and that is Romney at his best, using the power of "creative destruction" to cut some jobs and eventually create new ones. Without those hard decisions, Romney's own job might not have survived, and he very likely would not have had a future in politics. "The goal of the investor in Bain Capital is to make absolute returns," said Howard Anderson, the MIT professor who has also been a Bain investor. "When they do well, Bain does well. When Bain does well, they do well. It is essentially capitalism at its finest—and its worst."

Romney was nine years into his fifteen-year career at Bain Capital when he began having thoughts that he would later describe as "irrational" but that in retrospect seem wholly predictable. It was 1993, and he was feeling both fulfilled and restless. He had earned huge sums—but not yet the hundreds of millions of dollars that later deals would bring—and seemed ready for something new. Everyone who knew Mitt well, from his fellow missionaries to his college classmates to his business partners, knew that he seemed to have his heart set on someday moving from business to politics. The only question was when.

He minutely analyzed his options, as always. The numbers looked good; his tally of accomplishments was impressive. But still something seemed missing. What he had achieved to date seemed to him more a means than an end. There had always been another goal held in reserve, one he had carefully safeguarded through the years. At many stages of his life, he had confided to colleagues: I need to be careful about this; I might run for public office one day. The thought was always there. Now he talked it over with Ann and then with his father, the author of his life's ambitions. In deciding when to exit the business of making money, George Romney's example was vivid. Mitt often recalled how his father had "walked away from success" and run for governor of Michigan. His father's business experience had been as pure as it got; George took over a failing company, stayed long enough to set it on a course for success, and left on his own terms. As Mitt later wrote, "Work was never just a way to make a buck to my dad. There was a calling and purpose to it. It was about making life better for people."

And so, in the security of his boardroom at Copley Place or at his family manse in Belmont, Romney kept asking himself "Do I really want to stay at Bain Capital for the rest of my life? Do I want to make it even more successful, make even more money? Why?" The answer, when it came, made his choice seem obvious: "I thought of my dad," he said. He would follow the family standard and enter the world of public service. And, like his father, he had it in mind to start near the top rung. Indeed, he had a big target in mind, with huge risks and also huge potential. Victory could put bring him national notice and put him on the path to the White House, the path on which George Romney had stumbled badly; defeat—well, it was too soon to think about that.

SEVEN

TAKING ON AN ICON

---

*I was getting ready for this guy that was going to be kind of a
doddering old fool. I'd be able to crush him like a grape.*

—MITT ROMNEY ON TED KENNEDY

He came from Welsh coal-mining stock and built his own
American dream: a job at General Motors; a successful engineering firm hatched in his basement; three children
who found a spiritual home within the Mormon church; and a bevy of
grandkids who knew him as "Pops." But those were better days. Edward
Roderick Davies, in his midseventies and living with his daughter and
son-in-law, Ann and Mitt Romney, was now fighting a losing battle
with prostate cancer. One afternoon in the early 1990s, as Ann was putting dishes away in their kitchen, he turned grave. "I'm so mad that I'm
dying," he said.

Davies wasn't looking for pity. Decades earlier, his own father,
having nearly been killed by a runaway coal cart, had brought his
family from South Wales to the United States in search of something better. Now confronting his own mortality, Davies wanted his
daughter to make the most of the opportunities she and her husband
had been given. "He said, 'Ann, you've got so much living to do.
Think of the exciting things that will happen in the world. I'm so

jealous of all the wonders you're going to see in your lifetime,'" Ann later recalled.

Edward Davies died not long after. But his words stuck. Ann suddenly couldn't bear the thought of looking back over her life with regret. Losing both her parents within a year had sparked a slew of existential questions: "Who am I? What am I doing? And what's life about?" The conversation with her father still resonated in 1993, when Mitt and Ann began talking about taking on Edward M. Kennedy in the upcoming Senate race. Mitt resented Kennedy's rakish behavior and rejected his liberal ideas. "I've been living here for twenty-three years, and I've been saying, 'Somebody ought to go after that guy,'" he later told a group of business leaders. Besides, Kennedy had been in the Senate since 1962, when Mitt was a geeky fifteen-year-old. Ann remembered thinking, "Are we going to die some day and then say, 'Mitt, you never did it? You never tried?'"

So she brought it up as they were lying in bed one morning that summer, as Ann and Mitt tell it. They'd been blessed, Ann told her husband, and now they should share that blessing. Invoking his family's political legacy, Ann told Mitt it was time he took on Kennedy himself. "You can gripe and gripe and gripe all you want about how upset you are about the direction the country's going," she recalled telling him. "But if you don't stand up and do something about it, then, you know, shut up and stop bothering me." Mitt pulled the covers over his head. "No! No! I don't want to do it," he said. "I just about fell out of bed," he would say later. But he couldn't deny that she had a point. He was, at forty-six, firmly established in his business career and contented with his life at home and church. He seemed to have it all. But that just left him with a question: what was left to do? So following a period of reflection, polling, and soundings with influential Republicans, Romney convinced himself that his wife was right. That October, after they secluded themselves, fasted, and prayed, Ann and Mitt Romney made up their minds: he would launch a campaign to oust Ted Kennedy, one of the United States' great liberal fixtures, from the U.S. Senate.

It was a bold proposition. This was Massachusetts, where Democrats reigned and the Kennedys were royalty. Over more than three decades in Washington, Ted Kennedy had established himself as one of the preeminent voices on the left, a champion of civil rights, education, health care, and the working class. And he was a political titan: no opponent had ever given him a serious scare. By 1994, however, the political climate had turned, and not in his favor. Around the country, voters had grown weary of incumbents. Republicans had begun, after years of political near irrelevance, to prove they could win in Massachusetts. And Kennedy himself had just emerged from a period of reckless personal behavior, further eroding his public standing.

In the decade between his divorce from Joan Kennedy in 1982 and his second marriage, to Victoria Reggie, in July 1992, Kennedy's reputation for womanizing, drinking, and partying had become the stuff of lore—and the shadow of Chappaquiddick still fell over his head. It surely didn't help his image with voters that in 1991, he had taken his younger son, Patrick, and a nephew, William Kennedy Smith, to a bar in Palm Beach, Florida, and that Smith had been charged with raping a woman he'd brought back to the Kennedy estate. Smith was later acquitted, but for Kennedy, whose testimony at the trial was nationally televised, it was a low point. He took criticism even from friends and friendly media. His efficacy in the Senate was diminished, which was especially evident during the Supreme Court confirmation hearings for Clarence Thomas. Kennedy, usually a forceful presence, played only a secondary role. Ultimately, he was forced to turn a planned speech at Harvard University into a public apology, a promise to mend his ways. "I recognize my own shortcomings—the faults in the conduct of my private life," Kennedy said. "I realize that I alone am responsible for them, and I am the one who must confront them." All in all, it was hardly the profile of a man invulnerable to political challenge.

Early internal polls confirmed a growing fear inside the Kennedy camp: "Kennedy fatigue" had set in. Many Massachusetts voters viewed him unfavorably. "There was, from the beginning, a sense of urgency about the campaign that had nothing to do with who was

on the other side," said one former Kennedy staffer. "It was dangerous territory." Nor did Kennedy, who was going on sixty-two, look or
sound especially good. He had gained weight. His face was mottled.
He spoke with labored breaths. "People said to me, 'You know, he's
getting older, he's over the hill. He's not coherent anymore,'" Romney
recalled. "I was getting ready for this guy that was going to be kind of
a doddering old fool. I'd be able to crush him like a grape."

In Romney, Republicans had found a perfect foil for Kennedy—in
appearance, worldview, background, and lifestyle. He was a fit, up-
standing Mormon with thick, winning hair and an unassailable home
life. He was brainy, well spoken, and enormously successful. All the
ingredients seemed to be in place. And with Kennedy faltering, it
seemed just the right moment. "This," Romney would later say, "is
the chance of a lifetime." But first Romney had a lot of questions to
answer. It would be his first foray into politics, after all, the first time,
really, that he'd been forced to ponder what he stood for outside business, family, and faith. What were his issues? What did he believe?
Sure, he was against Kennedy, but what was he for? In other words,
who was Mitt Romney?

He was, technically speaking, a first-time candidate. But he entered
the political arena with a seasoned self-confidence, a legacy of his parents. He'd learned something of politics at the feet of his father and
watched his mother, Lenore, wage her unsuccessful U.S. Senate bid in
1970. For the Romneys, as for the Kennedys, public service seemed to
be almost a prewritten chapter. And like the Kennedys, the Romneys
had the money to enter the fray at an immediate financial advantage.
Young Mitt had absorbed many lessons at his parents' side. He knew
the arc of a campaign. He knew the importance of communicating a
clear message and the pitfalls of straying from it. He knew the loss
of privacy that came with public life and that he must be wary of
the press—whose questions and scrutiny had helped bring his father's
presidential hopes crashing down. "He knew what the game was,"
said one veteran Massachusetts GOP operative. What he didn't know

much about was the state Republican electorate, whose support he would need—and quickly—to fend off primary challengers and build toward a showdown with Kennedy in the fall. He knew almost none of the key players, but it was, as it happened, a fortuitous time to be a fresh Republican face. After a long political winter, the Massachusetts GOP was on the rebound, in large part because of William F. Weld, who, campaigning as a fiscal conservative and social liberal, had captured the governor's office in 1990. Weld had modernized the traditional, moderate mold of Massachusetts Republicanism, and the formula seemed to hold great promise.

Even before all that, state Republican leaders, desperate to rebuild the party behind new names, had tried to recruit Romney into politics. "You ought to consider running for office—we could use a guy like you," Joseph D. Malone, then the executive director of the Massachusetts Republican Party, said he told Romney over lunch in the late 1980s. "You'd be a perfect fit for a run for governor or U.S. senator." Romney just chuckled and said, "Someday." So he stayed in the background, sending occasional checks and showing up at fund-raisers. He didn't even join the Republican Party until October 1993, switching his registration from unenrolled in preparation for his Senate run. He had given money to Democratic congressional candidates and had voted for Paul Tsongas, the iconoclastic liberal, in the 1992 Democratic presidential primary. He was, the veteran Republican operative said, "a very attractive unknown quantity."

Romney's enigmatic political identity would become a liability as he got deeper into the race. Some of his positions seemed to be calibrated for voter approval, not necessarily reflective of personal convictions. Strategy trumped ideology: what kind of candidate did he need to be to win? One conservative columnist complained that Romney was simply "philosophically vacuous." Over the top as that assessment may have been, it was indicative of Romney's challenge, even among those who should have been his natural allies. One thing, however, was clear from the beginning: change was the campaign's watchword, the theme around it they would build everything else. Kennedy, Rom-

ney felt, had simply lost touch with Massachusetts. "That was his entire focus," said Seth Weinroth, a lawyer whom Romney tapped to run
his state convention effort.

Romney opened his campaign headquarters near Fresh Pond in
North Cambridge, a short drive from his Belmont home. As in most
campaigns, factions developed; there were the Washington consultants who handled TV ads; the local team led by the chief strategist,
Charles Manning, and campaign manager Robert Marsh; the family
and friends who volunteered; the fund-raising team; and some of his
colleagues from Bain. At least in the beginning, they worked in concert. The team staged a campaign kickoff at the Copley Plaza in February 1994. Romney promised two hundred supporters that he would
go to Washington and "tame the monster" of government, saying, "It's
time to come home, Ted."

His bravado masked the practical concerns that lay ahead. Romney
needed the support of the Republican rank and file—many of whom
wouldn't have been able to pick him out of a police lineup—to get
past his GOP competitors, including John Lakian, a businessman and
unsuccessful candidate for governor in 1982. Early in 1994, Lakian
and Romney had lunch in downtown Boston at what is now the Lang-
ham Hotel. Romney urged him to drop his Senate bid and run instead
for the House against Gerry Studds, a Democrat from the South Shore.
Romney's rationale was that Kennedy would be hard enough to beat
without having to weather a primary battle. Lakian thought about
it but called Romney a few days later and told him no. Romney also
had to win the backing of at least 15 percent of the delegates at the
spring state Republican convention to qualify for the primary ballot
in September. So the Romneys set out for Republican meetings and
caucuses across the state, splitting up to woo delegates from different
regions on the same night. Mitt would go to one town, while Ann,
George, and Mitt's eldest son, Tagg, then twenty-four and just finished
with his degree at Brigham Young University, would each represent

the campaign in others. "We knew nobody," Ann said later. "We did not know a single Republican activist." George, who was eighty-six, was a particularly popular attraction, given his political résumé and reputation for brash honesty.

The campaign's media team, Greg Stevens and Rick Reed, meanwhile, was putting together Romney's first TV ad, a sixty-second biographical spot designed to introduce him to voters. In it, Romney earnestly described how he and Ann, with Tagg as a baby, had first come east on the Massachusetts Turnpike in a Ryder truck, deciding to make a life there. The early spots worked. "He just made great strides. The more he was known, the better he did," Reed said. Romney was also proving that he could raise money, above and beyond whatever contributions he might make from his own bank account. His fundraising success helped establish him as the GOP front-runner. The campaign organized a team of lawyers to carefully vet each contribution, to avoid accepting checks from unsavory donors.

In those early days, Romney was getting a rousing reception, and his team was in high spirits. He was David, cheered on for having the temerity to go after the lurching Goliath, and that aroused a kind of underdog spirit. "Clearly everybody understood that this was the tallest order and the most daunting challenge in American politics at the time," Weinroth said. Inside the campaign, the mood was serious but playful and informal. At headquarters, family and friends mixed easily with staff. They ordered so often from a nearby Domino's that pizza boxes littered the place. "To this day," a former staffer said, "I don't think I can eat Domino's pizza."

Ann assumed the role of den mother, supplying the troops with M&Ms, popcorn, and cookies; making sure the office was stocked with supplies; and coordinating volunteers. "She was looking out over everybody," said one former aide. Ann also expanded her role as a sounding board for her husband, speaking out when a phrase or message didn't sound right or a piece of clothing was unflattering—"Don't wear that shirt," she'd say. And she could dish out prickly retorts. After Janet Jeghelian, a former talk-show host and one of Romney's GOP rivals, called Romney an "empty suit," Ann fired back, "How would she know? When's the last time she checked?"

Romney, often in open-collar shirts and slacks, was heavily engaged in day-to-day campaign operations, former staffers said, but he let people do the jobs for which he'd hired them. "He doesn't sit in a campaign office and micromanage," said Michael Sununu, who oversaw research and policy. That had also been his way at Bain: gather smart people, focus on the target, and let them go to work. Often the Romneys' kitchen in Belmont, which opened up into a family room space, was a site for campaign meetings. Aides would spread their work out on the counters as the Romney boys ran into and out of the kitchen for food. "You kind of are immediately absorbed into this family atmosphere," one former adviser recalled.

On May 14, 1994, after several months of frenzied preparation, Romney faced his first political test. State Republican convention delegates gathered in a Springfield civic center to anoint a slate of candidates for that fall's races. Romney's game plan was to win big, knock out as many rivals as possible, and shred any doubt that he was the Republicans' sole hope against Kennedy in November. The campaign ran its operations from a motor home parked in the belly of the convention hall. Inside the command center, for most of the day, were four people: Seth Weinroth; Weinroth's assistant; and Romney's parents, George and Lenore. "Here was this major political heavyweight who had been elected governor of Michigan, ran for president, and served in Nixon's cabinet," Weinroth said. "I was expecting that he would be there all day, telling me how to do my job."

It was the opposite. Weinroth said George told him, "You're the guy. You're running the show. You just tell me what you need from me." So Weinroth did, employing George to twist the arms of wavering delegates. He sent George, escorted by an aide or volunteer, onto the convention floor for surgical strikes, and the patriarch would make a forceful case for his son. On at least one occasion, his tactics backfired. Penny Reid, a Republican state committeewoman who was backing Lakian, was so offended by George's pushiness that she told him she would work for Kennedy if his son won the nomination

But Weinroth's convention strategy—from courting the delegates to distributing foam baseball gloves saying "I'm with Mitt"—worked. Romney cleaned up, winning 68 percent of the vote on the first ballot. Lakian, squeaking by with 16 percent, was the only other candidate to meet the threshold, setting up a September primary battle between the two. And the prize of winning the GOP nomination glistened brighter by the day: *The Boston Globe* published a poll that weekend suggesting that a majority of Massachusetts voters no longer felt that Kennedy deserved another term.

Romney used his convention speech to attack what he called the "failed big brother liberalism" of the thirty-two years Kennedy had been in office, highlighting increased crime and welfare dependency. "I will not embarrass you," he told the crowd. "I will carry out with all my energies an attack, a resurgence of the principles you find dear." Lakian, who vowed to press on, found encouragement from an unlikely source. After his speech, he was shaking hands when George Romney approached him. "He came out and said, 'You gave the best convention speech, including my son,'" Lakian recalled. That evening, there was a private reception for staff nearby at a Sheraton. Romney had never been one to linger at parties, and this one was no different. As the staff, eager to blow off some steam after a grueling campaign stretch, celebrated over drinks, he made brief remarks, shook a few hands, and was gone.

———————

On paper, it was now Romney versus Lakian. In practice, the race between Romney and Kennedy was already well under way. Few doubted that Romney would win the Republican nomination. Some called on Lakian to drop out, which he refused to do. Instead he cast both Romney and Kennedy as children of privilege who could never understand the middle class. As the primary campaign unfolded, Lakian began to charge that Romney was more conservative than he was letting on, particularly on social issues, a hint of what would follow in the months ahead. It was the first time Romney had been pressed for his views on abortion, gay rights, guns, and other issues. "Ideologically, I'm not sure he knew where he was until he got into the campaign," Lakian said. "I think he was filling in the blanks as to what he did believe in."

When Michael Sununu, the campaign's research and policy guru, would sit down with Romney and talk through key issues like welfare reform, Romney, true to his Bain training, wanted to drill down into the details: Who supports this? Are there other alternatives? What does the national Republican leadership say about it? Less natural to him was the question "What do I think?" "There were a number of issues that were put before us that we responded to that I hadn't really given a lot of thought to," he recalled. He did his best to keep his focus on Kennedy. In August, the campaign distributed a slick videotape to voters, in which Romney, dressed in a blue-and-white-plaid shirt and addressing the camera directly, celebrated what he called "a real opportunity in Massachusetts to replace one of the most liberal members of the United States Senate." He urged people to show support for his campaign by dialing 1-800-TEDS-OUT.

Kennedy and his advisers, meanwhile, were furiously trying to ramp up a moribund campaign apparatus. They launched a registration drive targeting minority voters. They prepared an unusual summer ad blitz to repair Kennedy's image and remind voters of his accomplishments in the Senate. They hired additional advisers and even contracted with a Washington investigative firm to probe Romney's background. One former Kennedy aide described a summer gathering of Kennedy's statewide political organization in a ballroom in Hyannis, near the Kennedys' Cape Cod compound. It looked to the aide more like a Bingo convention. "I'm looking out over this crowd of six hundred or seven hundred people, and they're basically geriatrics," the aide said. "I remember thinking 'This is what happens when you're a senator and you really don't have a campaign, and you get reelected automatically every six years.'" That, the aide said, "was the point where I was most worried."

It only got worse. As Labor Day came and went, polls began showing Romney just behind Kennedy, then pulling even with him, thanks in part to an effective TV ad Romney aired challenging Kennedy's record on crime. With this Republican upstart suddenly level with Ted Kennedy, the race drew national, even international, attention. Kennedy and his advisers acted as if all this was to be expected—"polls do go up and down," Michael Kennedy, the senator's nephew and campaign manager, said at one point—but they were rattled. On September 20, Romney crushed Lakian in the Republican primary, fueling his momentum. Appearing before some five hundred supporters at the Sheraton Boston, Romney, flanked by Ann, their five sons, and his mother and father, cast himself as an agent of change, saying, "Now we ignite the final stage of this rocket, and the next stop is going to be the U.S. Senate." It was, for the Romney team, the high-water mark. "I remember the feeling," one former aide said, recalling the buzz among the staff, the volunteers, and the supporters. "It was a great night."

Romney said he told colleagues when he first joined the race that he thought his chance of winning was about one in twenty, although Ann was more optimistic. Charles Manning, according to Romney's account, was more blunt. "There's just no way you can win," Romney said Manning told him early on. Now, with the primary behind him, the wind at his back, and Kennedy seemingly on the run, Romney had changed his assessment. This was the race he'd wanted. New face against old. Change versus clout. And he liked his prospects. "After the primary," Romney recalled, "I began to think, 'Wow . . . I'm sort of catching on here. Maybe I could actually win this thing.'"

If Romney was beaming from behind the podium on primary night, the smile slipped from his face the moment he descended the stage. A TV reporter, nudged by the Kennedy camp, immediately challenged Romney on his record at Bain. Hadn't the firm slashed some jobs? "You saw the flash of anger," said one former Kennedy aide, describing Romney's reaction. It was something of an epiphany for the Democratic campaign: Romney seemed to have a glass jaw. "I walked in to Michael Kennedy and said, 'All we have to do is keep the pressure on this guy, and we can beat him.'" It was the first strike in what would become the Kennedy team's principal method of halting Romney's ascent: tagging him as a corporate raider who had made millions at workers' expense. For Romney, who claimed credit for having helped create some ten thousand jobs, the line of attack hit where it hurt. It was a soft spot he would never effectively address. "We didn't want to let him take a deep breath," the former Kennedy aide said.

Two days before the GOP primary, Kennedy; his wife, Vicki; and his top political advisers gathered at his Back Bay condominium for a Sunday-night strategy session. The mood was tense. Kennedy pollster Tom Kiley presented his newest numbers, which confirmed Romney's surge. The race was deadlocked. The room overflowed with aggressive personalities who clashed over what to do, while Kennedy ate from a bucket of Kentucky Fried Chicken that a maid brought him. "It was one of the most sort of strange 1960s French film–type moments, where you can't believe what's going on around you," recalled one adviser who was present. Kennedy agreed that they would have to raise and spend a lot more money than they initially planned. He knew his field organization would have to be rebuilt, and fast. Robert Shrum, one of Kennedy's closest advisers, read scripts from TV ads he had produced, which portrayed Romney as a heartless businessman. Shrum recommended that they begin airing the spots immediately. Kennedy was uncomfortable going negative on Romney, a campaign tactic he had never before been forced to use. But he was persuaded by the gravity of the situation, by the fact that Romney had been attacking him, and by Vicki, who understood the political trouble her new husband was in. Vicki's message was "This is real." So Kennedy signed off. They would go after Romney, and hard.

For months, Kennedy researchers had been quietly mining Romney's business record for political vulnerabilities. One recent deal caught their eye. A company called Ampad Corporation, which Romney's firm, Bain Capital, had acquired in 1992, had just purchased a paper products plant in Marion, Indiana, from SCM Office Supplies.

The day Ampad bought the factory, SCM fired the workers. Many were rehired, but at lesser wages and reduced benefits. The notice that workers received upon returning from their July Fourth weekend made clear that the layoffs were integral to the deal. It read, "The assets of SCM Office Supplies Inc. are being sold to Ampad Corporation. Therefore as of 3:00 p.m. today . . . your employment will end."

Having taken a leave of absence from Bain six months earlier, Romney was not directly involved in the firings. But it was still his company. One former Bain executive who sat on Ampad's board later said that Romney had the authority to resolve the dispute but instead let Bain managers focus on maximizing the company's investment. Romney was a private-equity man and a coolly brilliant one; he was not yet thinking as a candidate. The story was political gold for the Kennedy campaign, neatly aligning with its overarching goal of casting Romney as a self-interested capitalist. "It was devastating," said one former Kennedy staffer. But aides to Kennedy, thinking the senator would look weak coming out so aggressively against a comparative unknown, were leery of letting on about their involvement. So they encouraged the notion that the gift had fallen into their laps.

On September 1, 266 members of the United Paperworkers International Union went on strike at the plant to protest the deal. A union official called Kennedy's campaign, which arranged for the union to tell the Indiana press about the Romney connection, allowing Kennedy aides to suggest that the story had broken independently. In the meantime, the Kennedy team was making big plans. On September 26, the campaign sent a crew to Indiana to film the workers for TV ads. Tad Devine, a partner of Shrum, went to Marion with a script for the workers to read on camera. But Devine quickly scrapped it when he realized how much better the workers' own words were. Kennedy's campaign tested those and other spots with a focus group. The ads featuring the Ampad workers were the clear winners.

Three days after the filming, on September 29, Kennedy went up with a series of six thirty-second TV spots featuring nine Ampad workers. They were withering. The workers—angry, plagued by economic uncertainty, and as authentic-sounding as they come—were more than happy to lay the blame at Romney's feet. "He has cut our wages to put money in his pocket," one worker said. Another, Sharon Alter, a packer who had been laid off after twenty-nine years with the firm, said, "I would like to say to Mitt Romney, 'If you think you'd make such a good senator, come out here to Marion, Indiana, and see what your company has done to these people.'" Robert White, a fellow Bain executive who had joined Romney's campaign for the general-election push, complained that the firm had invested in dozens of companies, the vast majority of which had been successful. "Wouldn't it be nice if Kennedy would run some ads about the forty success stories instead of the handful that weren't?" But White was no expert in politics. As Tad Devine said in defending the Ampad spots, "I don't think we are under an obligation to tell Mitt Romney's side of the story."

As the TV onslaught continued, the Romney campaign was caught flat-footed. At first, Romney justified layoffs as sometimes being necessary to revive a business. "This is not fantasy land," he said. "This is the real world. And in the real world, there is nothing wrong with companies trying to compete, trying to stay alive, trying to make money." Romney's pollster Linda DiVall started noticing that the Ampad ads were hurting, because voters, in open-ended questions, began describing Romney as coldhearted. Still, his favorability rating remained high, so the campaign didn't panic. Romney did not air a TV ad giving his side of the story. Instead his campaign tried to change the subject, airing a spot criticizing Kennedy over his opposition to the death penalty. The paperworkers' union then doubled down, sending a group of striking workers to Massachusetts as a "truth squad" to hound Romney on his home turf. They distributed leaflets blasting Bain and depicting Romney dressed as an executioner. On October 7, with reporters and cameras on hand to capture it all, strikers and local union officials got into an angry confrontation with Romney strategist Charles Manning outside Bain headquarters, shouting "union buster" as a deal to meet with Romney fell through. Two days later, the workers confronted Romney at the start of a Columbus Day parade. "I'd love to help and I'll do my best," Romney told them. "But there's a separate management team running the company Ampad. I don't work there." One worker fired back, "I don't anymore, either." That night, Romney sat down with the workers at a hotel in suburban Newton to try to persuade them that the layoffs weren't his fault.

But the damage was done. The Kennedy ad campaign would prove pivotal. Romney's poll numbers began to slip. Despite his strategic mind, he had failed to adequately prepare for a most predictable line of attack—criticism of his business career. After he took leave from Bain, the campaign had not monitored the firm's business deals for potential political fallout. "It's something we should have anticipated, maybe not Ampad specifically, but an attack on his business record," Robert Marsh, Romney's campaign manager, said later. "I do blame myself for not reading the signals." The whole thing began to wear on Romney. Lining up for the East Boston Columbus Day parade, Joseph Malone, the state treasurer, came upon Romney outside Santarpio's, a legendary pizza joint. "You can tell by looking at a candidate's eyes whether they're weathering the fight well or whether it's negatively impacting him," Malone said. "He looked like he'd been through the wringer." Malone remembered telling Romney to keep plugging away, that voters would give him a fair shot. "He goes, 'I don't know. What I'm hearing is these negative ads take their toll. And it sure does feel that way.'"

The Ampad episode taught Romney several lessons, some of which would inform his future adventures in politics. One, holding your fire when under attack is not an effective strategy. Two, the Kennedy political machine, whatever Kennedy's weaknesses might have been, was still a dangerous thing to rouse. And three, it was difficult to go from being a CEO—accustomed to controlling information, operating in a quantitative climate, and calling the shots—to being a candidate in a high-wattage political campaign. Being a candidate also meant being handled, leaving certain decisions to the experts. "Letting go of that was really hard for him at first," one former staffer said. Romney was, after all, a child of Detroit, descended from a long line of pragmatic, take-charge men. He liked to drive his own cars, change his own oil. He liked holding the reins. As the Ampad ads went up, Romney—along with his father—wanted to hit back hard and to respond in kind on TV. But the Washington consultants believed it would be a waste of money. "He felt almost powerless that he couldn't make that decision," the fomer aide recalled.

At the same time, Romney resisted pleas from others in the party to attack Kennedy on his personal life. "You'd get Republicans who would come up and say, 'You gotta go at him on Chappaquiddick! You gotta go at him on his lifestyle!'" one former Romney aide said. It might have been an appealing idea to a man who believed that political leaders' personal shortcomings threaten the nation's values. But Romney would politely ignore the suggestions. It wasn't his style, and, perhaps more important, it wasn't something he thought would advance his campaign.

The race against Kennedy was, for Romney, his first opportunity to define himself in the public eye. From the beginning, his campaign made a deliberate effort to highlight his economic conservatism but present him as an acceptable choice on social issues to independents, wayward Democrats, and especially women. Mindful of Weld's success in 1990 as a socially liberal Republican—and of Kennedy's legacy as a defender of civil rights—Romney set out to prove that he would be as good as, or even better than, Kennedy in advancing those causes in Congress. That moderate, nuanced political profile, akin to the one established by his father, seemed to be in his blood. But it was also a strategy conceived to win over the left-leaning state electorate. One Massachusetts Republican leader recalled a conversation with Charles Manning early on in which Manning said, "We've got this thing mapped out, and Mitt has bought into the idea that the key to victory here is that he's a Bill Weld Republican." No one, the thinking went, would be able to paint Romney as a right-winger. "That was really the foundation of the campaign," the Republican said.

Romney's willingness to embrace socially moderate, even liberal, positions—Romney himself preferred the term "socially innovative"—made him an attractive candidate for groups such as the Log Cabin Republicans, a grassroots GOP gay and lesbian organization. In September, as Romney was seeking the group's endorsement, he sat down with Richard Tafel, the group's founder, and a local leader, Mark Goshko, and received a primer on gay rights issues, from antidiscrimination legislation and the military's "Don't ask, don't tell" policy to the scourge of AIDS. Romney was deeply engaged, asked probing questions, and noted that he had gay employees at Bain. "I'd met with businessmen and politicians, and this felt like a business meeting. It felt much more pragmatic," Tafel said. Romney's approach was "What do I need to do here? How do I get this done?" One Massachusetts Republican who has known Romney for years summed up his approach this way: "In Mitt's mind, it doesn't matter what my positions are. I'm someone who solves problems." Toward the end of the meeting, Romney turned to Tafel and said, "Now on the Boy Scouts, you wouldn't want gay Scout leaders, would you?" Romney was on the executive board of the Boy Scouts, which banned homosexuals from participating. Tafel, who, as an openly gay appointee, had previously directed the state's adolescent health program under Weld, explained why that question was offensive and why he believed the Scouts' policy was so wrong. "I'm with you on this stuff," Tafel recalled Romney saying. "I'll be better than Ted Kennedy."

The Log Cabin leaders told Romney they wanted his commitment in writing before the organization would make an endorsement. So not long afterward, Romney wrote a letter thanking the group for the meeting and asserting, "I am more convinced than ever before that as we seek to establish full equality for America's gay and lesbian citizens, I will provide more effective leadership than my opponent." Same-sex marriage was not on the table at that point, so he made no mention of it. Romney had previously said that, like Weld, he did not believe gay marriage was "appropriate at this time," but he was willing to follow Weld's lead. Romney, in the letter, did promise to cosponsor a federal employment nondiscrimination act to protect gays and lesbians in the workplace, expressed concern about suicides among gay and lesbian youth, and said he believed that "Don't ask, don't tell" was "the first in a number of steps that will ultimately lead to gays and lesbians being able to serve openly and honestly in our nation's military." He continued, "That goal will only be reached when preventing discrimination against gays and lesbians is a mainstream concern, which is a goal we share."

All this was music to the ears of the Log Cabin Republicans, who saw great value in having a high-profile Republican voice speaking out for their cause. Every indication was that Romney, like Weld before him, genuinely cared about gay rights and gay people. Even after the group endorsed Romney, his campaign would sometimes call and ask for guidance on things. A couple of years later, a gay man who had worked on Romney's Senate bid, Eduardo Paez-Carrillo, was dying of AIDS. He was deeply touched by receiving a phone call from Romney.

Abortion burned even hotter as a front burner issue in 1994. Romney established himself as a passionate supporter of abortion rights early on in the campaign, despite his personal opposition to abortion. In fact, his professed views would grow more liberal over the course of the race. Romney initially said he opposed Medicaid funding for abortion. He later softened that position to say he favored leaving the question of coverage up to the states. Romney also endorsed the legalization of RU-486, the abortion-inducing drug, and appeared in June at a fund-raiser for Planned Parenthood. Ann Romney gave the group $150.

Though the Church of Jesus Christ of Latter-day Saints was firmly opposed to abortion in all but a few circumstances, Romney asserted that his family had supported a woman's right to a safe, legal abortion ever since the October 1963 death of his brother-in-law's sister, Ann Hartman Keenan, from complications following an illegal abortion. Keenan's death—she was twenty-one—was devastating to her family and to the Romneys, with whom she was very close. "Ann was like a part of our family," said Romney's sister Jane. "Since that time, my mother and my family have been committed to the belief that we can believe as we want, but we will not force our beliefs on others on that matter," Romney said in an October debate, pushing back against Kennedy's charge that he was "multiple choice" on abortion. "And you will not see not see me wavering on that."

It's true that Lenore Romney had openly said during her 1970 Senate campaign that she supported the liberalizing of abortion laws. At the time—this was three years before the Supreme Court legalized abortion in its landmark *Roe v. Wade* decision—abortions were illegal in the state of Michigan. But it is overstating things to say, as Romney has since, that she truly "championed a woman's right to choose." In June 2005, *Boston Globe* columnist Eileen McNamara quoted two former Republican heavyweights close to Lenore's Senate campaign, Elly Peterson and William Milliken, who said they had no memory of Lenore campaigning on abortion rights. Nor does David Plawecki, a twenty-two-year-old prolife Democrat running for the Michigan Senate that year against a Republican, N. Lorraine Beebe, who actively supported abortion rights. Abortion had been a defining issue in his own successful race, Plawecki said, but not in Lenore's. "I cannot recall that being any part of her central theme at all," he said. Lenore's comments in a May 1970 story in the *Owosso Argus-Press*, a newspaper in Owosso, Michigan, perhaps best captured her ambivalence. Asked for her views on abortion, she was quoted as saying, "I think we need to reevaluate this, but do not feel it is as simple as having an appendectomy." She went on to say, "I'm so tired of hearing the argument that a woman should have the final word on what happens to her own body. This is a life."

As the 1994 race went on, voters had increasing difficulty discerning who Mitt Romney was. His outreach to social moderates was undermined by contrary things he was reported to have said in private settings. And as the race matured, his political definition—some amalgam of his core values, what his church taught, and what Massachusetts voters had come to expect from Republican hopefuls—never came into sharp relief.

That summer there was a report that Romney had told a Mormon gathering the previous fall that he was alarmed by reports of homosexuality in the congregation, denouncing it as "perverse." Four people at the gathering confirmed what he had said, though Romney, who at the time of the alleged statements was president of the church's Boston stake, denied making them. Then, in the fall, Romney acknowledged that as a Mormon leader he had counseled women not to have abortions except in rare instances where the church says they may be justifiable: rape, incest, if the mother's life is at risk, or if the fetus has a severe abnormality. That revelation followed the anonymous account in Exponent II, the local Mormon feminist journal, by the mother who said Romney pressured her not to have an abortion, despite her precarious medical condition. Romney maintained throughout that he was representing the tenets of his church as a private citizen, which he said would have no bearing on his official duties as senator. Further muddying his pitch to moderates was an endorsement of sorts from a leader of Massachusetts Citizens for Life, an antiabortion organization. The rationale was simply that Romney would be better than Kennedy, even if only marginally. Romney's campaign accepted the nod but quickly distanced itself from the group, saying it had never met with Massachusetts Citizens for Life or filled out a questionnaire and that the organization had misrepresented his views.

Indeed, Romney was groping for a comfortable perch within the GOP. He pushed some reliably Republican themes, including requiring welfare recipients to work, cracking down on crime, and creating private-sector jobs. But he often strayed from the party plank as he sought to broaden his base of support. He supported raising the minimum wage and tying it to inflation, echoing Kennedy on one of the senator's main causes and putting himself at odds with many business leaders. He rejected a national Republican proposal to trim the overall tax rate on capital gains. He backed two gun-control measures that were strongly opposed by the National Rifle Association: the Brady Law, which imposed a five-day waiting period on gun sales, and a ban on certain assault weapons, saying, "I think they will help." Romney also discouraged the Christian Right from contributing money or airing anti-Kennedy ads on his behalf. And he distanced himself from the "Contract with America," the political blueprint of Republican leaders in Washington, dismissing it as too partisan. "I'm not going to Washington to toe the line," he said in one debate.

And then there were times when Romney was simply hard to pin down. He expressed lukewarm support for John Lakian's proposal for a flat income tax. He was critical of federal spending but also questioned Kennedy's effectiveness in bringing government dollars home to Massachusetts. Late in the race, Romney told an interviewer that he would have reluctantly voted for a universal health care plan pushed by John Chafee, the moderate Republican senator from Rhode Island. A linchpin of Chafee's plan was a federal mandate that individuals buy health insurance, a principle that years later Romney would embrace on a statewide scale but vigorously oppose as national policy. "I told people exactly what I believed," Romney said of the 1994 race. Kennedy, too, seemed to be moderating his views as the political sands shifted beneath him. He tried to retain his liberal credentials while keeping pace with a new, more centrist orientation in Democratic politics, represented, at times awkwardly, by President Bill Clinton. Kennedy came out for "workfare"—requiring able welfare recipients to find work in exchange for benefits—after rejecting similar bargains in the past. And he tried to cast himself as a crime fighter—a hard sell.

The most delicate issue in the race, though, had nothing to do with social policy, economics, or the direction of the country. It was much more personal. Romney was forced into the uncomfortable position of having to defend something he felt had no place in the campaign: his faith.

Everyone was there—print reporters, photographers, TV cameras. This was a press conference one didn't want to miss. It was September 27. Romney's campaign had summoned the media to its headquarters on Moulton Street in Cambridge. Questions about his faith had reached the point where Romney felt he had to address them head-on. Several days earlier, Kennedy's nephew Joseph P. Kennedy II, the self-styled "pit bull" of his uncle's campaign, had criticized the Church of Jesus Christ of Latter-day Saints over its exclusion of blacks and women from leadership positions. He later apologized, saying he didn't know that the church had let blacks join the priesthood in 1978. But then, on September 26, Senator Kennedy—despite previously insisting that religion had no place in the campaign—said Romney should be asked where he stood on the Mormons' racially exclusive policies in the past. "Where is Mr. Romney on those issues in terms of equality of race prior to 1978 and other kinds of issues in question?" the senator said.

So Romney took to the podium and invoked the words of Kennedy's own brother, John F. Kennedy, who confronted skepticism about his Catholic faith in 1960 by saying, "I do not speak for my church on public matters, and the church does not speak for me." Romney said that John F. Kennedy's presidential victory that year "was not for just forty million Americans who were born Catholic, it was for all Americans of all faiths. And I am sad to say that Ted Kennedy is trying to take away his brother's victory." As Romney was talking, George Romney, who had been circling the press gaggle, shaking his head and deeply agitated, suddenly couldn't keep quiet any longer. "George Romney literally dove into the scrum," a former Romney aide recalled. "He shoved his arms in between reporters like the parting of the Red Sea." Everyone turned to him. "I think it is absolutely wrong to keep hammering on the religious issues," said George, who had experienced anti-Mormon bias in his own life. "And what Ted is trying to do is bring it into the picture." Romney, who had come to expect his father to do things like that, made light of the outburst and then retook control of the news conference.

Romney's faith posed an unusual quandary—not only for him but for Kennedy and his aides, for the press, and for voters. Was it fair game in a political context? Was it off-limits completely? Or did it fall somewhere in the middle, especially as it informed his social views? They weren't always easy questions to answer. And they surely weren't helpful to Romney. Christopher Crowley, who took over as Romney's research director, said at the time, "Every time it's brought up, it makes women think, 'Is he really prochoice?'" Romney's media team even produced a TV ad in which Romney, addressing the camera in short sleeves from his driveway, said that John F. Kennedy's election should have ended "religious bigotry" in politics. "All my life I've been guided by a set of strongly held beliefs. One is my religion. Another is tolerance of others," he said. "Unfortunately, some in this campaign have chosen to use religion against me." But his campaign never aired the spot. Kennedy backed off on Romney's religion, and Romney and his aides decided that running the ad was unnecessary. His bitterness at the whole episode, though, would not fade easily.

———————

Life on the Romney campaign, as the press conference had shown, was never dull with George around. And one never quite knew when George would be around. One former aide recalled, "I would ask Ann, 'So what is George Romney's schedule for the next week?' She goes, 'Oh, we never know.'" George's schedule was his own, but he was deeply invested in his son's success. He was also in his late eighties at that point but irrepressibly defying his age. He would fly in to Boston's Logan Airport from Detroit, take the bus to the subway, catch the Blue Line into the city, pick up the Red Line outbound to Cambridge, take it to the end of the line, and then walk, wheeling his bag, the mile or so to campaign headquarters. Then he'd walk upstairs and say simply, "Is Mitt here?" Sometimes he'd arrive with his thoughts about the campaign sketched out. "He would get all kinds of ideas going, because he was on the plane," one aide remembered. "George Romney had a sense that Mitt had to be his own man, his own candidate, but wanted to share his own strategies." Eventually, he and Lenore moved into the guest suite above Mitt and Ann's garage. Campaigning for his son one day at a nursing home in Salem, north of Boston, George was asked if Mitt wasn't very much like himself back in the day. "He's better than a chip off the old block," George replied. "No. 1, he's got a better education. No. 2, he's turned around dozens of companies while I turned around only one. And No. 3, he's made a lot more money than I have."

On the campaign trail, Kennedy and Romney had distinct styles and strengths. Romney was a far more polished speaker—authoritative, confident, and cogent in his arguments. But Kennedy, having honed his street skills over more than three decades, was a more natural campaigner, always finding a way to connect with the person on the other end of a handshake. Making such fleeting personal connections stick was hard for Romney; sometimes, indeed, he seemed painfully awkward. A series of campaign appearances filmed by C-SPAN captured the difference between the men: Kennedy, leaving a Rotary Club event in Salem, comes across a man with a cigar dangling from his mouth. "I used to smoke those myself," Kennedy tells him, eliciting a knowing laugh. Romney, shaking hands in Waltham, just west of Boston, approaches a woman outside a convenience store, who turns from him. "Don't run," he says. "I'll shake your hand anyway." The woman stops and puts her hand to her face. "I know, you haven't got your makeup on yet," Romney says good-naturedly. "I do! I do!" the woman protests, mildly offended. "You do! You do!" Romney responds with a nervous laugh. "Good to see you!"

One side of Romney that voters rarely saw on the campaign trail was his tremendous charitable impulse when people needed help. They were often, but not always, people he encountered in his church. After Kennedy's assault on his business career set him off, Romney did complain publicly that no one saw all his compassionate work as a private citizen, a theme he would revisit in a televised debate. But he largely kept it out of the spotlight. There was the time he took an entire day off from campaigning so he and his sons could help a single mother move, because she couldn't afford movers. Or the time he called Rick Reed's mother in the hospital when she was dying of cancer. Romney made a point of telling her how proud he was of her son. "It was something I'll never forget," Reed said. But the story that perhaps best illustrates Romney's charitable streak in those days came late in the campaign. It began with a gaffe.

About a week before election day, Romney spent a half hour courting voters at the New England Shelter for Homeless Veterans, a regular stop on the political trail near Boston's City Hall. After giving his pitch, Romney was talking to the center's director, Ken Smith. Romney asked him what his biggest problem was. Smith said that just that morning, he'd met with the guy in charge of food services for the shelter and learned that the high price of milk was killing their budget.

And they went through a lot of milk, some one thousand pints a day. Romney attempted a joke: why don't you just teach the veterans how to milk cows? Then he was out the door. Smith was stunned. "One of the press guys said, 'Did he just tell you to have veterans milk cows?' I said, 'That's what I heard,'" Smith recalled. "It did rub everyone the wrong way."

But then, several days later, Smith got a phone call. It was Romney. "He personally called and said something to the effect of 'I want you to know that I truly do support American veterans, regardless of what has been said or how this was portrayed.'" Romney told Smith he wanted to cover part of the shelter's milk costs, and he didn't want any publicity for it. "He said something like 'I'm looking to do it because it's the right thing,'" Smith said. He didn't know exactly how Romney had done it—he figured Romney had arranged something with one of the shelter's milk suppliers. But now, instead of paying for a thousand pints a day, the shelter was paying for just five hundred. And it wasn't just some political stratagem. "It wasn't a short-term 'Let me stroke you a check," he said. "It happened not once, not twice, but for a long period of time." In fact, Smith said he understood that Romney was still supporting the shelter when Smith left in 1996.

———————

With Kennedy pulling ahead in the closing weeks of the race, Romney had but one hope left to arrest his own slide. The two candidates, after furious negotiations, agreed to meet for two debates. The first, scheduled for October 25 at Boston's historic Faneuil Hall, promised to be an especially high-stakes clash. With a commanding performance, Romney could show everyone on live television why he, and not Kennedy, belonged in the United States Senate. The day of the Boston debate, a poll published in the *Boston Herald* underscored the pressure on Romney for a game-changing night: Kennedy was up by eighteen points. That evening, Kennedy supporters, including many union members, were out in force around Faneuil Hall. Many had been drinking. They were rowdy, even menacing. Tussles broke out. "I saw little old ladies getting kicked over by guys in hard hats," recalled one former Romney aide. There was, the aide said, "palpable hostility toward Romney." Even though he had many supporters on the streets, too, Romney didn't linger long, slipping into the building soon after he pulled up. Kennedy basked in the show of support, however unruly, clasping hands and pumping his fist, and he and Vicki entered the hall.

Kennedy had the benefit of low expectations—peerless with a script, he could be halting and incoherent as an extemporaneous speaker. Aides worried that a major stumble could be damaging. But his campaign did score one important aesthetic victory by persuading Romney's advisers to agree to larger podiums. The giant wooden boxes made both candidates look smaller, which was bad for Romney but good for Kennedy, because, as his aides intended, it concealed his ample girth. "You could stay overnight in this podium," Kennedy cracked at one point beforehand.

Faneuil Hall was electric that night, packed with supporters of the candidates, opinion leaders and journalists from around the country, and a lively audience speckled with boldfaced names. For Romney, it was one of the biggest moments of his life. He started strong. He did not shrink from Kennedy's stature. He tagged the senator as a relic of the 1960s. He offered a full-throated defense of his beliefs and values, criticizing the Kennedys for drawing attention to his Mormon faith. He accused Kennedy of misleading attacks on his business record. But as the debate unfolded, Romney made crucial missteps while Kennedy gained steam, particularly after a memorable, rehearsed retort to Romney's charge that he had benefited from a no-bid real estate deal. "Mr. Romney," Kennedy thundered, "the Kennedys are not in public service to make money."

When Kennedy asked Romney what health care plan he favored, Romney answered with a solid summation of his beliefs. But after Kennedy followed up by saying, "What is the cost of your program?" Romney said, "Uh, I don't have a cost to my program, Senator Kennedy." "You don't have a cost?" Kennedy replied with exaggerated incredulity. Romney complained that, unlike Kennedy, he didn't have Congressional Budget Office number crunchers at his disposal. But it sounded like an excuse, and Kennedy, in his best patronizing tone, succeeded in painting Romney as an amateur who didn't understand lawmaking. Romney had rehearsed an answer to that very question in debate prep—something along the lines of "Senator, since when did you care about the cost of programs?" But he didn't use it.

Not long after, in defending himself against Kennedy's charge that he was a political clone of Ronald Reagan, Romney said, "Look, I was an independent during the time of Reagan-Bush. I'm not trying to return to Reagan-Bush." It was at that moment that some of Romney's fellow Republicans felt the race slip from his grasp. John Lakian, who was doing commentary for one of the local TV stations, looked over at his Democratic counterpart and said, "I think the election is over." Reagan had won Massachusetts twice, and Romney needed those voters. His answer also revived the same pesky question that had dogged him for months: what did Mitt Romney believe? "When you deny Ronald Reagan when you're a Republican, then nobody knows where you are," Lakian said. "You just become another person running against Kennedy."

And then, toward the end of the night, Romney fumbled an appeal to working women. "Women are concerned about the glass ceiling," he declared, attempting solidarity. He asked Kennedy what the senator would do to break it. Kennedy had been around long enough to recognize the softball question that it was. Almost smiling as he recited his legislative record, Kennedy said, "You're not going to find a member of the United States Senate that's a stronger supporter." Robert Shrum, Kennedy's senior adviser, said at the time, "That question was like asking Babe Ruth how to hit a ball." Romney overall didn't perform badly, especially for a novice. But Kennedy fared better, and with some three million people watching. With less than two weeks to election day, a fresh *Boston Globe* poll put Kennedy up by twenty points.

A few weeks earlier, a Kennedy sign holder in a run-down part of Boston's Dorchester neighborhood had warned Romney to steer clear of "Kennedy country." Romney figured he could turn that phrase on its head, using it as shorthand for the appalling urban blight that had persisted through years of Democratic control of the state. It would neatly capture, he thought, the failure of Kennedy's liberal policies. Romney and his wife discussed making it a campaign slogan, but advisers dissuaded them. Then, in his final debate with Kennedy, Romney used the phrase anyway, going on to make it the centerpiece of his closing message to voters. He launched a tour of economically depressed neighborhoods and aired a late TV ad bemoaning the "legacy of Kennedy country—a legacy of failed programs, welfare dependency, and rising crime."

The Kennedy camp seized on the chance to cast Romney as a suburbanite who didn't know his way around a city. "What do you expect from a candidate from Belmont?" Boston Mayor Thomas M. Menino said at a late-October event arranged by Kennedy aides. On the Southeast Expressway, one of the main highways into and out of Boston, a union local put on its electronic signboard KENNEDY COUNTRY, AND PROUD OF IT. Ann Romney hadn't helped things with an interview she'd given in which she had said that she and Mitt had struggled in college because their only money had come from his selling of stock. And she seemed not to know that West Roxbury, a wealthier, more suburblike, and mostly white neighborhood of Boston, was distinct from Roxbury, which was largely black and poor. Those who already thought the Romneys lived a fairy-tale life weren't about to change their minds.

The campaign spent $100,000 to air an unusual half-hour infomercial on Boston TV stations beginning the Friday night before election day. Romney toured the state by bus and by rail, flanked by relatives who had descended on Massachusetts. No fewer than sixty-eight Romneys joined the candidate on a whistle-stop train tour the final weekend, decked out in sweatshirts reading ROMNEYS FOR MITT and TKO SQUAD, short for TED KENNEDY OUT. In the days leading up to the vote, Romney's staff saw the writing on the wall but held out hope for an unlikely victory. Romney himself knew how it would end, but it wasn't his nature to just fold up the tent. He remained sunny and optimistic, wanting each person on the campaign team to take something positive away from the race. "Everybody felt they fought a good fight," one former Romney aide said. Romney had given Kennedy a run, the best he'd ever have. But in the end, it would not be nearly enough.

———————

At about 8:45 a.m. on election day, Romney, joined by his wife, sons, and father, strode into Belmont Town Hall and cast a ballot for himself. He knew that most of the state would be voting differently that day and was already being asked about his next run for office. George and Ann Romney seemed to be taking it harder than Mitt was. "There wasn't a thing he said about my son that was true," George complained of Kennedy. After he voted, Romney sat and watched his boys play touch football on the lawn, a pastime long associated with the political dynasty that was on its way to another victory at the polls. That night, Romney put on a brave face before cheering supporters at the Westin Hotel in Boston, appearing onstage shortly after 10 p.m. to concede. Kennedy won by a comfortable seventeen points, 58 to 41 percent. But Republicans took heart that they had swept into power on Capitol Hill, a wave that Romney celebrated in his speech. "Sometimes in isolated little pockets there may be some sandbags of clout and pork and seniority that are able to hold back that tide," he said. "But let me tell you the tide keeps rising and we are going to make it keep rising."

Ann Romney, worn out by the brutal campaign, descended from the stage and, once out of the TV lights, burst into tears. "It was heart-wrenching," a former aide recalled. "They gave it their all." It was Ann who'd gotten them into this, and now it was Ann who was vowing never to return to politics. "You couldn't pay me to do this again," she said toward the end of the race. But she also left an opening, saying on election day that if one asks any new mother right after delivering a baby whether she will have another child, "She will always say no."

As the Romney team watched Kennedy deliver his victory speech, they saw him revert back to what one former Romney adviser called "the old Teddy"—the one who often had difficulty stringing sentences together. "Why couldn't he have done that last week?" Tagg asked in exasperation, the former adviser said. One irony was that Kennedy had spent the campaign trumpeting his outsize influence. That very night, with Republicans taking control of the U.S. Senate, Kennedy and his Democratic counterparts were instantly marginalized.

For Romney, it was the first major defeat of his life, and he didn't swallow it easily. He had sunk $3 million of his own money into the race and was disgusted with how effective Kennedy's attacks had been. In future years, Romney would develop a productive relationship with Kennedy and come to respect his political chops. Asked two weeks after his loss if he harbored any ill will toward the man, Romney said, "I wouldn't say ill will, but I was surprised he went as negative on my character and I was very surprised he brought up my church. I was surprised he went after my business record with such variance with the facts." The mistakes and missed opportunities of the campaign, he said, kept him up at night. It wasn't just the attacks that did Romney in, though, and he knew it. He had failed to make a compelling enough case for himself, failed in crafting a narrative of his character and convictions that could move voters. "I think people were searching for something else, as much as they liked the Kennedys," Rick Reed said. "We were never able to capitalize on it." Or, in the words of one longtime Republican, "His main cause appeared to be himself." Over dinner not long after the race, Romney told a fellow party member that one thing really ate at him: that one couldn't sum up in a sentence why he had run. "After all the weeks and months of that campaign, if you ask, 'Why did Mitt Romney run for U.S. Senate, and what did he stand for?' most people had no clue," the Republican recalled Romney as saying. "We didn't do a good job getting that message across."

Not that Romney was just sitting around Belmont marinating in self-pity. The morning after the election, he waltzed back into the Bain office as if he'd only been away on a short Caribbean cruise. He called a meeting, dished out assignments, and reviewed forthcoming deals. "I'm sure I ruffled a lot of feathers," he wrote in *Turnaround*. "But Bain Capital was my baby, and I was back in town." Yet Romney could not deny that even with the bitter taste of defeat still fresh in his mouth, the pull of public leadership remained strong. At that moment, he may have been comforted by returning to the familiar embrace of private equity. But the restlessness would return soon enough.

EIGHT

THE TORCH IS LIT

---

*The best way I can describe what he faced is trying
to rebuild an airplane while it's flying.*

—AN ALLY, DESCRIBING ROMNEY'S EFFORTS
TO SAVE THE 2002 OLYMPICS

In their sixty-four years of marriage, George Romney was said to have brought his wife, Lenore, a rose every day. At the dawn of their courtship, he had gone to tremendous lengths to win her hand—trailing her when she ventured out with other suitors, buying her a piece of cake each day in the school cafeteria, and once actually yanking her off a dance floor when he thought she'd been out there too long with another boy. And after they married his devotion continued, inexhaustible.

They built quite a life together: four children, six years as Michigan's first family, and many adventures in politics, business, and faith. They had their disagreements—as their grandchildren would later attest. But Lenore was a partner in George's many successes. The roses were a daily ritual of love and thanksgiving. A little after 9 a.m. on July 26, 1995, Lenore awoke to find no rose. She knew right away: he was gone. She found him collapsed on a treadmill in the exercise room of their home in Bloomfield Hills, Michigan. The man who for so many years had seemed immortal was dead of heart failure at eighty-eight.

On his birthday two weeks earlier, George had been in Los Angeles to see his younger daughter, Jane, star in the opening performance of the play *Shadowlands* at the Tracy Roberts Theatre in Beverly Hills. Her father, Jane said later, grasped her love of acting that night. "He got why I was an actor in that performance," she said. "He had just given me that gift." But George Romney had, over many decades, given Mitt Romney much more. From the day he was born, Mitt enjoyed the love of a father who never seemed to lose his excitement at this youngest son, the son whose very existence the doctors had said was improbable. George became more than just a role model or mentor to Mitt. He was a pathfinder, showing the way in their Mormon faith, through the thickets of politics, in family life, and in character. Through his achievements and his mistakes, George had bestowed many lessons, and Mitt soaked them up. "His whole life was following a pattern which had been laid out by his dad," said John Wright, a close family friend.

Some twelve hundred mourners came to bid George farewell at his funeral service. All four children spoke. Mitt's older brother, Scott, said his father liked to say that the family was descended from mules. "The Romneys tried to do the right thing, and everyone else was stubborn," he said to laughs. Mitt, too, leavened the service with humor, saying that his father was probably aware of what was happening. "He got here good and early," Mitt said. "His eyes are closed, but I'm sure he's listening. And he'll be the first to leave." After the service, George was buried in a plot he had chosen in the city of Brighton, roughly halfway between Bloomfield Hills and Lansing, the state capital. As she said her final good-bye, Lenore returned her husband's gesture of devotion, laying a single red rose on his casket.

Mitt, having just lost the Senate race, was already wrestling with what to do next. The campaign had transformed him. "The experience of walking into diners and onto construction sites and hearing people tell me their problems was not easily put aside," he would later write. "I had the bug of wanting to be more involved." This sense of duty was almost written into his genes. "I don't know where it comes from, and I don't know how far back I'd have to look to see it in my ancestry," he said. "But there's a sense of obligation to help my country, to help my family, to give back in some respects." But where was he needed? What would he do? Those questions would have been hard to answer anyway. And now he was, for the first time, confronting life without his father, his North Star. The family legacy of public service now rested in Mitt Romney's hands. The torch was his to carry, alone.

There are good years in private equity, and then there are really good years in private equity. After his frustrating loss to Kennedy, Romney returned to Bain Capital just in time to ride the booming economy of the late 1990s to unimaginable heights. The next five years were extremely profitable, with Bain scoring some of its biggest deals ever, deals that yielded sensational returns. Romney, the cautious, skeptical investor, had become a confident, comfortable deal maker, reaping millions for himself and his partners. "I was in the investment business during the most robust years in the history of investments," he would later say. Indeed, the 1990s further established Bain Capital as one of the country's elite leveraged buyout firms.

But as lucrative as those years were, Romney's business career couldn't hold his attention for too long. He began dabbling in politics as he searched anew for a path to public service. In 1996, he was so concerned about Republican presidential candidate Steve Forbes's proposal for a flat tax that he spent $50,000 attacking it in a series of newspaper ads during GOP primary contests in Iowa, New Hampshire, and Massachusetts. Forbes's plan, Romney said then, was dangerously tilted toward the rich. "It's a tax cut for fat cats," he said. A couple years later, Romney steamed over President Bill Clinton's affair with the intern Monica Lewinsky, outraged by Clinton's moral and political failure, one longtime Romney associate remembered. Though Romney never said it directly, the takeaway was clear: this would never happen if Mitt Romney were in the White House. Romney also inquired about playing a significant role in the inchoate presidential campaign of a Texas governor named George W. Bush. He flew down to Austin to meet with Bush and explored taking time off and campaigning for him full-time ahead of the 2000 race, said one Republican close to Romney. "He felt he had been really lucky in life," the Republican said. "And [he] felt that the next president was important."

Around the same time, Romney helped a longtime dream of many local Mormons come to pass. For years, area Mormons had had to travel to the temple near Washington, D.C., for sacred rites such as baptisms and marriages. It was the closest temple to Boston, and many felt that Boston, with its growing number of congregations, deserved one of its own. Years earlier, Romney, along with other local church leaders, had been active in acquiring a roughly seventeen-acre plot of land on Belmont Hill. It was where the church had, after overcoming objections from neighbors, built its Belmont chapel in the 1980s. But there was a lot of room left over, and Romney and other leaders had long hoped that they could persuade Salt Lake City to build a temple there, too. In the mid-1990s, they finally got their wish. Gordon B. Hinckley, then the president and prophet of the worldwide Mormon church, had wanted to erect a temple in the Northeast anyway. After a visit to the property, and after Romney and other local leaders advocated through various church channels, Hinckley was sold. The church's hundredth temple would be built there on a rocky ledge in Belmont, high above a major highway and visible from miles away. Though neighborhood opposition was again fierce, the groundbreaking for the $30 million project came in 1997, and the temple, built of striking white Sardinian granite, opened in 2000. Hinckley called it "a reminder to the world."

The new temple was a beacon beckoning the faithful and a statement asserting the church's growing influence in New England. For Romney, it was also a fitting capstone to all the work he had done to promote the church. "It feels great to have a temple closer to home," he said at the time. "It makes you feel proud that the membership of the church has grown large enough to merit a temple being constructed so close." But there was one glaring omission. Unlike every other Mormon temple, this one lacked a spire with a gold-leafed statue of Moroni, the angel who Mormons believe led Joseph Smith to a set of gold plates that Smith translated into the Book of Mormon. That aspect of the sacred building—the plans called for a 139-foot steeple—was still being challenged in court.

An unlikely booster came to the Mormons' defense: Ted Kennedy, Romney's bitter rival only a few years earlier, who had once suggested that his opponent's religion was a legitimate political target. One afternoon in September 2000, Romney led Kennedy on a personal tour of the new temple, an important step in their gradual rapprochement. The two former rivals downplayed their past battles. "As individuals, we certainly respect each other," Romney said. "I wish he was a Democrat," Kennedy said. "I'm glad he's not running against me this time." On the tour, Kennedy praised what he called the "magnificent" temple architecture, and he said the Mormons should be allowed to build their steeple. "If other churches are going to have their expressions in terms of spires, this one should as well," Kennedy said. "What's fair is fair." The following spring, the state's highest court agreed, to the delight of the local Mormon community, and Moroni got his rightful home atop the towering spire.

The more tests the doctor did, the further their hearts sank. She couldn't feel pinpricks in her foot. She couldn't keep her balance. She couldn't touch her nose with her eyes closed. Something was seriously wrong with Ann Romney, and she and Mitt were scared.

Ann, fit and not yet fifty years old, had first noticed the symptoms in 1997, and they had gotten progressively worse. Numbness in her right leg, which spread up her right side. A hard time climbing stairs. Difficulty swallowing. Initially they'd thought that maybe she had a virus or a pinched nerve. But their family physician recommended that she see a neurologist, so they made an appointment with Dr. John Stakes at Massachusetts General Hospital.

Mitt and Ann didn't know what to expect when they arrived at Stakes's office that day in 1998. In the waiting room, seeing pamphlets on Lou Gehrig's disease and multiple sclerosis, they quickly realized that the news wasn't going to be good. "There weren't many other brochures there; that was basically what he was dealing with," Romney said. "And of course, it was very frightening." Stakes brought them in and performed a series of tests. That's when it became clear: this was no virus. "She's failing test after test after test," Romney recalled. They were crushed. When the doctor left the office, Ann and Mitt broke down. "We hugged each other, and I reminded her that as long as this was not fatal, we could deal with it," Romney said. It was, he would later say, the worst day of his life.

After an MRI, Stakes made his diagnosis: Ann Romney had multiple sclerosis, a life-changing, chronic disease of the central nervous system whose course is difficult to predict. It was a devastating moment for a family that, by almost any measure, had lived a charmed life. "When I heard that she had MS, I didn't believe it," Tagg Romney said. "Like my dad was Superman, she was Superwoman to me. I just couldn't accept it." Mitt had known a few hard moments: the fatal accident in France as a missionary, the bruising defeat by Kennedy, and, more recently, the loss of his father. But for the most part he'd had it pretty good, his golden childhood largely unimpeded. "It continued for Mitt," said his sister Jane. That had not been the case for his siblings, who had suffered divorces and other setbacks.

Ann's diagnosis changed everything. Now the love of Mitt's life was facing a potentially crippling disease. Their future was in doubt, his political aspirations in serious question. "It subdued him," Jane said. The diagnosis also came as Mitt was mourning the loss of his mother, Lenore, who had died at the age of eighty-nine after suffering a stroke at home in July 1998. Ann and Mitt's youngest son, Craig, who was seventeen, cried after a friend told him the disease would kill his mother. "Craig, I'm not going to die from this," Ann said she told him. But early on, she felt as though she wanted to. The uncertainty was too much to bear. "I frankly would have rather died than be the way I was," she once said. "At the time, I wished I had cancer instead of this, something more tangible. We didn't know what was in store for me. This would take bits and pieces of you." The months after her diagnosis were trying. Ann would improve, raising everyone's hopes, only to dash them by suffering relapses. Even limited physical activity could exhaust her. She was so weak she could barely take care of herself. She received intravenous steroid treatments at first, but she stopped because the drugs made her sick, according to Tagg.

The Romneys, having given help to so many over the years, now required it themselves. Close friends from church offered prayers, emotional support, meals—whatever was needed. "A lot of people were pulling for them," John Wright said. But Ann was relatively private about her affliction early on, and she also had five able sons at her side. For a time, Mitt and Ann wondered how her illness would affect his career. Grant Bennett asked him that question once, and Mitt's immediate response was clear: family came first. "I remember to this day, it wasn't 'Maybe this will slow us down,'" Bennett said. "It was 'Of course Ann's my first priority.'"

Eventually, Ann hit upon an effective assortment of treatments, including yoga, Pilates, reflexology, acupuncture, and a controlled diet. Perhaps most important, she rekindled her childhood love of horses, improving both her mobility and spirit. "Riding exhilarated me; it gave me a joy and a purpose. It jump-started my healing," she said later. "When I was so fatigued that I couldn't move, the excitement of going to the barn and getting my foot in the stirrup would make me crawl out of bed." Tagg said that his dad tried to relieve some pressure on Ann by urging her to focus on herself instead of everyone else, as she had for years. Mitt told her, "You don't have to be perfect—having meals cooked for everyone, getting presents for the grandkids," Tagg said. That, he said, gave Ann license to say, "It's okay if I deal with me and get myself better." She gradually began to see that living with multiple sclerosis didn't have to mean a future in a wheelchair. "I am very strong right now, but I have been in a deep, dark hole," she would say several years after her diagnosis. "And I have crawled out inch by inch."

The lesson that she did not have to be captive to her disease came not in Belmont, where she had made her home for nearly thirty years, but in Utah, where she found equestrian companions and a retired guru of reflexology by the name of Fritz Blietschau, who, although he was nearly eighty, agreed to take her on as a patient. By early 1999, Ann and Mitt were based there as well. That's because when the call

came offering Mitt Romney the opportunity of a lifetime, it came
with a Utah address.

———————————

Around the time the Romneys were reeling from Ann's diagnosis, they
got a phone call from Kem Gardner, an old family friend and fellow Mormon. Gardner, who was now a prominent developer and civic
leader in Utah, had a wild proposal that he knew Mitt would reject
out of hand. So he went first to Ann.

The 2002 Winter Olympics, which Salt Lake City would host in
a few years, were in deep trouble, severely damaged by an influence-
peddling scandal. The Salt Lake Organizing Committee, swept up in
a culture of corruption that permeated the high-stakes international
competition to get the Games, had embarrassed itself and the state
of Utah by lavishing gifts on international Olympics executives. The
shame of the scandal—in which ten members of the International
Olympic Committee had resigned or been expelled for accepting gifts
from the Salt Lake committee—had sent corporate sponsors fleeing,
leaving the budget for the Games woefully underfunded. Relations
between the international and U.S. Olympic committees deteriorated
badly. Morale among Salt Lake staffers nose-dived. The competition
would go on; there was little doubt about that, with three years to go
before the Games were set to begin. But there was a palpable sense of
peril, particularly among vital supporters and sponsors. The Games
needed a turnaround artist, and fast.

Utah's political elite launched an urgent search for a leader who
could revive their flickering Olympic flame. Robert Garff, the chairman of the Salt Lake Organizing Committee, quickly submitted
Romney's name to Governor Michael Leavitt. Garff had known Romney since childhood, their fathers having been close friends since they
had attended Latter-day Saints High School in Salt Lake City in the
1920s. In the fifty years since Romney and Garff had first met at a
family gathering, Garff had established himself among Utah's most
influential leaders by helping to expand his father's automotive empire,

becoming speaker of the Utah House of Representatives, and rising to the upper echelons of the Mormon church.

Romney's own ties to Utah ran deeper than his ancestral roots. After visiting the state regularly as a child, he and Ann had married at the Salt Lake City temple, and he had graduated from BYU, where two of his sons were enrolled as Utah prepared for the Olympics. The Romneys had also just built a magnificent mountainside retreat near Park City, on the aptly named Rising Star Lane, thirty miles from the organizing committee's office in downtown Salt Lake City. To Garff, Romney was the perfect candidate, the "white knight" they so desperately needed. "Mitt had his father's charisma, his mother's good looks, his own intellect, and his wife's supporting hand," he said. Romney also had a keen sense of the Mormon hierarchy's role in Utah's social and political culture. "It would have been a disaster if we just picked a stranger and they didn't understand the mores of this community," Garff said.

The only other serious candidates—Jon Huntsman, Jr., one of Utah's most prominent and politically connected figures as the son of the billionaire industrialist Jon Huntsman; and Dave Checketts, the chief executive of Madison Square Garden—were also Mormons with Utah roots. But Garff wanted Romney. So did Leavitt. "I was looking for a businessman who had a good political instinct," Leavitt said. The younger Huntsman, then a top executive of his father's chemical conglomerate and a former U.S. ambassador to Singapore, grew disillusioned with the search process in its final days. With Romney emerging as the likely winner, Huntsman responded by withdrawing his name from consideration, then rebuffing an invitation to serve on Romney's management committee. "A search was never fully carried out," Huntsman complained. "If I was not able to support the process which they were employing in bringing in new leadership, I shouldn't be serving in a position like that." Huntsman would later assist Romney's bid for the 2008 Republican presidential nomination, only to then defect to his main rival, Senator John McCain of Arizona. Four years later, Huntsman would himself compete against Romney, joining a field of candidates for the 2012 GOP presidential nomination.

Leavitt, in trying to sell the job to Romney, appealed to his desire to serve. "Based on the fact that he had run for office, I figured his public service gene might be lighted," he said. Gardner's strategy of starting with Ann, meanwhile, had paid off. She helped her husband flick the switch. "Just think about it," she told him. "If there's any one person ideally suited for this job, it's you." Mitt said he had resisted, but "the more I protested, the less crazy the idea seemed." As was his nature, he also found the challenge difficult to turn down.

To Romney, a major civic achievement would be good for his soul—and his political résumé. He understood that if he could turn the Games around, it would be the perfect bridge to elective office. One of his campaign consultants from 1994, Rick Reed, told him as much in a letter urging him to take the Olympics post. In a return letter, Reed said, Romney told him that the advice had been integral to his decision to go for it. "It was serendipitous," Garff said. "Mitt wanted to leapfrog from the world of business to public service, and this was a perfect opportunity for him to propel himself into the national spot-light, which I believe was all part of his overarching plan of his life."

His departure from Bain Capital, though, was not so neat. The partners squabbled over how the firm would operate without him. A power struggle ensued. Several partners made plans to leave. Sud-denly, a company that relied on loyalty, long-term relationships, and Romney's personal courtship of investors seemed to be at risk. And such a breakup could be messy. A Bain meltdown might mean law-suits with tens of millions of dollars at stake. The potential existed for embarrassing disclosures of how much money Romney had made on certain deals. "It would have been a circus, and circuses over money are not good for politicians," one Romney associate said. Romney grew worried that the company he had worked so hard to build would be destroyed. The anxiety escalated until finally, one Sunday afternoon, Romney and one of his fellow Mormons at Bain Capital, Bob Gay, knelt on the floor together and prayed for its survival. "We were fac-ing a crucial event that threatened the very existence [of] our firm's partnership," Gay said later. In the end, the crisis abated. Romney left the firm, retaining a financial interest in it, and Bain Capital continued to thrive.

David D'Alessandro's office on the fifty-ninth floor of Boston's Hancock Tower was decked with pricey artifacts: original images of Abraham Lincoln's funeral, Marilyn Monroe's silver compact, Muhammad Ali's boxing trunks. Romney, who had come seeking the counsel of D'Alessandro, an insurance executive and one of the Olympic movement's most influential corporate voices, was especially drawn to one item: a memoir authored and signed by Winston Churchill, the British statesman who had epitomized resolve in the face of grave danger. Romney viewed Churchill's memoir as a source of inspiration. He understood that taking over the Olympics would likely be his personal crucible.

Romney knew the perils of squandering the Salt Lake opportunity. Failure would scar the United States' pride, further embarrass Utah and the Mormon church, and derail Romney's effort to position himself as a leader for the new millennium. "If this doesn't work," Romney told D'Alessandro that day, "I can come back to private life, but I won't be anything anymore in public life." D'Alessandro thought to himself that the damage could extend to Romney's business career, rooted as it was in his reputation for high competence and probity. "I knew that more than his public reputation was at stake," D'Alessandro said. "If he failed to put on a quality Olympic Games, his reputation as a private-equity guy would have been significantly damaged." D'Alessandro later said he was surprised that Romney had taken the job. "I think he took it because he felt that the Mormons were in trouble. He never said that, but I think he saw the scandal as a stain on his religion. But I don't believe he understood what he was getting into." He might not have anticipated all the challenges he would encounter, but he knew a calamity when he saw one. And that was the appeal. "He loves emergencies and catastrophes," Ann Romney said the day her husband took the reins. "He would never have considered doing it if it wasn't a big mess."

And it was. In Romney's first weeks in charge of the Winter Games—he was formally introduced on February 11, 1999, as chief executive officer of the Salt Lake Organizing Committee—he grasped the enormity of his challenge. "We were in a psychological zombieland; the whole community was walking around dazed," he said of the early days. "Here they were, jubilant that the Games had been won, and then their reputation was now tagged with scandal—that this was the center of Olympic scandal. It made people sick." Romney left no doubt when he started the job that he would take full command. A century and a half after his pioneer ancestors had trudged through Emigration Canyon to help make the Great Salt Lake Valley a land of the righteous, Romney arrived that February morning in a cheerless ballroom in a Salt Lake City hotel to open a new chapter in his family's history. He began by issuing a stern warning to the remaining trustees of the scandal-tainted organizing committee about restoring dignity to the mission. "There is no justification for compromising integrity," he informed the fifty-three-member board. "I will expect that if, as the investigations continue, any of you casts a shadow on the Games, even where no wrong may have been done, you will stand immediately aside."

Much of the damage seemed to stem from decisions made by two previous executives on the Salt Lake Organizing Committee: president Thomas K. Welch and vice president David R. Johnson. Welch and Johnson had allegedly embraced some of the seamy practices that had long greased the international selection process for Olympic sites. After Salt Lake City had lost its bid for the 1998 Games to Nagano, Japan, Welch and Johnson had concluded that they had failed to lavish enough largesse on members of the International Olympic Committee, which chooses host cities. Whereas Nagano had plied IOC members with about $540,000 worth of souvenirs, including laptop computers and video cameras, Salt Lake City had handed out little more than cowboy hats and saltwater taffy. Vowing not to be defeated again, Welch and Johnson funneled, through the committee, more than $1 million in gifts to numerous IOC delegates for the 2002 Games—a

ABOVE LEFT: Miles Park Romney, Mitt's great-grandfather, was born in the Mormon base of Nauvoo, Illinois, moved to Utah, and then followed the church's instruction to undertake multiple moves and establish a polygamous colony in Mexico. *(Special Collections Dept., J. Willard Marriott Library, University of Utah)* ABOVE RIGHT: Hannah Hood Hill Romney, Mitt's great-grandmother, was the first of five wives taken by Miles. She kept the polygamous family together as Miles fled U.S. marshals and went to Mexico, where she eventually joined him. *(Special Collections Dept., J. Willard Marriott Library, University of Utah)*

Gaskell Romney, Mitt's grandfather, fled a revolution in Mexico and moved the family, including five-year-old George Romney, Mitt's father, to the United States. *(Special Collections Dept., J. Willard Marriott Library, University of Utah)*

Mitt Romney (*right*) idolized his father, George, who served as governor of Michigan and unsuccessfully sought the presidency. *(Courtesy of Romney family)*

Mitt hopped behind the wheel of his dad's Rambler. His father, as chairman of American Motors Corporation, championed smaller cars. *(Courtesy of Romney family)*

As a nineteen-year-old student at Stanford University in May 1966, Romney picketed antiwar protesters who were holding a sit-in at the office of the university president. *(George Romney Collection, Bentley Historical Library, University of Michigan)*

Romney was nearly killed in June 1968 when the car he was driving through France as a Mormon missionary *(at left)* was hit head-on by an oncoming vehicle. Leola Anderson, the wife of the French mission president, died from her injuries. The fateful accident quickly eroded Romney's sense of youthful invulnerability. *(Courtesy of Richard B. Anderson)*

High school sweethearts Mitt Romney and Ann Davies were married in a two-part ceremony in March 1969, the first at her parents' home and the second in the Mormon temple in Salt Lake City. Ann's parents, because they were not Mormon, were forbidden to witness the temple ceremony. *(Courtesy of Romney family)*

Armed with degrees from an elite graduate program at Harvard University, Romney helped build his private equity firm, Bain Capital, into a powerhouse. *(Boston Globe/David L. Ryan)*

The Romneys arrived in the Boston area in 1971 with a one-year-old son, Taggart. Over the next decade, Mitt and Ann had four more boys. *From left, with their parents:* Tagg, Ben, Matt, Craig, and Josh. Mitt said that being a part of that household "was so much fun, because of the jokes, the laughter, bathroom humor, the physical, you know, fisticuffs, wrestling, games." *(Courtesy of Romney family)*

In 1994 Romney launched an improbable bid to oust Edward M. Kennedy from the U.S. Senate. It was the strongest challenge Kennedy had ever faced. But in the end, Romney, after failing to articulate a clear campaign message, came up short, losing by a wide margin to the Democratic stalwart. *(Boston Globe/Mark Wilson)*

Romney took over the 2002 Winter Olympics in Salt Lake City at a time of crisis, with the Olympic community and much of Utah reeling from an influence-peddling scandal. Many credit Romney for turning things around and leading a solemn, successful games just months after the 2001 terrorist attacks. *(AFP/George Frey)*

Eight years after his bruising loss to Edward Kennedy, Romney returned to Massachusetts and fulfilled his lifelong dream of winning public office like his father. Romney, shown here with his lieutenant governor, Kerry Healey, won the 2002 election at fifty-five, the same age his father was when he was first elected governor of Michigan. *(Boston Globe/Jim Davis)*

Massachusetts's groundbreaking universal health care law, which Romney signed in an elaborate ceremony at Boston's Faneuil Hall in April 2006, was his biggest achievement as governor. Though the law has succeeded in covering nearly everyone in the state, its controversial requirement that residents purchase insurance has become a political burden for Romney. (*Boston Globe/ David L. Ryan*)

In his first bid for president, Romney tried to win early contests in Iowa and New Hampshire by showering local political leaders with campaign contributions and shaking hands at countless events like this, the September 2007 Labor Day parade in Milford, New Hampshire. *(Boston Globe/Dina Rudick)*

Mitt and Ann kicked off his 2012 presidential campaign by serving her homemade chicken-and-bean chili to supporters on a farm in Stratham, New Hampshire. The June 2011 announcement, in keeping with Romney's new approach, was far more low-key than his presidential launch four years earlier. *(Boston Globe/Jonathan Wiggs)*

stunning trove of booty that included cash, college tuition, medical
care payments, jobs, lodging, beds and bedding, bathroom fixtures,
Indian rugs, draperies, doorknobs, dogs, leather boots and belts, per-
fume, Nintendo games, Lego toys, shotguns, a violin, and trips to ski
resorts, Las Vegas, and a Super Bowl in Miami.

Then came the comeuppance. After Salt Lake City won the Games,
a local ABC affiliate, KTVX, received a news tip about college tuition
payments the organizing committee had made for the daughter of an
IOC delegate from Cameroon. Soon investigations were launched by
the Justice Department, Congress, Utah's attorney general, the United
States Olympic Committee, the IOC, and the Salt Lake committee.
There was scandal in the land of the righteous, and some critics began
wearing T-shirts that depicted the five Olympic rings fashioned as
handcuffs. The challenge Romney faced was clear to D'Alessandro.
"The best way I can describe what he faced," he said, "is trying to
rebuild an airplane while it's flying."

———————————

Within months of Romney taking the job, his Mormonism and the
church itself became sources of contention. Romney had been chosen
in part because of his Mormon credentials, but some people in Utah—
most notably a prominent Mormon, Jon Huntsman, Sr.—complained
that the church was too actively involved in the secular event. After all,
the church had made no secret of its desire for Salt Lake City to host
the Games. In a state where nearly every political leader and a majority
of the population was Mormon, church president Gordon B. Hinckley
had delivered the message that he viewed the Games as a vehicle to
fulfill pioneer Brigham Young's prophesy that Salt Lake City would
"become the great highway of the nations." "Kings and emperors and
the noble and wise of the earth will visit us here," Hinckley said, quoting Young. Church leaders had traveled the world with the Salt Lake
Organizing Committee. Documents in Garff's archives at the University of Utah show that church officials recommended employees to the
organizers, commented on committee policies, and sought direct pub-

lic relations benefits from the Games. A Mormon leader, Elder Robert Hales, met privately with an NBC executive in New York to offer the church's cooperation in the television presentation of the Games.

Romney's initial decisions exacerbated concerns about the level of church participation. He had requested an additional $8 million—on top of the $5 million it had already committed—in lent property and cash from the church, among other contributions, as he tried to strengthen the Games' financial position. And he hired Fraser Bullock, another prominent Mormon who had been a partner at Bain, as his chief operating officer. The moves prompted the elder Huntsman to assail Garff and Romney for exploiting their ties to the Church of Jesus Christ of Latter-day Saints. "We've got a chairman who is active LDS, now we've got a present CEO who is active LDS," Huntsman said. "They claim they're going out [to] really scour the world to find the best person, so Mitt brings in one of his cronies to be the COO. Another broken promise. Because we've got three LDS folks who are all cronies. Cronyism at its peak. . . . These are not the Mormon Games."

Garff, Leavitt, and Romney quickly paid a series of visits to Huntsman, who cooled his opposition to their management and made peace with all three. The elder Huntsman later helped fund Romney's political campaigns and served on his national finance committee for the 2008 presidential race. But church leaders, wary of appearing to exert too much influence on the Games, asked Romney to scale back his requests for aid, which he did. They also curbed their ambitions to use the Olympics to promote Mormonism—with a few exceptions, among them a book that one of the church's publishing companies released for the Games, *Why I Believe*, in which the Romneys joined about fifty other prominent Mormons in expressing their faith. "It all worked out beautifully after the church backed off and the prophet [Hinckley] said we won't have any missionaries on the streets proselytizing," Garff said.

Yet Romney's funding problem persisted. Three years shy of the opening ceremonies at the University of Utah's hillside stadium, Romney had inherited a $1.45 billion budget. After he and Fraser Bullock completed an analysis, they concluded that the organizing committee needed to raise at least $400 million more to meet expenses. Romney later joked that at the rate they'd been going, the Olympic cauldron "was going to be a couple of Weber grills welded together and hauled up a flagpole." He cast the challenge as a monumental crisis, one that would require extraordinary efforts from everyone involved. "We could be liquidated by our creditors, and shamed," Romney told Leavitt.

Romney rapidly trimmed the budget to $1.32 billion, launched marketing campaigns, and established an austerity program, which included an end to free lunches for the trustees. No more grand, multicourse buffets such as committee members had enjoyed until Romney's first meal with them. After that he ordered pizza and required board members to pay $1 a slice, plus $1 per soda. No one resisted the change, although one financially strapped trustee considered the new meal plan part of her sacrifice to public service. Joan Guetschow, a former winter Olympic biathlete who served on the board as an athlete representative, said, "Mitt was very adamant about being squeaky clean on every single rule. I was a volunteer. I didn't have a lot of money. I had to take vacation time from work to go to meetings. I thought, 'Can I at least get a free piece of pizza?'" Romney's culinary austerity extended through the Games. Olympic sponsors and other VIPs had grown accustomed through the years to organizing committees' providing lavish buffets. Yet their main staple at the Salt Lake Games was not exactly haute cuisine. "I ate more chili in those two weeks than I've eaten in my entire life," D'Alessandro said. When a local paper ran a cartoon depicting Romney saying, "We're cheap!" in front of "MittFrugal's," his discount Winter Games, Romney was so proud he had it framed.

Though most of Romney's belt-tightening moves—both symbolic and substantive—energized the organizing committee, some people familiar with the budget insisted that his dire forecasts were overstated. A review of archived records shows that the organizing committee had already secured commitments of nearly $1 billion in revenues, including $445 million as its share of NBC's broadcasting contract, and nearly $450 million in sponsorship deals before Romney arrived. In addition, the Utah State Legislature had already loaned $59 million in sales tax revenue to the Games and Congress was prepared to provide hundreds of millions of dollars in direct support, despite protests from Arizona Senator John McCain. Direct federal aid for the Games ultimately totaled $382 million. "Most of the federal money was already in place before Mitt came on," said Robert Bennett, a Republican senator from Utah who served as point man for the federal funding for the Salt Lake Games. "The Clinton administration was completely supportive in saying these are America's Games, we will do whatever we can to make sure they are successful. The one concern I had was whether we would get the same degree of support from the Bush administration, which we did."

Even some key members of the organizing committee, including the friend who had recommended him for the job, didn't believe the budget was in peril. "Yes, we were out of balance, but we had [three] years to organize that," Robert Garff said. "In my mind, there was no sense of panic." But Fraser Bullock said the hole was real. "I was the chief financial officer, so all these other people who say it wasn't that dire, they needed to be in my shoes," he said. Romney has since touted his economic rescue of the Games as a hallmark of leadership, saying, "The tsunami of financial, banking, legal, government, morale, and sponsor problems following the revelation of the bid scandal swamped the organization. It was the most troubled turnaround I had ever seen."

Romney did have to work quickly to revive corporate support, the lifeblood of the Olympics. Fund-raising had stalled as companies feared becoming stained by the scandal. Sponsors had been scared off in part by D'Alessandro, who was highly critical of Olympic leaders for bringing on the assorted investigations. D'Alessandro had standing. After all, his company, John Hancock Mutual Life Insurance, had committed $50 million to sponsor the Games. "The I.O.C.'s sponsorships have become radioactive," D'Alessandro said before Romney took over. "They've got to find a way to make sponsorships safe again." Romney cast himself as the clean businessman who would restore integrity to the Games. He began by personally winning over D'Alessandro. That cleared the way for Romney to assume the role of chief salesman, criss-crossing the country to help reap more than $300 million in additional sponsorships. His entrepreneurial abilities were on full display. He was prohibited from soliciting companies in marketing categories that already had been claimed—Pepsi could not sponsor the Games, for instance, because Coca-Cola had already committed. So he created more than twenty new categories, including an official furniture supplier (Herman Miller) and an official online job recruiter (Monster .com). He also struck a deal for an official meat (certified Angus beef), which explained why the free buffet for VIPs was heavy on chili and hot dogs.

His salesmanship, which involved raising an additional $100 million from companies and individuals in Utah, extended to NBC headquarters at Rockefeller Plaza in New York. The company had paid $555 million to broadcast the Games, and network executives would not abide having their investment damaged by subpar management. Had Romney botched the assignment, he would have earned the wrath of Dick Ebersol, who was chairman of NBC Sports and Olympics and had been identified by *Sporting News* in 1996 as "the most powerful person in sports." But Romney impressed Ebersol. "I have no doubt whatsoever, as the representative of the chief investor in the Salt Lake City Olympics, that Mitt Romney was single-handedly responsible for those Games being the immense success they were," Ebersol later said. "The list of people who could have pulled it off began and ended with Mitt Romney."

Part of Romney's appeal was his insistence on high ethical standards. Yet Romney himself risked the appearance of conflicts by soliciting sponsorships from companies such as Staples and Marriott International on whose boards of directors he served. As a director, he was responsible for protecting the companies' interests; as CEO of the organizing committee, he had promised to get the best deal for the Olympics. Romney dismissed the notion of any conflict. "That's not a conflict of interest," he said. "That's trying to encourage people who I knew to get with the Games." In Garff's view, Romney mitigated any risk of a possible conflict by citing his business affiliations on an ethics statement. To be sure, Romney's entanglements were nothing like those of the trustees who resigned when he took over, including Alan Layton, whose construction company had received a $29 million contract from the organizing committee, and Earl Holding, whose ski area had signed a $13.8 million deal with the committee. Romney did allow some trustees whose companies engaged in relatively small business with the committee to remain on the board. "They removed the people who were the poster children for conflicts of interest," said Glenn Bailey, a leader of Salt Lake Impact 2002 and Beyond, a coalition of community groups. "But they still had conflicts."

Even if the finances were fixable, there were real doubts about how to repair the reputation of the Games, Salt Lake City, and Utah. Romney acted quickly to remove the taint of the scandal, partly by laying the blame on Welch and Johnson. Romney joined Leavitt in casting the two men as rogue members of the organizing committee who had betrayed their Olympic cause. Until the case went to trial, Romney supported federal prosecutors, who alleged that Welch and Johnson had participated in defrauding the committee of more than $1 million by doling out gifts to IOC delegates. Facing bribery charges punishable by up to seventy-five years in prison, Welch asserted that everyone involved in the process, including Governor Leavitt, had known that favors were being given to members of the international selection committee. "We amassed significant, undeniable information that everybody involved in the process was knowledgeable about what was going on, all the way to the governor's office," said Max Wheeler, one of the defense attorneys. Leavitt denied knowing anything.

Welch's view was widely shared in the Salt Lake community. "If you're going to fault Tom Welch for anything, you can fault him for having tunnel vision," said Zianibeth Shattuck-Owen, who served on the organizing committee. "He was given the directive to get the Olympic Games. He's not a dumb man. He saw what he was up against and played to win. Is he a devil or a bad man? No, not to me." Garff, too, took exception to criticism of Welch, Johnson, and the committee for some of their actions before Romney arrived. "All we did was do what other people did," Garff said. "The IOC expected and wanted to be pampered." But Romney barred Welch's and Johnson's names from appearing on a list of more than 20,000 other committee staffers and volunteers on a Wall of Honor at the city's Olympic Legacy Plaza. And he went so far as to encourage Welch to accept a plea bargain for the good of the Games. One of Welch's friends, Sydney Fonnesbeck, said that Romney had urged her to persuade Welch to plead guilty to a lesser charge. "Mitt called and said he thought it would be best for the Olympics and for everyone's benefit," said Fonnesbeck, a former Salt Lake city councillor who helped organize the Games as a member of the state's Sports Advisory Council. "He said they would slap Tom's hands and it would be over."

She recalled Romney saying, "You never know what could happen if he goes to trial. He could end up going to jail." "Well, I'll send him some books," Fonnesbeck replied. She said that Romney was civil throughout their conversation, but she resented his urging her to intervene. "I don't know if it was legally inappropriate, but I felt it would have been incredibly inappropriate to do what he asked me to do," she said. Welch's defense team agreed. "Tom was represented by counsel, and it was inappropriate for [Romney] to be talking to a defendant in a criminal case about pleading," Wheeler said.

Romney's request looked even worse when a federal judge threw out all fifteen felony charges against Welch and Johnson for insufficient evidence and praised their contributions to the Games. "I can only imagine the heartache, the disappointment, the sorrow that you and your loved ones suffered through this terrible ordeal," U.S. District Judge David Sam told Welch and Johnson. Yet even after the charges were dismissed, Romney continued to express doubt about Welch's and Johnson's innocence, blaming the acquittal in part on ineffective prosecutors. "Of course, not being convicted of a crime isn't vindication of wrongdoing, and not all unethical behavior is criminal," he contended. "Even when criminal conduct occurs, it may be difficult to prove—and that's with effective prosecutors. I believe those who pursued Welch and Johnson were inept."

"Mitt's objective was to look as good as he could, to wear the white hat," Welch said. "The more critical he could be of what was there before him, he was. He didn't have to do it, but he chose to. He viewed everything in terms of how he could promote himself and his legacy, even at the expense of others. He showed a mean side, as well as a competent side." The government's chief prosecutor, Richard Wiedis, believed he had lost the case primarily because several members of the organizing committee who had agreed to testify against Welch and Johnson "essentially went south during the trial" by supporting the defendants. "Mitt Romney, as far as I know, was never in the courtroom, didn't review any of the evidence, and never asked the prosecutors for a summary of their case," Wiedis said. "I don't see how he was in a position to make a judgment as to the competence of the prosecution team."

To that point in his life, the tests of Mitt Romney's capacity for leadership had been conducted largely in private, his skills and shortcomings visible only to business associates and fellow Mormons. The Senate campaign had brought him to the fore, but it had ended in failure. It wasn't until he took over the Olympics that his ability to lead was put on intensely public display. The overall picture that emerges from those who worked with and observed him in Salt Lake City is of a man focused on the task at hand with laserlike intensity. To some, that was an inspiring thing. His many admirers viewed him as ethically pristine, amiable, and self-effacing. But he developed another image among a group of dissenters: as petty, vindictive, and self-aggrandizing. Romney was also dismissive of the few trustees who aggressively questioned his practices.

Romney's chief foe was Ken Bullock, who was no relation to Fraser Bullock. Ken Bullock, who served on the organizing committee as executive director of the Utah League of Cities and Towns, believed his professional position made him an official watchdog for the state's $59 million investment in the Games. "He tried very hard to build an image of himself as a savior, the great white hope," Bullock said of Romney. "He was very good at characterizing and castigating people and putting himself on a pedestal." Bullock was among those bemused by Romney's efforts to promote himself, which seemed to run against the grain of his buttoned-down business persona. Romney became the first Olympics executive to approve a series of commemorative pins bearing his likeness. One pin depicted his face under a heart with the words: HEY MITT . . . WE LOVE YOU!

Romney and Bullock clashed often, never more publicly than when they went nose to nose at the Utah state capitol after Bullock failed to support Romney's request to defer repaying the state its $59 million.

"You don't want me as an enemy," Romney said in the corridor outside a conference room, according to Bullock. "Ted Kennedy and I get along. Why can't you and I?"

"I'm doing my job," Bullock replied.

To which Romney repeated, "You don't want me as an enemy."

Salt Lake City Mayor Rocky Anderson, a Democrat who also served on the organizing committee and has remained a Romney friend, said that Bullock had long played a "very destructive" role in the Olympic movement. "We were all running out of patience and were pretty proud of Mitt that he finally put Ken in his place," he said. Yet Garff, the organizing committee chairman whose association with Romney went back to childhood, believed that Romney had inappropriately tried to silence Bullock. "Mitt saw him as an agitator," Garff said, "and I saw him as a watchdog who needed to be heard." Romney showed little sympathy for another trustee who criticized his stewardship. Lillian Taylor, a small-business consultant, questioned why the organizing committee continued to retain a pricey, well-connected law firm that claimed to have lost documents related to the scandal. "I wanted to know why we were spending millions of dollars for a law firm that was expected to keep the records and then told us, 'Somebody ate the homework,'" she said. "I thought I was asking a legitimate question." Romney, she said, offered no support, sitting silently while the board's attorney aggressively dismissed her complaint. "Shame on Mitt for that one," Taylor said. "He didn't stand up and protect me, and they just dropped it like a hot potato."

On other occasions, Romney artfully defused tension by reaching out to leading critics of the Games. The most vocal was Stephen Pace, the head of a group called Utahns for Responsible Public Spending. A business consultant, Pace had made great sport of ridiculing Romney's predecessors in an effort to cast the Olympics as an exercise in wasted tax dollars. Pace's group produced a line of T-shirts mocking the Games, and Pace had taken to standing in front of television cameras wearing a shirt that said "Slalom & Gomorrah." Romney wasted no time trying to disarm him. "His first day in Utah, he called me and started blowing in my ear," Pace said. "It was very clear what he was doing, but it was a very smart gesture after the people before him had treated us very contemptuously."

Romney also impressed Guetschow, the former Olympian on the organizing committee, by demonstrating a measure of respect for her as a lesbian. Geutschow recalled the first meeting of the new committee members after Romney's arrival. It was at the governor's mansion. Garff, after presiding over an opening prayer, began by asking the members to stand and introduce their spouses. Guetschow, who had brought her partner, went last. She recalled saying to herself, "What am I going to do?" Many of the trustees were members of the Mormon church, which considers homosexuality sinful. When Guetschow's turn came, she said, "This is my friend; I guess that's a safe way to put it.'" Everyone, she said, "was a little horrified." Soon, Guetschow herself was horrified when the organizing committee proposed an antidiscrimination employment policy that did not include a provision for sexual orientation. "They skipped over my minority, and I was too shy to speak up," Guetschow said. Instead, she spoke to Lillian Taylor, who served on the board's human resources committee. Taylor conveyed the omission to Romney, who approved an amended policy that covered homosexuality. Romney later reached out to Salt Lake's gay community as part of the committee's effort to enhance diversity in the Olympic workforce. "He treated me well, and I think he genuinely believes that all people should be treated well," Guetschow said.

The most publicized moment of controversy during Romney's Olympics stint came later, during the Games, when he clashed with Utah police after they alleged that he had twice used the F-word in berating a teenage student who was directing snarled traffic at an Olympic venue, the Snowbasin resort. Police were angry that Romney had denied shouting the expletive. "Both the Job Corps student and a sergeant who witnessed the scene related the same story," said Weber County Sheriff's Captain Terry Shaw, who was in charge of security at Snowbasin. "There was no reason to indicate they weren't telling the truth." Romney denied using the obscenity and said that two other witnesses—a Secret Service agent and an Olympic aide, Spencer Zwick—corroborated his denial. "I have not used that word since college, all right? Or since high school," he said. Law enforcement authorities were further miffed that Romney offered a partial apology to the police but not to the student. "There were a lot of people in public safety who were extremely angry," said Peter Dawson, who was serving as an intern in the Olympics communications center at the time. "The general consensus was 'I hope he doesn't need any help from us because we aren't going to respond very quickly.'"

The warm morning air was thick with black smoke as Mitt Romney and an aide raced away from Capitol Hill. They'd been preparing for a day of lobbying the federal government for Olympic aid. Then came reports of planes hitting the World Trade Center and the Pentagon. The acrid smoke billowed down on their BMW convertible as they fled. It smelled, Romney would say later, "like war." Of all Romney's challenges as head of the 2002 Winter Olympics, this was the most grave: the terrorist attacks of September 11, 2001, had changed everything. The blueprint had to be rewritten with the Games just five months away, a joyous celebration now set in a wounded land. Security became paramount. Romney expected calls for the Games to be canceled. Delegations, teams, or certain athletes might refuse to come. "I think Mitt wondered inwardly whether we could even hold the Games. He couldn't say it publicly, but that was his nagging fear," Ann Romney said at the time. "For a few days, like the rest of the country, he was floating in this sense of gloom and doom. But when he realized that the Olympics might help things, he became extra-determined. It was: We're holding these Games no matter what, even if it's just the athletes."

When he returned to Utah, Romney gathered hundreds of staffers and volunteers in an outdoor plaza and delivered a speech that several described as the most presidential moment of his Olympic tenure. While he addressed the fears many harbored of terrorists striking again during the Games, Romney invoked the glory of patriotism, public service, and facing down danger. "By the end, he had everybody singing 'God Bless America,' but not in a 'Kumbaya' kind of way," said Zianibeth Shattuck-Owen, who had served as a trustee and later as a luge manager. "It was leadership." Romney delivered a similar message in an e-mail to the staff. "In the annals of Olympism and the history of Utah, this may stand as one of the defining hours," he wrote. "I am confident we will perform with honor."

The attacks required tightened security for the Games—and a new infusion of federal funding to pay for it. The Salt Lake Games were designated as a national special security event, with the FBI, Secret Service, and Federal Emergency Management Agency overseeing an effort that also involved the Central Intelligence Agency and numerous other international, national, state, and local military and law enforcement agencies. Congress had already earmarked about $200 million to try to make the Games safe. After the attacks, it fell largely to Senator Robert Bennett, a member of the Senate Appropriations Committee, to secure an additional $34.4 million, and Bennett enlisted Romney to help lobby key legislators. "It was very easy to make the case with Mitt because he had the credibility," Bennett said. "Whenever he was questioned, Mitt had his homework done. It made my job a lot easier to have him as the salesman for all of this."

Romney's team displayed less tact in handling another sensitive matter related to the terror attacks. Trouble began when his executive assistant, Donna Tillery, twice rejected requests to provide free or discounted tickets to widows and orphans of firefighters who had died at the World Trade Center. Tillery sent e-mails to a former Salt Lake City firefighter, A. J. Barto, in which she explained the denial by citing a policy barring ticket giveaways. That made Romney, who professed not to know about the requests to Tillery, appear callous six weeks later when he offered a hundred surplus tickets, valued at $885 each, free to Utah legislators. "I was outraged at the hypocrisy," Barto said. "In less than two months, he went from saying, 'We're going to run a tight ship' to throwing out free tickets to a group of people who could help him politically."

As the Games neared, Romney and his Olympic colleagues wrestled with what tone to strike at the ceremonies. "It's been much more somber since September 11, so I don't know that we will have the same kind of exuberant, celebratory feeling for the Olympics that other cities have enjoyed," he said at the time. "But we are what we are, and the nation is experiencing the mood it's experiencing. In some respects, the Olympics take on a more profound meaning now than they might have in a more giddy time."

It was, in the end, a solemn display of resiliency that became the most unforgettable moment of the Games. Romney and Sandy Baldwin of the United States Olympic Committee persuaded the IOC to allow the U.S. team to carry into the opening ceremonies a tattered American flag found in the rubble of the World Trade Center. IOC officials had argued that the gesture would be viewed as too political, but they ultimately relented, clearing the way for eight American athletes, accompanied by New York firefighters and police officers, to carefully walk the flag into the Olympic stadium, to silence.

With September 11 still on his mind and Winston Churchill as his inspiration, Romney was prone, as the Games approached, to grand historical analogy. "Dwight Eisenhower was quoted as saying that as D-day got closer, he realized there wasn't much more he could do than salute the soldiers as they went off," he said. "But I'll still be in the main operations center during the Games. I'll still be on the radio and the phones constantly. I doubt I'll sit down and just watch events." In taking on the job, Romney had "gone through all the usual Mitt stages," said Charles Manning, his former political strategist. "First it's Mr. Worrywart: 'This will never happen; it's going to be a disaster.' Then it's: 'Let's take it all apart, then put it back together.' Now it's: 'Let's make this the greatest success we can.'" Indeed, Romney, true to his nature, fretted and fussed and fidgeted over his Olympics until the flame in Salt Lake City was finally extinguished.

His tight grip paid off. Over seventeen days in February 2002, 2.1 billion people around the globe watched the competition, with the United States winning thirty-four medals, and representatives of nearly every nation expressing satisfaction with the event. In the end, Romney helped generate nearly a $100 million budget surplus and a trove of political goodwill for his next endeavor. "The people who say he is given too much credit for restoring the reputation of the Games," David D'Alessandro said, "I don't think they understand what he was up against."

Ann Romney, after a difficult few years following her multiple sclerosis diagnosis, enjoyed a special lift. Before each Winter or Summer Games, a host country typically stages a torch relay, in which the Olympic flame is passed from runner to runner en route to its final destination in a stadium cauldron for the opening ceremonies. Many of the runners are everyday people—so-called personal heroes nominated for their courage, acts of charity, or some other attribute. Mitt nominated Ann in 2002, which afforded her the privilege of helping run the Olympic torch into Salt Lake City. "It was an amazing thing," she would say years later, "to be too weak to barely walk when we got out there, and to be strong enough after three years, to have my children helping me to hold my arm up, and my husband was at my elbow, running with me and running the torch into Salt Lake City, as his hero."

Had he been around to witness it, George Romney would have no doubt been proud at his youngest son's achievement. It was a classic

Romney project, and the son had been true to his name, in the land where his ancestors had overcome their own long odds a century before. Mitt Romney had helped to clear Brigham Young's "great highway of the nations" for the kings and queens, presidents and nobles, elite athletes, and everyday people who converged on Salt Lake City. As a missionary, businessman, church leader, and friend, he had repaired many things in his life. None, however, had been nearly as big as this.

NINE

THE CEO GOVERNOR

I don't think that government is about doing favors for people.
I think it's doing the right thing for the folks we represent.

—MITT ROMNEY, 2007

A year before the torch was lit for the Winter Olympics, Mitt Romney's leadership was already earning him mentions as a potential candidate for governor in two states. With his 1994 loss to Ted Kennedy now but a bad memory, Romney was eager to make another run at public office. But would he stay in Utah, where some pundits suggested his politics aligned more with the Democratic Party, or would he return home to try again in Massachusetts? As with other major decisions in his life, this one would not be driven by impulse, but by a careful strategic assessment of the political landscape. "If politics is part of the mix, which it may be," Romney said at one point, "a lot of that depends on where the opportunity is."

Friends and his former chief political adviser figured Romney was destined for Washington, for a cabinet or other high-level position under President George W. Bush. Speculation had also grown about a run in conservative Utah, however, after Romney objected to a July 2001 story in *The Salt Lake Tribune* describing him as "pro-choice" on abortion. "I do not wish to be labeled pro-choice," he wrote in a letter to the editor. But later in 2001, a column in Salt Lake City's *Deseret*

News, citing sources close to Romney, suggested that he wanted a perch with enough national exposure to enable him to run for president. And Massachusetts was clearly the bigger launching pad. There was just one problem: Jane Swift, elevated to acting governor of Massachusetts when her boss, Paul Cellucci, had become U.S. ambassador to Canada, was preparing to anchor the GOP ticket in 2002. Romney had said there was little chance that he would challenge his fellow Republican.

Swift, though, was struggling politically. Panic-stricken Republicans feared losing the governorship, their only prize in the lopsidedly Democratic state. As one Republican town chairman put it at the time, "People are just sort of ready to jump ship and line up with somebody they think that can carry the ball in November." Barbara Anderson, Massachusetts's best-known antitax activist, recalled leaving this message on Romney's answering machine: "I know you're really busy now with the Olympics, but when you're finished, please come back and save Massachusetts." The state Republican Party's new chairwoman, Kerry Healey, discreetly flew to Salt Lake City to gauge Romney's intentions. He was noncommittal. Romney was telling other leading Republicans he wasn't planning to run.

But then the Olympics went off with poignant beauty, and Romney was widely credited with rescuing the Games from financial ruin and scandal. "The guy looks like he walks on water," Dan Jones, whose Utah poll registered Romney's approval rating at 87 percent, said at the time. In Massachusetts, Romney commissioned his own poll, which showed he would be a viable candidate against any Democrat. His agents on the ground were quietly signing up staff and consultants and scheduling a formal announcement—whether Swift was in or out. One Massachusetts Republican close to Romney said that Romney had hated having to muscle Swift aside, but the allure of the job, and the pressure from party leaders, had been too great to resist. "It was clear Jane could not win. It was clear he could," the Republican said. Romney never called Swift to say that he had changed his mind and would run for governor after all. Swift found out from her staff, which had gotten wind that Romney had rented a big hotel ballroom. This was consistent with what is, by several accounts, Romney's aversion to confrontation if he can help it.

Ann Romney's health was a factor in his decision to run. She said she had "huge qualms" about going back east, because her multiple sclerosis symptoms had abated during three years in Utah. "I've been healthy out here," she said. But the next day, March 17, the Romneys, decked out in matching jackets with an Olympic theme, flew to Massachusetts. They were met at the airport by reporters and a fresh poll indicating that Romney would crush Swift in a race for the GOP nomination. Within forty-eight hours, Swift, the first woman to lead Massachusetts, pulled out of the race at an emotional press conference. The mother of three young children said it would be impossible to balance the demands of her family and her work as governor and fend off a primary fight from Romney. That afternoon, Romney was gracious toward Swift—he'd even canceled the big ballroom announcement for a more subdued kickoff. But he made clear that the race was now his. "Lest there be any doubt," he told reporters gathered on the driveway of his Belmont home, "I'm in."

The Mitt Romney who hit the campaign trail in 2002 was a different man from the political neophyte of 1994. He was, at fifty-five, now something of a media phenomenon—*People* magazine would soon name him one of the fifty Most Beautiful People in the World. With a high-profile public achievement now under his belt, he was that much more self-confident. After the bruising he had endured in his race against Senator Kennedy, he was wiser in the ways of politics, and as a result, he was tougher. Within days, his campaign produced television ads designed to preemptively beat back any Democratic attacks. It was a lesson learned from the Kennedy onslaught eight years earlier, which had typecast him as a heartless corporate raider. This time Romney would define himself, instead of letting his opponent do it for him. "I learned in my race against Senator Kennedy, don't be so naïve as to just sit back and say, 'Oh, I'm going to be positive while they're negative,'" Romney told a TV interviewer. "That doesn't make sense."

Soon after jumping into the race, Romney deposited $75,000 in a new campaign account, the first installment of $6.3 million of his own money he would eventually spend on the campaign, then the most expensive in state history. But Romney's money—he was now considerably wealthier than he'd been during the Senate race—was both an asset and a liability. His aides feared fresh criticism that he was out of touch with middle-class voters. They knew how effective that line of attack had been the last time around. Romney's campaign was especially worried that another wealthy businessman—a former state GOP chairman named James Rappaport—might win the Republican primary for lieutenant governor. Romney declared that he would remain neutral in the lieutenant governor's race, but, as had happened with Swift, political necessity changed his mind. Robert White, his longtime wingman from Bain, and Michael Murphy, his chief political strategist, began working to avert a Romney-Rappaport ticket of two rich white men. After evaluating a list of potential alternatives, including African Americans and women, they settled on Kerry Healey, a bright but little-known figure who had two failed campaigns for state representative behind her and a wealthy husband who could neutralize Rappaport's self-financed effort. Healey won the primary to become Romney's running mate, but the expedience of their alliance was evident at the outset, when he called his new partner "Sherry" on a radio show.

During the same interview, Romney also displayed a weak grasp of Massachusetts politics, using the wrong first name of a past governor and misstating the hometown of a current Democratic candidate for the office. Listeners were perhaps willing to forgive him—he'd been away from the state for a few years, after all. But Democrats, sensing an opening, had other ideas. The Massachusetts Democratic Party launched a challenge to Romney's eligibility for office, contending that his three years in Utah disqualified him from running. (The state constitution requires seven years of state residency before one can run for governor.) Romney was incensed. "Any effort to try to remove me by hook and crook and trick and legal machinations is going to end up failing," he said.

Romney's campaign initially insisted that he had filed income taxes as a resident of Massachusetts. Soon after, he acknowledged he had filed as a Utah resident for two years and had amended those tax returns, after announcing his candidacy, to show Massachusetts as his home. Over three days in June 2002, the State Ballot Law Commission heard evidence about his tax returns and the fact that the Romneys' Utah home had been classified as his "primary residence," giving him an $18,000 annual property-tax break. Romney attributed the mistakes to his accountant and the local tax assessor in Utah, who acknowledged the error under oath and, after the commission proceedings, sent him a new bill to recoup $54,587. But if the tax filings suggested that Romney was hedging his political bets, the evidence also showed that he had always maintained his Belmont voting address—choosing George W. Bush over John McCain in the 2000 Republican presidential primary, he would say later—and had returned from Utah for special occasions, maintaining ties to Bay State boards and organizations. The ballot commission, deeming Romney's testimony "credible in all respects," rejected the Democrats' challenge in a unanimous forty-one-page decision.

The residency fight proved to be a good tune-up for the race against his Democratic opponent, Shannon P. O'Brien, a feisty Yale-educated former legislator and the state's first female treasurer. Her family had been active in Massachusetts politics for four generations. She had won a rugged four-way primary but drained her campaign account in the process. At the outset, she was the aggressor, attacking Romney incessantly on trustworthiness and needling him on issues where he seemed vulnerable. As Kennedy had in 1994, O'Brien homed in on abortion, hoping to exploit uncertainty about Romney's position and undercut his appeal to women and independent voters. It was hardly a pressing issue at the statehouse, and O'Brien herself had opposed abortion in a previous political campaign. But as a litmus test, it seemed to hold promise, and O'Brien's campaign knew it.

"My position has not changed." It was an unusually defensive assertion for a convention speech, but Romney felt obliged to make it. He was speaking to a Massachusetts Republican Party gathering that April. But he was also addressing a broader concern: growing doubts about his views on key social issues. Romney's assertion signaled that he was the same man who had run against Kennedy, a fiscal conservative with moderate views on issues such as abortion, gay rights, guns, and the environment. But there were confusing signals, too, and Romney was again forced to persuade a skeptical electorate of where his heart truly lay. His 2001 letter to the editor in the Utah newspaper, in which he said he no longer wanted to be considered prochoice, was one source of political friction, leading opinion makers and abortion rights advocates to question his stance. Romney sought to answer those questions definitively at the convention, declaring, "I respect and will fully protect a woman's right to choose. That choice is a deeply personal one, and the women of our state should make it based on their beliefs, not mine and not the government's." Ann Romney also tried to assuage the concerns of women voters, saying in a joint TV interview, "I think [they] may be more nervous about him on social issues. They shouldn't be, because he's going to be just fine." Mitt cut in and said that when he's asked whether he will "preserve and protect" a woman's right to abortion, "I make an unequivocal answer: yes."

His answers to an April 2002 questionnaire from the local NARAL chapter were similarly unequivocal. Romney said he would oppose attempts to change state law, either by adding new restrictions on abortion or by easing existing ones. He expressed support for Medicaid funding for the procedure, efforts to expand access to emergency contraception, and the restoration of state funding for family-planning and teen pregnancy prevention programs. He also said he supported comprehensive sex education in public schools and would oppose "'abstinence-only' sexuality education programs." "The truth is, no candidate in the governor's race in either party would deny women abortion rights," he wrote. "So let's end an argument that does not exist and stop the cynical, divisive attacks made only for political gain."

Later that year, Romney picked up the endorsement of the Republican Pro-Choice Coalition, a group that backed GOP candidates nationwide who supported abortion rights. "We applaud your commitment to family planning and protecting a woman's right to choose," the group's national director wrote Romney in October. "You set a wonderful example." On flyers Romney printed to court women voters, the first bullet point assured them that he had "promised to protect a woman's right to choose."

But abortion, although an ideological barometer, was not really much of a focus for state politicians. The bigger culture battles in 2002 were being fought over gay rights. The political landscape had shifted following Vermont's pioneering decision in 2000 to legalize civil unions. The decision spooked gay marriage opponents in Massachusetts, who organized a push for a constitutional amendment restricting marriage to heterosexual unions. Ann and Tagg Romney alarmed Romney's gay supporters—the many he'd won over in his 1994 campaign—by signing a petition to put the question on the election ballot. But Romney quickly distanced himself from his family's decision, saying he did not support the proposed ban. The proposal, which later died, would also have withheld domestic-partner benefits such as bereavement leave and health care coverage from the gay and lesbian partners of public employees.

Romney did not support same-sex marriage, declaring in a 2002 questionnaire for *Bay Windows*, New England's leading gay and lesbian newspaper, "I believe that marriage is a union between a man and a woman." He also said he opposed civil unions, believing they were too close to marriage. But at the same time, he was assuring gays and lesbians—publicly and privately—that he would not crusade against them. Plus he was voicing support for domestic partner benefits that sounded an awful lot like civil unions. Richard Babson, a board member of the state chapter of the Log Cabin Republicans, the Republican gay rights advocacy group, recalled an event at a bar in Boston's Bay Village neighborhood where Romney spoke to gay voters. Babson came away from the meeting—and a more intimate question-and-answer session that immediately followed—believing Romney to be reasonably progressive. "He said, 'I support everything that it calls for in terms of recognizing unions between people. But just don't use the M-word,'" Babson recalled.

In addition, Romney said, as he had in 1994, that he would be especially effective as a Republican voice supportive of gay rights, and he vowed to use his bully pulpit to stand up to the powerful, culturally conservative Democratic House speaker, Thomas Finneran. At Boston's gay pride parade his campaign distributed pink flyers that said, "Mitt and Kerry Wish You a Great Pride Weekend! All citizens deserve equal rights, regardless of their sexual preference." On the *Bay Windows* questionnaire, he expressed support for strengthening laws against hate crimes and expanding funding for AIDS treatment and prevention and said he would look to both "protect already established rights and extend basic civil rights to domestic partnerships." All in all, he was again sending signals that, though he wasn't willing to go as far as endorsing marriage, he was sympathetic to gay rights and would advance the cause once in office. As it had in 1994, that stance earned him the endorsement of the Log Cabin Republicans.

In the closing weeks of the race, Romney sat down for a private meeting with Bill Saltonstall, a former Republican state senator whose family had helped establish the Massachusetts brand of socially moderate Republicanism. Saltonstall had a gay daughter living in Alaska with adopted children and wanted assurance from Romney that his assertions on gay rights came from the heart. Romney, according to one participant in the meeting, assured Saltonstall that he would promote tolerance and fight discrimination. He also, according to the participant, proposed a thorough review of state laws to see where life-long gay and lesbian relationships were negatively affected, and how the state could change its practices to make them nondiscriminatory. Saltonstall's mind, for the moment, was eased.

Romney credited George and Lenore with giving him his civil rights values. "At a very young age, my parents taught me important lessons about tolerance and respect," he said. "I have carried these lessons with me throughout my life and will bring them with me if I am fortunate enough to be elected governor."

In an echo of his 1994 platform, Romney positioned himself as an agent of change, vowing to "clean up the mess on Beacon Hill," the seat of state government. And there was plenty to clean up: government was still rife with patronage and waste. Romney exuded competence and promised fresh thinking, an especially appealing profile to voters who had for years elected Republican governors to check the power of the dominant Democratic establishment. Even though the Republican Party had long owned the corner office, Romney had no ties to government, the GOP's own cronyism, or its oversight of the granddaddy of all messes—the cost overruns and mismanagement of the $15 billion Big Dig highway project in Boston.

Romney debuted a sophisticated "microtargeting" program to drill deep into voter behavior, seeking to identify supporters through their voting history and other personal information. He pitched himself squarely to independents, who made up half the Massachusetts electorate. Unlike O'Brien, he supported the death penalty, which had long been abolished, and an initiative petition on that year's ballot to replace bilingual education with English immersion. With the state losing jobs and facing a ballooning budget deficit, Romney said his business experience had been the perfect proving ground. He insisted he could cut $1 billion from the state's $23 billion budget by eliminating the fat—all while rolling back taxes over several years. Neither candidate took a no-new-taxes pledge; Romney's campaign spokesman dubbed it "government by gimmickry," though in a future campaign he would embrace just such a pledge. Romney vowed to fight any tax increase, although he did propose raising excise taxes on vehicles such as trucks and sport-utility vehicles that got poor gas mileage. That idea unnerved his political advisers, given that Romney needed truck and SUV owners to stand with him at the polls.

If Romney's message was smartly framed, the packaging—that of the financier with crisp shirts—needed work. One of the first things his campaign did was to make sure his BMW remained stowed out of sight. The image of Romney behind the wheel of a fancy car—and a European one, at that—was not exactly what advisers wanted in voters' minds. Later in the campaign, Romney engaged in a number of made-for-TV-news "workdays," in which he took turns performing blue-collar jobs—an auto mechanic, a road paver, and a garbage man, among others—doffing his suit and tie for tradesmen's garb. O'Brien, in her victory speech on primary night, got some good mileage out of it. "Massachusetts," she said, "doesn't need a governor who thinks getting in touch with working people is a costume party."

Romney was indeed struggling to connect with the people. One of his TV ads, designed to humanize him, backfired, instead underscoring the disconnect between his seemingly storybook life and the lives of many voters. Called "Ann," the spot featured the Romneys talking about their high school romance and family. Mitt professed his love for his wife: "Ann's just good to the core." It closed with video of Mitt in a bathing suit horsing around with his sons on a raft at the lake. His poll numbers tumbled. Three weeks into the seven-week general-election campaign, his own internal polls had him trailing O'Brien by ten points. Within the campaign, some advisers thought the cloying tone of the ad had encased the candidate in plastic. "I think it made him appear to be too perfect," recalled Murphy, the irreverent strategist and media consultant who crafted the ad, insisting that it would have worked in any other state. "There's a certain cynicism in the Massachusetts electorate, and it locked into that."

Panic was setting in at Romney headquarters in North Cambridge. Some advisers grew resentful of Murphy, who had developed a big national reputation after leading John McCain's valiant fight against George W. Bush for the 2000 Republican presidential nomination. After a weekend meeting of senior campaign aides at Romney's home, the candidate stood by his chief strategist. "Mike came and said, 'It's not working. My strategy is wrong, and you either ought to change the strategy dramatically as I'm going to suggest, or you ought to fire me,'" Romney later recalled. Romney considered his options but decided to keep Murphy and try the new tack. Instead of selling voters on Romney, the campaign would work to sour them on O'Brien—round two for the tough new Mitt.

It was October 2, the morning after Romney and O'Brien had met for their second debate, at a technical college in the central Massachusetts city of Worcester. Many thought O'Brien had won. But Romney and his advisers sat in the family room of his home in Belmont that day and made a shrewd judgment: O'Brien's aggressive style, they predicted, would not wear well. So after initially resisting additional debates between the two candidates, Romney did an about-face and called for more. The strategy paid off. The subsequent clashes allowed Romney to perfect his attacks on O'Brien and showcase himself as a confident, focused executive whom voters could easily imagine running the state.

He perhaps went too far in trying to highlight O'Brien's aggressive tone in the final debate, calling her tactics "unbecoming." To O'Brien and her supporters, the comment smacked of patronizing sexism, and it earned Romney some unfavorable headlines. But O'Brien probably did herself more damage, in part by endorsing lowering, to sixteen, the age at which girls could get abortions without parental consent. Together with her recent shift to supporting gay marriage—a year before the state's highest court ordered it—the abortion answer eroded her centrist appeal. "That last debate was probably the point where things really began breaking my way," Romney told reporters while campaigning in Boston the night before the election. "That was the time when the independents and the undecideds really started to move."

Murphy, meanwhile, produced two devastating ads that featured a "watchdog"—Duncan the basset hound—snoozing as men removed bags of money from the state treasury. Using humor, the ads fused loosely related issues—including $7 billion in state pension fund losses in the stock market crash on O'Brien's watch and the lobbying work of O'Brien's husband, a former state representative—to shake her credentials as a manager and cast her as an insider. O'Brien called the ad "as disgraceful as it is inaccurate," but it worked. At the same time, O'Brien's advertisements exhumed the corporate-raider attack Kennedy had used so effectively on Romney in 1994. One featured a steelworker who had lost a job in the shutdown of a Kansas City plant bought by Bain Capital. "It was the Ted Kennedy punch, but it was gone," said Michael Travaglini, O'Brien's deputy treasurer at the time. "The Olympics had made Mitt a real celebrity. That carried significant weight, and it made him more credible."

After trailing O'Brien by double digits, Romney had climbed back into the race. By November 1, the contest was deadlocked. Romney's team had counted on O'Brien turning people off, and that seemed to be happening. Voters indicated that they liked her less than they had a month before and liked Romney more. Four days later, on November 5, the Romney-Healey ticket romped, defeating O'Brien and her running mate, Christopher Gabrieli, by 106,000 votes, or 5 percent, of 2.2 million cast. The Republicans held their tiny base, limited Democratic margins in the urban strongholds, and, most crucially, swept the independent-rich suburbs and exurbs in Greater Boston. Romney's obsession with building campaign strategy around voter data had worked. "We took on an entrenched machine and we won," he declared to supporters at the Boston Park Plaza Hotel. "Tonight we sent a loud and clear message. That message is that it's time for a new era." Romney, while celebrating his victory, took a congratulatory phone call from President George W. Bush from aboard Air Force One.

There was much to celebrate. Forty years after George Romney had won his first election as governor of Michigan, his son had fulfilled a lifelong goal: following his father into public office. Both were fifty-five when first elected. Both turned their business backgrounds into political assets. And both won by running as moderate Republicans who appealed to the political center. Heartening as well, especially for Massachusetts Mormons, was that Romney's religion had never become an issue in the 2002 campaign. It felt good, at least for the moment, to be past all that. And for Romney, it felt even better to exorcise the demons of 1994 once and for all. The bitter disappointment at Kennedy's easy victory, the frustration he felt afterward for having failed to articulate a clear message and for having offered a phlegmatic response to attacks on his business record—all of that was now gone, lost in the glare of his success and growing national notice. Mitt Romney would soon be the seventieth governor of the Commonwealth of Massachusetts.

———————

He followed in the footsteps of some famous men: three signers of the Declaration of Independence and a future U.S. president. Also, of course, some rogues. But Romney was the state's first self-styled CEO governor, and he brought a new approach to the top job. A few weeks after being sworn in, he was visiting with leading state lawmakers in the Senate Reception Room at the Massachusetts statehouse, an ornate sanctuary with tall windows and a barrel-vaulted ceiling that sits just off the Senate chamber. Campaign season was over. Now it was time to cool the partisan rhetoric and figure out how to govern together. Romney, a Republican in a sea of Democrats, gamely offered his hellos, shook some hands, and then gave brief remarks.

Romney told his new colleagues that he'd led Bain Capital, held leadership positions in large companies, and was now looking forward to working with all of them in doing the people's business. "My usual approach," one former lawmaker recalled him as saying, "has been to set the strategic vision for the enterprise and then work with executive vice presidents to implement that strategy." Andrea F. Nuciforo, Jr., then a state senator from western Massachusetts, said he had leaned over to a fellow legislator and sighed, "This is going to be a long four years." Romney seemed to be suggesting that he was the boss and state lawmakers his deputies. "My take on it was, here is a person who is well-intentioned and competent but unclear on the basic concept," Nuciforo remarked.

Romney had long wanted a high-profile job in politics, and now he had it. A lifetime of study, preparation, and hard work seemed to be paying off. But he also had a state to manage, and that wasn't something he knew much about. Nor was he familiar with the clubby quarters of the statehouse, where Democrats had run the show for decades, often from back rooms. He had staked his candidacy on his outsider status—the Bain consultant who would come into state government, throw open the hood, and make it run better. Now the outsider had come inside and found it a foreign place, with rituals and customs all its own. Every time he spouted a "Holy cow!" or dropped a "Golly!"—not exactly vernacular in the world of politics—he seemed born of a different species.

In an early showdown with Romney over the state budget, lawmakers were eager to teach the new governor a lesson in how things worked. After the House and Senate passed the final version of their joint spending plan, Romney issued a litany of vetoes. The legislature responded by holding one of its marathon sessions, taking up the vetoes one by one, and promptly reversing scores of them. Around midnight, a Senate leader stood up and proposed that the body adjourn "in memory of Mitt Romney." As Nuciforo put it, "The legislature had really killed this notion that he was going to run the place and we were there to execute his wishes." It was a glimmer of the hurdles that lay ahead, as Romney sought to apply his corporate methods and cool analytical acumen to a political culture that was deeply entrenched and, many believed, unaccountable and ill suited to carry Massachusetts into the next millennium. Yet, from the beginning, Romney showed he could drive a bargain, thwarting an attempt by legislative leaders to raise the pay of their lieutenants and forcing a budget that ultimately avoided a broad-based tax hike.

Willing to look beyond the Republican Party for top talent, Romney turned to high-powered figures such as Robert C. Pozen, a former top executive at Fidelity Investments; and Douglas I. Foy, the longtime president of the nonprofit Conservation Law Foundation. Over twenty-five years, Foy had been a relentless force at the law foundation, which had won numerous landmark lawsuits to protect the quality of air, water, and open space in New England, including litigation preventing a second reactor at the Seabrook, New Hampshire, nuclear plant and oil drilling in the Georges Bank fishing grounds. Pozen had been vice chairman of Fidelity Investments, practiced and taught law, and been associate general counsel of the Securities and Exchange

Commission. They were the biggest names in a strong lineup, and Romney gave both broad authority over new "supersecretariats" that oversaw economic development, housing, the environment, and more. This had long been a Romney credo: surround yourself with smart, aggressive players and let them go to work.

Once Romney had his team in place, he managed the way he knew how: he wanted piles of data and robust, free-flowing debate. One of Pozen's major tasks was to rescue the state's nearly insolvent unemployment insurance fund, which was draining quickly because of the state's generous benefits during a time of high joblessness. He focused on a plan that would reward firms with stable employment and require greater contributions from employers that often cut their workforce. "We probably presented him with ten or twenty different [computer] runs," he recalled of his dealings with Romney. "He would go through them and really absorb the data and try to figure it out."

Foy came from an environmental background, Romney from a business one. But they found common ground in wanting to minimize sprawl, using the state's leverage to promote density in development, and maximize open space. As a candidate, Romney had called sprawl "the most important quality-of-life issue facing Massachusetts." Romney threw himself behind the notion of "smart growth," as it was known, rewarding communities that clustered new residential and commercial projects, particularly around mass transit hubs. This view of economic development also manifested itself in what became known as a "fix-it-first" transportation policy, in which Romney's administration would focus on repairing existing roads and bridges instead of building new highway systems at the request of local politicians.

On these issues or any other, Romney wanted to know every argument for and against a certain course of action—he was "a decision maker by devil's advocacy," said Eric A. Kriss, with whom Romney had worked at Bain for years before tapping him to be state budget chief. "He would take conflicting facts and different perspectives and put them all on the table and have everyone argue all sides. There were times when I knew privately the outcome he favored, but he would actually take the opposite view of his own opinion, so that he could not only try it on for size but also know what the pushback would be."

———————————

From the moment they got into office, Romney and his team faced an urgent task. The budget was in meltdown. The state faced a potential $650 million deficit. Worse, his analysts discovered that the projected shortfall for the following year was exploding—from $2 billion to $3 billion in a $23 billion budget. Romney and the legislature settled on an austere plan to close nearly all the projected gap. His leadership in steering the state through the fiscal maelstrom is unquestionably one of his key achievements. In the retelling, however, he and his aides have often overstated the accomplishment. The budget gap never became as wide as Romney said, because tax collections came in more than $1.2 billion above preliminary estimates. He also understated the side effects of the rescue plan: big fee increases, more tax dollars wrested out of business, and increased pressure on local property taxes.

The plan raised at least $331 million in new revenue through increased fees for permits, licenses, and services—about a 45 percent jump. Massachusetts residents suddenly found it more expensive to get a driver's license, marry, or buy a home. In addition, it raised $128 million from tax code tweakings to close what Romney called loopholes for businesses. Some eliminated schemes by corporations to avoid state taxes by sheltering income in shell companies. Many business leaders considered the changes tax increases, even if Romney didn't call them that. Another $181 million in so-called loophole closings followed in the next two years.

The effects of the deep spending cuts were also felt in cities and towns, which absorbed hundreds of millions of dollars worth of reductions in state assistance over the first eighteen months of Romney's term. When Romney left office, Massachusetts communities—in a twenty-five-year high—relied on property taxes to cover 53 percent of their budgets, up from 49 percent before Romney took office. Many communities cut services and raised fees on residents as a result. Eric

Fehrnstrom, Romney's longtime spokesman, said that local officials had exaggerated the impact of the cuts, which had been made necessary by a plunge in state tax revenues. "Cities and towns share in the state's revenues when times are good, but they also must share in the belt-tightening when times are bad," he said. The budget cuts were often deep and painful, but Barbara Anderson, the antitax activist who had begged Romney to run for governor, identified a trade-off. "What people don't credit him for," she said, "is what he prevented from happening . . . a [broad-based] tax increase."

From the outset, Romney's team maintained tight control of information and displayed a disciplined, centralized management style. The administration spoke with one voice, usually Romney's. Assisted by Boston-based management consultants, they scoured the bureaucracy, culling data in a search for inefficiency, waste, and savings. That work provided the basis for what Romney would tout as "the most significant restructuring of state government in half a century." Sensing a mandate, the new governor pursued bold goals. He aimed to overhaul the sprawling human services system, a court network beset by legislative meddling and patronage, and the twenty-nine-campus public higher education system. At the same time, he wanted to loosen the teachers unions' influence on public education and fix a troubled state program that helped cities and towns finance school building and repair projects. Some changes Romney could make himself, by executive order. His cuts to the state bureaucracy, however, turned out to be fairly modest. After four years, he reduced the payroll of agencies under his direct control by 603 jobs, according to his administration's tally. By contrast, a Republican predecessor, William F. Weld, had closed state hospitals, privatized services, and slashed about 7,700 jobs during his first term, though the numbers had later increased when the economy improved.

Many other moves to reform state government required legislative support, and there Romney often faltered. That was partly because he showed little interest in the lawmakers themselves. He and his brainy, idealistic staff seemed too often blind to the fact that sweeping reforms, even if they made great sense in a white paper, counted for nothing unless the spadework had been done to cultivate legislative support. Romney, never a backslapper, invested little in building such ties—or even in getting to know the players. And so his court consolidation plan went nowhere, his vision for higher education vanished almost without a trace. Only pieces of his grand plan became law. A frequent complaint was that Romney, unlike previous Republican governors, rarely made an effort to develop meaningful relationships with the rank and file. Weld, for example, had once traded his support for a legislative pay raise in return for a promise by legislative leaders to cut the capital gains tax. "Weld had a genuine curiosity about the people in the building and what made them tick, and how to develop functional relationships that proved to be productive in the clinch," said Thomas Finneran, a Democrat who was House speaker for the first twenty-one months of Romney's term. "Romney was considerably more reserved."

A fellow Democratic lawmaker put it more pointedly: "You remember Richard Nixon and the imperial presidency? Well, this was the imperial governor." There were the ropes that often curtailed access to Romney and his chambers. The elevator settings that restricted access to his office. The tape on the floor that told people exactly where to stand during events. This was the controlled environment that Romney created. His orbit was his own. "We always would talk about how, among the legislators, he had no idea what our names were—none," the lawmaker said. "Because he was so far removed from the day-to-day operations of state government."

Even as some of his sweeping initiatives ran headlong into political reality, Romney remained certain about his path. And there was nothing he was surer of than his vision for economic growth. "My program for creating jobs is second to none in the entire history of this state," he said during the gubernatorial campaign. But he inherited a brittle economy; few states had been hit harder by the collapse of the tech bubble than Massachusetts. The state lost about 200,000 jobs, or nearly 6 percent of its workforce, between February 2001 and December 2003, the end of his first year in office. Then the climate began to gradually improve. Through it all, Romney was heavily involved in trying to sell business leaders on Massachusetts. A spokesman said he met each year with an average of about fifty chief executive officers who were considering expanding or locating in the state. Still, by the end of Romney's term, fewer than 40,000 net new jobs had been generated statewide, about a 1 percent increase. It was the fourth weakest rate of job growth of all states over the same period—a testament, ultimately, to the gap between what elected leaders say they will do about job growth and what they can actually do.

Nonetheless, Romney gets credit in some quarters for improving the state's competitiveness. His administration streamlined the public approval process to help businesses expand and revived an agency charged with recruiting businesses to Massachusetts. Under Ranch C. Kimball, a self-described "Romney Democrat" who succeeded Robert Pozen as secretary of economic development in 2004, the number of companies in the Massachusetts development pipeline jumped from 13 to 288 in three years, though much of that was driven by the recovering economy and the rise of the biotechnology and life sciences sector. In 2006, Bristol-Myers Squibb chose an eighty-nine-acre site northwest of Boston over one in North Carolina for a $750 million complex, which, as of the summer of 2011, employed more than three hundred people. The deal required tax credits, other state assistance, and an unusual show of teamwork by two reluctant playmates, Romney and the legislature.

Through it all, Romney remained something of a reluctant political negotiator. Instead of pushing his agenda through closed-door appeals to legislators, he favored well-orchestrated media events to generate public pressure. They were often heavy on stagecraft and carefully choreographed by his aides. "His theory of government was 'I'm going to the bully pulpit, which is the press, and beat you up so you succumb to my position,'" said Salvatore DiMasi, a Democrat who served as House speaker during most of Romney's term. (DiMasi would later become a symbol of the kind of deal making Romney sought to avoid; he was sentenced to an eight-year prison term after a corruption conviction in 2011.)

Romney's gifts as a communicator produced some triumphs. One was a bill on repeat drunken drivers that he and his lieutenant governor, Kerry Healey, prodded lawmakers to toughen in 2005 by enlisting family members of victims to make emotional televised appeals. Another came when Romney refused to sign a retroactive increase in the state's capital gains tax approved by the legislature in 2005 that would have affected 48,000 taxpayers. Romney's objections lit up the radio talk shows, and lawmakers backed down. He possessed, too, an instinct for the grand gesture, as he demonstrated after Hurricane Katrina ravaged New Orleans and the Gulf Coast in August 2005. Romney was unusually critical of the country's fumbled response under President Bush, calling it "an embarrassment," and he offered to take in thousands of evacuees on Cape Cod, where he set up an entire makeshift town on a military base. In the end, only 235 people were sent to Massachusetts. But Romney committed the state to helping anyone it could. "They're going to find it warm here, and hospitable," he said, "and they're going to find the people of Massachusetts have great big hearts."

One of his longest-running battles with the legislature was over the Big Dig, the beleaguered Boston highway project that became a symbol of patronage and mismanagement. Romney had twice sought control of the Massachusetts Turnpike Authority, the quasi-independent agency that oversaw the project, but had been rebuffed by the legislature, where the authority's chairman, former Republican state senator Matthew J. Amorello, had important friends. Then a tragedy changed everything. On the night of July 10, 2006, heavy concrete ceiling panels fell onto a car driving through a Big Dig tunnel. Milena Del Valle, a thirty-eight-year-old mother of three from Boston, was killed.

Within hours, Amorello was at the scene, where he remained as morning broke, briefing investigators, reviewing the accident with the state attorney general, and coordinating with state police. Romney aides had called Amorello's office, saying the governor wanted to see him. But before Amorello could make his way to the statehouse, Romney lost his patience, incensed that Amorello hadn't shown up yet. So Romney, his blood boiling, went to Amorello. He arrived at the scene and darted toward Amorello with an outstretched hand. In the bizarre seconds that followed, Romney, visibly agitated, gave Amorello a handshake, grabbed his shoulder with one hand, and then slapped him on the chest with the other. "You're too big for the governor?" Romney said, according to one witness to the exchange. "You're too big for the governor?"

Amorello and others around them were taken aback. "I'm standing there in shock," the witness said. "It was really something to see." As Romney and Amorello started walking awkwardly together down into the tunnel, the witness said, Romney's rant continued, before Amorello tried to calm him down by reminding him that a woman had just been killed. Asked later about their heated exchange, which had been caught from afar by TV cameras, Romney said he had expressed "disappointment" with Amorello at his snub. "And so, if you will, the mountain went to Mohammed," Romney said.

For years, the public had been troubled by cost overruns and design problems on the megaproject. Now the abstractions turned into a real threat: commuters were, for a time, afraid to venture into the warren of tunnels underneath the city. Romney had public opinion on his side. Three days after the accident, the legislature handed him emergency powers over the tunnel project. Amorello soon resigned. Immediately, Romney became a commanding and reassuring presence. The legendary quick study was on the case, demonstrating a stunning mastery of complicated engineering details. He unveiled plans for inspections and repairs. He vowed to restore public confidence. This, even many critics had to admit, was the take-charge CEO Massachusetts voters had elected in 2002. "At a moment of crisis, he exercised significant leadership," said David Luberoff, the coauthor of a book about the Big Dig and other megaprojects. Romney seemed to relish his role. When the accident happened, he was at his vacation home in New Hampshire and had to return to Boston. Tony Kimball, Romney's former colleague in local Mormon leadership, remembered running into Romney's middle son, Josh, at some point afterward. "Josh said, 'That's the first time I've ever seen my dad not really mad to have to go back,'" Kimball recalled.

He was, in almost every way, Mitt Romney's perfect foil. William M. Bulger, a former president of the state Senate, was a cunning and erudite Democratic pol from South Boston. A powerful figure during forty-two years in state government, Bulger enjoyed support among Democrats in the legislature, but his reputation for arrogance and Boston parochialism did not wear well with the general public. He was also weakened by embarrassing disclosures about his contact with his brother, the mobster James "Whitey" Bulger, who had been accused of nineteen murders and, at the time, was still a fugitive.

In his first year as governor, Romney zeroed in on Bulger, trying to break up the five-campus University of Massachusetts system and eliminating the office of president, which Bulger held. Major industries in the state objected, however, and the legislature thwarted the move. But Romney had a Plan B. Before Romney took office, Bulger had invoked his Fifth Amendment privilege before a congressional committee investigating the FBI's use of informants, his brother "Whitey" being among the most notorious. In June 2003, William Bulger faced a climactic second congressional appearance, testifying under a grant of immunity. Days before the hearing, State Attorney General Thomas F. Reilly broke ranks with Democrats and called for Bulger to resign from his state post. The next day, Romney, who resented Bulger's lack of cooperation with authorities, said he might call on the University of Massachusetts trustees to remove him.

After trustees rebuffed Romney, praising Bulger's job performance, Romney began to frame the conflict in moral terms, saying that Bulger, as a public university leader, should be held to "a much higher standard" even if he had committed no crime. The chairwoman of the trustees, Grace K. Fey, Bulger, and other backers on the board worried that the higher education system would suffer retaliation by Romney if Bulger survived. Lawyers negotiated an expensive buyout of the remaining four years on his contract, and he resigned. It was a triumph of the Beacon Hill newcomer over the ultimate insider. To Romney, that was what accountability looked like.

Romney had no personal relationship with Bulger as he pushed him out the door. William P. Monahan was a different story, one that illustrates Romney's allergy to controversy and willingness to cut loose even loyal associates if they threaten to sully his reputation. Three weeks after Bulger's exit, Monahan's long personal and political relationship with Romney ended abruptly with a thirteen-minute phone call. Romney forced Monahan out as chairman of the state Civil Service Commission just a month after he appointed him. His hasty ouster was engineered by aides who feared that the governor would be embarrassed by a *Boston Globe* story about Monahan's purchase of property from Boston organized crime figures twenty-three years earlier. From his lake house in Wolfeboro, New Hampshire, Romney called Monahan, who recalled Romney saying, "Bill, my stomach is turning. . . . My senior staff is unanimous that I have to ask for your resignation. I don't want to do this, but I am outvoted." Embittered, the lawyer and former Belmont town leader filed a lawsuit in U.S. District Court against Romney and others, seeking reinstatement. "He threw me under the bus," recalled Monahan, who had been a backer of Romney's political campaigns and a leading supporter of efforts by the Mormon church to build the Belmont temple. "When he needed me, I was always there."

Citing the litigation, Romney in 2007 declined to discuss the case in detail. He said that, had he known the full scope of Monahan's business dealings, he would never have appointed him in the first place. A federal court judge found in favor of Romney and his aides in September 2009.

Romney's relative coldness toward state lawmakers while in the State House did not extend to James Vallee. At least not initially. Vallee, a Democrat who chaired a key House committee, was one of the few legislators with whom Romney cultivated a relationship. Vallee had fought for high-profile bills that the governor backed. "The first couple of years, he'd call me on my cell phone and we met maybe a dozen times," Vallee said. "He supported me when I was going against the grain of my own colleagues."

But then Vallee lost his chairmanship, he said, and Romney stopped calling. Still, a few days before he left office, Romney thanked Vallee and asked whether there was anything he could do for him. The legislature had passed a measure awarding Vallee's hometown an additional liquor license, and all it needed was Romney's signature. So Vallee asked for it. "I'll take care of it before I leave," he quoted Romney as saying, though Romney would later contend that he did not recall the conversation. When Romney departed, the petition lay in a stack of last-minute bills left unsigned on his desk. "It was not a two-way street," Vallee said of the relationship.

This is consistent with how others who have worked with or watched Romney closely describe him: as a utilitarian who sometimes views others purely in terms of their value to him and his goals. "Mitt is always the star," said one fellow Republican. "And everybody else is a bit player." Indeed, Romney was, in many ways, a solo act, and his obsession with staying above the fray irked many State House regulars. They were accustomed to the transactional culture that has long permeated Massachusetts politics. Votes were traded. Political supporters were awarded state jobs. Family members of politicians were appointed to lucrative positions. And pension deals rewarded the well connected. That, quite literally, wasn't a language Romney spoke. When he used words like "poophead" and "gosh" in place of coarser constructions, hard-bitten political veterans rolled their eyes, as if to say, *Who is this guy?*

Over the course of Romney's term, Democratic lawmakers came to understand that it wasn't even worth approaching him for a favor. "I never asked the governor for anything political, never," said Robert Travaglini, the Senate president during Romney's tenure. "I've observed him, and never once did he demonstrate to me that that was part of his tool set." But although Romney's relative detachment was, at times, a hindrance, it also presented a healthy challenge to a stagnant, insular political culture. "He forced all of us to bring our A game to the table," Travaglini said as Romney ended his term. "Say what you will about the man, to some degree he initiated the action and direction on reform. . . . He brought out the best of us here in the Senate."

One change Romney made was sanitizing the judicial selection process, requiring a nominating panel to conduct an initial blind review of candidates without knowing their names, gender, or references. "The review process was completely apolitical," said Ralph C. Martin II, who chaired the state Judicial Nominating Commission for half of Romney's term. A July 2005 review of Romney's judicial picks by *The Boston Globe* detected no philosophical or partisan pattern. Indeed, Romney, who had all the information about judicial candidates when making his final selections, had filled three-quarters of thirty judicial vacancies with registered Democrats or independents, including two gay lawyers who had supported expanded same-sex rights. He similarly showed no evident preference for candidates who had contributed to his campaign.

Romney's distaste for crony politics was perfectly in character, another manifestation of his determined personal rectitude. But it was something very different in the State House, producing an administration that was virtually scandal-free and comparatively restrained in the exercise of patronage. "He never sent me anyone he wanted hired and never said, 'This is a major donor, see what you can do for him,'" Foy recalled. Romney gave a similar message to Daniel B. Winslow, who became his legal counsel. "He said, 'You will have lots of people calling you up to get their Uncle Oscar a job on the legal team,'" recalled Winslow, who served about two years under Romney. "'Don't do that,' he said. 'I need the best and brightest lawyers without regard to politics.'" It was not, as with many public figures, an image-buffing quote aimed at posterity. It was what Romney actually expected.

As governor, Romney did give jobs to many of his own campaign workers, but he was aggressive in ousting longtime operatives of his own Republican Party, including David Balfour, the head of the Metropolitan District Commission, a patronage haven that Romney would fold into another state agency. Much later, Romney rebuffed requests that he appoint Brian P. Lees, the Republican leader in the Senate, to the open job of clerk-magistrate of a district court in western Massachusetts. "I wanted to change the environment in Massachusetts from one of patronage to one of people getting jobs on the merit," Romney explained. "I don't think that government is about doing favors for people. I think it's doing the right thing for the folks we represent."

Romney's hiring policies were, however, flexible enough to tolerate some people with political ties. He found a job for Angelo R. Buonopane, a veteran of prior GOP administrations and a key liaison with labor and Italian-American voters. In April 2005, Buonopane resigned his $108,000-a-year job as state labor director after *The Boston Globe* reported that his post had no obvious duties and reporters observed him working an average of less than three hours on eight different days. And in his final months as governor, Romney filled more than two hundred slots on boards and commissions with party loyalists, state employees, and others. Eric Fehrnstrom, his communications director, was named to the part-time board of a town housing authority. But after the *Globe* reported that the appointment would help Fehrnstrom qualify for a large state pension, he resigned, protesting "unwarranted political attacks" on Romney.

By the start of his second year in office, Romney understood—better than he wanted to—the make-or-break power the Democratic legislature held over his agenda. If at the outset he had been fuzzy on that point, it had now become painfully clear. Frustrated by lawmakers' resistance to his ideas, he decided to go over their heads and right to the people.

The outlines of what would become a frontal assault on the Democrats in the 2004 elections emerged in his State of the State address that January. Using the word "reform" at least ten times, Romney laid out his challenge. "Quite simply, reform is about putting people first," Romney said. "We have to put people we don't know ahead of political friends we do know, schoolchildren ahead of teacher unions, and taxpayers ahead of special interests." The message may have been sound, but it was risky politics. In the past, Republican governors had tried to pick off a few legislative seats to build their numbers. Not Romney. He began an aggressive recruiting drive and in May unveiled a slate of 131 Republicans—the most in a decade—for the legislature's two hundred seats. All this in a year in which a favorite son, U.S. Senator John F. Kerry, was poised to win the Democratic nomination for president and be at the top of the November ballot.

Romney personally campaigned for more than forty Republican candidates, making almost seventy trips around the state. With his help, the state GOP raised $3 million and sent a blizzard of direct mail attacking Democrats for supporting in-state college tuition rates for illegal immigrants and being soft on sex offender laws. It was tough stuff, in many cases misleading, and Democrats were outraged. Raising the stakes further, Romney backed candidates who were challenging powerful Democratic committee chairs. But with so much advance warning, Democrats were ready for the fight. Many Republicans, meanwhile, weren't; the party had a famously short political bench in Massachusetts, and a number of its legislative hopefuls weren't ready for prime time. On election day, Kerry coasted to a nearly twenty-six-point win over George W. Bush in the state, providing long coattails for his party. "We could have had the twelve disciples running," said one Massachusetts Republican close to Romney.

For Romney, it was a train wreck. The party not only failed to pick up seats, it suffered a net loss of two in the House and one in the Senate. With only 21 of 160 seats in the House and 6 of 40 in the Senate, the party was down to its most diluted legislative presence since 1867. Romney's push may have protected a handful of GOP legislators, including an up-and-comer named Scott P. Brown, who would win election to the U.S. Senate several years later. But overall, Romney had invested immense political capital and now had nothing to show for it. "He put his personal reputation on the line," the state Democratic chairman crowed. "And he lost."

The resounding defeat spurred a resolution. "From now on," Romney said, "it's me, me, me." It was not long after the disappointing verdict at the polls, and the shift in Romney's mind-set was evident. No longer, he told the *Boston Globe* editorial board, would he spend so much time trying to build the Republican Party in Massachusetts. "The whole climate changed," Robert Travaglini said. "I think that's when they started looking elsewhere."

Even before the 2004 elections, Romney had been working to expand his national profile. That September, he earned a prime-time speaking slot at the Republican National Convention in New York, ripping into Kerry as a weather vane who went whatever way the wind blew. And he became an out-of-state surrogate for President George W. Bush's reelection campaign. Romney and his aides brushed aside questions about his long-term intentions, but an October appearance for Bush in Iowa ratcheted up the chatter. Romney, in fact, had been working quietly for months to build the scaffolding of a presidential campaign. In the summer of 2003, barely six months into his governorship, his old friend and confidant Robert White had huddled in Washington with two political strategists, Michael Murphy and Trent Wisecup, and a top GOP lawyer, Benjamin Ginsberg, to ponder Romney's next move. At Ginsberg's law office near Georgetown and later, over steak in a private room at Morton's, the seeds of Romney's eventual presidential campaign were planted. They conceived the Commonwealth PAC, a political action committee that would enable Romney to build a staff and travel the country with checkbook in hand, currying favor with Republican leaders by contributing to their campaigns and causes. Romney's advisers organized the PAC in an innovative way, setting up affiliates in six key states—including some states with no limits on contributions, which allowed Romney's wealthy associates, notably several Bain partners, to give five- and six-figure sums. The PAC went on to raise at least $8.8 million and dole out $1.3 million, much of it in key presidential-primary states.

He also put himself onto the leadership ladder of the Republican Governors Association, with the aim of becoming chairman in 2006. The move was designed to give him visibility, contact with Republican donors, and the chance to visit other states, and it worked. Romney's statehouse staff, meanwhile, began collecting biographical information to answer media inquiries about Romney's background, including his Michigan draft board records from the 1960s. "I didn't know if I wanted to run, I didn't know what would happen, I didn't know who the opposition would be," Romney said of those early preparations. "But I knew I didn't want to foreclose the possibility." That, according to a Republican close to Romney, had been the advice of Robert Bennett, then a Republican senator from Utah, whose family had long known Romney's. Timing was everything in politics, Bennett told Romney. If he thought he might want to run, he had to start now.

Romney's furious preparations for a possible campaign sparked controversy in October 2006, when *The Boston Globe* reported that his political team had privately consulted with Mormon church leaders on building a nationwide network of Mormon supporters. E-mails showed that Romney's political operatives, family members, and church officials had discussed building a grassroots political organization using alumni chapters of Brigham Young University's business school around the country. Representatives of BYU, which is run by the church, and Romney's political action committee had also begun soliciting help from prominent Mormons, including a well-known author suggested by Romney himself, to build the program, which his advisers had dubbed Mutual Values and Priorities, or MVP.

The e-mails also indicated that Jeffrey R. Holland, one of the twelve apostles who help run the worldwide church, was handling the initiative for the Mormon leadership and had hosted a September 19, 2006, meeting about it in his church office in Salt Lake City with Josh Romney, Mitt's middle son; Don Stirling, a consultant for Romney's Commonwealth PAC; and Kem Gardner, Romney's friend who had helped recruit him to the Olympics. Holland, a former BYU president, suggested using the BYU Management Society, an alumni and networking organization of the university's business school, to build the supporter base, the e-mails indicated. Romney would then have an established infrastructure—the group at the time had 5,500 members in about forty U.S. chapters—to help raise money and provide political support.

Romney's team downplayed the MVP program and chalked it up to overzealous advisers and supporters. Stirling and Gardner took the blame. The risk for the church was running afoul of federal laws prohibiting political advocacy by tax-exempt organizations. For Romney, the risks were political, as Gardner acknowledged in an interview with *The Salt Lake Tribune*. "We know Mitt can't use the church," he said. "Nobody wants a Mormon presidential campaign. It would kill us with the evangelical groups."

Gradually, Romney drifted further away from his day job, traveling almost constantly to promote himself and the Republican Party—a tally at the end of 2006 found that he'd been out of Massachusetts more than 200 days that year. Ann Romney, who had maintained a very low profile as Massachusetts's first lady, began expanding her own visibility on the national campaign trail, helping establish her husband's sterling reputation as a husband and father. The same could not be said for several of Romney's competitors in the Republican presidential primary. All the while, Romney was doing far more to prepare for a possible presidential bid than raise money and network with party leaders. Indeed, none of that would mean anything if he couldn't sell himself to primary voters in states where people knew little about him. And so Romney, employing the same careful, methodical strategy that had served him so well in the past, set out to build a new political identity.

———

On November 9, 2004, a week after his bid to elect more Republicans failed, Mitt Romney and two aides met in his statehouse office with a renowned Harvard University stem cell researcher named Douglas A. Melton. In Romney's retelling, Melton coolly explained how his work relied on cloning human embryos. "He said, 'Look, you don't have to think about this stem cell research as a moral issue, because we kill the embryos after fourteen days,'" Romney would later say. Melton afterward vigorously denied Romney's characterization of the meeting, saying, "We didn't discuss killing or anything related to it." Melton said, "I explained my work to him, told him about my deeply held respect for life, and explained that my work focuses on improving the lives of those suffering from debilitating diseases."

But for Romney, it was a seminal day, triggering what he describes as an awakening on "life" issues after he had spent his entire political career espousing very different views. In the official account of Romney's rebirth as a social conservative, the meeting with Melton would become the Genesis story. On February 10, 2005, three months after his meeting, Romney came out strongly against the cloning technique, saying in a *New York Times* interview that the method breached an "ethical boundary." He vowed to press for legislation to criminalize the work. Romney's opposition stunned scientists, lawmakers, and observers because of his past statements endorsing, at least in general terms, embryonic stem cell research. Six months earlier, his wife, Ann, had publicly expressed hope that stem cells would hold a cure for her multiple sclerosis.

Some scientists wondered if he simply didn't fully understand stem cell research. So they held a meeting with his deputy chief of staff, Peter Flaherty. In that meeting, Flaherty made clear that they knew the issue cold, Leonard Zon, a stem cell scientist at Children's Hospital in Boston who participated, said later. "I felt that they had thought this through and that their reasons for making this decision either was that he was a true believer or that there were other things going on politically," Zon remarked. Many critics accused Romney of political expedience. "There's evidence that he is clearly concerned with the national agenda," Robert Travaglini, who led efforts to pass a stem cell bill over Romney's objections, said at the time.

Romney has rejected suggestions that politics motivated his change of heart. "Changing my position was in line with an ongoing struggle that anyone has that is opposed to abortion personally, vehemently opposed to it, and yet says, 'Well, I'll let other people make that decision,'" he said. "And you say to yourself, 'But if you believe that you're taking innocent life, it's hard to justify letting other people make that decision.'

"You know, everybody's entitled to their own view," he continued. "I think there's some people who look at the issue of the beginning of life from the lens of their faith, of their religion, and say, 'When does the spirit enter the body?' That is not the lens that I think a secular leader should use. I look at this from a scientific standpoint. And I don't know when the spirit enters the body or the soul enters the body, and that's not something I would endeavor to find out." Romney said he arrived at his moral answer after pressing scientists on the cloning process. "When you create this clone, when you take the skin cell or the nucleus of a skin cell of a male and put it in an egg of the female, do you at that point have life?" Romney recalled asking. "And they said, 'No question, it is life. Once you put those together, you have life.' . . . That's all I need to know for when the definition of when human life begins."

For all the deep reflection Romney said he had engaged in, he was also acutely aware of the political significance his newfound views held. He turned Harvard into a useful foil on the conservative circuit, boasting in one fund-raising letter of his valiant fight against the liberal establishment over what he termed "human cloning."

The extent of Romney's shift became clearer in July 2005, when he wrote an op-ed in *The Boston Globe* saying that on abortion, his "convictions have evolved and deepened during my time as governor." He began calling himself "firmly prolife." On that basis, he had vetoed—and returned from a New Hampshire vacation to do it—a bill to make the so-called morning-after pill available over the counter at Massachusetts pharmacies and to require hospitals to make it available to rape victims. Supporters of the bill noted that the pill was categorized as a contraceptive, because it halted ovulation, fertilization, or implantation of a fertilized egg in the uterine wall but had no effect on a firmly implanted egg. But Romney said he believed it could function as "an abortion pill." Romney's veto, which the legislature eventually overrode, drew condemnation from reproductive rights advocates, who remembered that Romney, only three years earlier, had answered "yes" to a question on their questionnaire about whether he supported expanding access to emergency contraception. By 2006, Romney's aides were telling the national political press that he would have signed a controversial bill in South Dakota that outlawed abortion even in cases of rape or incest but then sought those exceptions through separate legislation.

As his term went on, Romney would make a series of shifts—in some cases wholesale reversals of past positions, in others significant changes in emphasis—on issues that had nothing to do with abortion or the question of when life begins. It was a sweeping recalibration that erased any doubt that he had set his sights on a bigger prize than Massachusetts. What made that possible was partly his gift for salesmanship. "He can come to a position or realization that's newly found and go out there an hour later and convey it with great zeal," one former adviser said. "A lot of candidates couldn't pull that off." After saying he opposed abstinence-only education, Romney adopted just such a program for twelve- to fourteen-year-olds, primarily in Hispanic and black communities. He backed out of a multistate plan to reduce greenhouse gases which Massachusetts had helped craft—and for which he had earlier voiced support. And perhaps most striking of all, he adopted a wholly new tone on gay rights, a break that produced a deep sense of betrayal among those who had taken him at his word for the previous ten years.

The decision, on November 18, 2003, hit like a lightning bolt. The Massachusetts Supreme Judicial Court, one of the nation's oldest appellate courts, had legalized same-sex marriage in a close ruling, making Massachusetts the first state where gays and lesbians could legally marry. Suddenly Massachusetts was on the vanguard of a burning civil rights issue. And its governor, Mitt Romney, who had already been pondering a national political career, was on the hot seat.

This was the man who had promised in 1994 to be more effective than Ted Kennedy at reaching "full equality" for gays and lesbians. He had never supported gay marriage, but at one point in 1994 he hadn't ruled it out, either. He had also said it should be left up to states to decide whether to sanction same-sex marriage and had criticized Republican "extremists" who imposed their views on the party. "People of integrity don't force their beliefs on others," he had said then. "They make sure that others can live by different beliefs they may have." But now, with the immensely controversial decision making gay marriage legal in his home state, social conservatives nationally were sounding the alarm. Romney made a fateful decision: he would become a leading crusader against gay marriage.

He railed against "activist judges," whom he accused of remaking American society by fiat, asserting that the Massachusetts justices had issued their landmark ruling to promote their liberal values and those of "their like-minded friends in the communities they socialize in." It was a sneer perhaps aimed at Cambridge, home to Harvard and to the chief justice of the court, Margaret H. Marshall. He pressed Congress to pass a federal gay-marriage ban and lobbied Massachusetts lawmakers hard on a proposed state ban, once making a show of distributing copies of the state constitution to legislators after the legislature resisted holding a vote on the measure. He warned at a nationally broadcast anti-gay-marriage rally at a Boston church about the threat from "the religion of secularism."

And he went around the country ridiculing the state he'd lived in for more than thirty years, reporting with an air of disgust what was transpiring back home. "Some are actually having children born to them," Romney said of gay couples before a nationally televised address to South Carolina Republicans in February 2005. On another occasion, he quipped that Massachusetts had become "San Francisco east." Gay marriage, he suggested, was a stain on his corner of the nation that would spread if it weren't stopped. But Romney didn't stop there. He sought to amend Massachusetts's antidiscrimination laws so a Catholic adoption agency could deny placements to gay couples. He backed away from his earlier advocacy of gays serving "openly and honestly" in the armed forces. And, after a flap over its mission, he eliminated the Governor's Commission on Gay and Lesbian Youth, a panel that funded programs for gay teens and their schools. William Weld had started the groundbreaking commission a decade earlier in response to troubling research into teen suicides, something Romney had said in 1994 was a concern he shared. "He dealt a death blow to a one-of-a-kind program in the nation," said David LaFontaine, who had started the commission under Weld and chaired it until 2000.

All of these moves signaled to Republican voters that Romney would be with them when it counted. "I stood at the center of the battlefield," he would tell a meeting of conservatives not long after leaving office. "On every major social issue, I fought to preserve our traditional values." That would become Romney's central defense when he was pressed on his long history of directly contrary views: don't look at what I said back then; look at what I've done as governor. "I know there are some that want to go back . . . and say, 'Yeah, but look at what you said in '94,'" Romney said. "I said, 'Well, yeah, but I was governor for four years. Look what I did as governor, and look what I ran on as governor." On gay rights in particular, it is not that his positions have changed, Romney suggested, but that his audience has. "Now someone will say, 'Yes, but look what you wrote in 1994 to the Log Cabin Club,'" he said. "Well, okay, let's look at that in the context of who it's being written to."

A few months after Romney, in December 2005, surprised absolutely no one by saying he would not seek a second term, he and other Massachusetts political leaders gathered for Boston's famous St. Patrick's Day breakfast, where, as the joke went, he was on the menu. The roasts at the annual breakfast frequently had bite, but Salvatore DiMasi's that morning seemed to have a little extra.

The liberal House speaker, who had often clashed with Romney, presented him with a collection of mock gifts, including a replica of an Academy Award. "For a great job of acting like you really enjoyed being governor of Massachusetts," DiMasi deadpanned. DiMasi then gave Romney a tiny piece of paper—"a list of all your accomplishments as governor"—and noted that the governor was not accepting a salary from the state. "I guess it's true what they say," DiMasi said. "You get what you pay for." It was caustic stuff, but Romney sat there and took it admirably. When it was his turn, he gave as good as he got.

Their political tussles aside, DiMasi's sharp barbs captured the depth of resentment felt by many in Massachusetts, whose governors always seemed to be on their way somewhere else. Romney's repackaging for national office had produced a range of reactions, few of them particularly positive. Some were furious at his reversals on issues that were personally important to them. Others felt his presidential ambitions had clouded his judgment about what was best for Massachusetts. "It's almost as though he had his eye on higher office very early into his governorship. I think it ended up hurting his performance as governor and the fortunes of the party in general," said Richard R. Tisei, a former Republican minority leader in the Massachusetts Senate, who later broke with many of his GOP legislative colleagues in endorsing a Romney rival in the 2008 presidential contest.

For many voters, for state lawmakers, for political reformers, for environmentalists, for business leaders, for his former gay supporters, and for countless others, seeing Romney turn so sharply from Massachusetts produced not just disappointment, frustration, or anger. There was a wistfulness, too, a lament about what could have been had Romney fully committed himself to the job voters had given him. The high expectations he had set and his obvious capacities only made the letdown that much harder. "I was intrigued when he became governor, because he was supposed to be a breath of fresh air. He was nonideological, looked at each issue individually, and did not play games," said one Democratic lawmaker. But, the lawmaker added, "Romney blew in and blew out of here." A majority of voters in a poll cosponsored by *The Boston Globe* toward the end of Romney's term expressed an unfavorable view of him. And when his lieutenant governor, Kerry Healey, ran her own unsuccessful campaign for governor in 2006, she often kept her distance from him.

As public resentment about his national ambitions grew, Romney swatted it away as best he could. How could liberal Massachusetts be expected to understand? The state didn't matter for Republican presidential candidates, anyway. Yet just when many expected him to disengage completely, Romney did the opposite, displaying a dedication and focus that people around him had never seen. Eager to notch a signature achievement before he left office, he took on a problem many others had tried, and failed, to solve. It was a puzzle he had worked over in his mind for years. No state had ever put all the pieces together. But now, Romney decided, Massachusetts would: every resident would have health care insurance. He'd find a way.

TEN

HEALTH CARE REVOLUTIONARY

An achievement like this comes once in a generation.

—MITT ROMNEY ON HIS HEALTH CARE PLAN

I t was a sunny October afternoon in 2008, and Mitt and Ann Romney were making a return visit to the Massachusetts State House to meet with the portrait artist Richard Whitney. Together they walked to the third-floor office Romney had once occupied, its broad windows offering expansive views of the Boston Common and bustling downtown. His successor as governor, Democrat Deval Patrick, had arranged for them to use it for a photo shoot. Whitney needed the snapshots to paint Romney's official portrait, an honor afforded every governor.

Romney had been clear about the image he wanted to convey for posterity. Wearing a blue suit, white shirt, and striped tie—the dress uniform of a businessman—he would be sitting on his desk in front of an American flag, next to symbols of two things he held dear. The first was a photo of his wife, the center of his personal universe. The second was the Massachusetts health care law, represented by an official-looking document with a caduceus—often used as a symbol of the medical profession—embossed in gold on the cover. Romney was deeply proud of the law and felt strongly that it should figure prominently in the portrait, which would hang alongside others dating back to the Colonial era. "As long as the symbol was there, that was important," Whitney said. "He wanted to be remembered for that."

Romney hoped the revolution in health care that he, more than anyone, had driven into law would redound to his benefit as a presidential candidate. Who else on the Republican side had tried to do anything as difficult or ambitious—much less gotten it done? But it wouldn't be that simple. Relentless attacks would follow from conservatives and rival candidates, with objections to its provisions, scope, and progeny: the reform plan would later serve as a template for the polarizing national universal health care law that President Barack Obama signed in 2010.

The story of how the Massachusetts law was conceived, crafted, negotiated, and passed showcased Romney in all his complexity, illuminating his remarkable skills, stubborn limitations, and ideological flexibility. He displayed a willingness to challenge convention with creativity and confidence, a gift for framing a problem and seeing a solution through, and the courage to disregard some political risks. But while his outsider's mind-set brought fresh energy to an old puzzle, his aversion to the basics of political engagement—cultivating allies, making personal connections, calling in chits—complicated the nitty-gritty of the negotiations. And, at the end, he showed more concern for his own political ambition than willingness to stand behind a bipartisan compromise. Indeed, his yearning for higher office turned out to be both a spur and a shackle, making him hungry for a deal but ever mindful of how it might play to Republicans around the country.

A push by one state to universalize access to health care—a push led from the right—was a novel idea and a high-stakes gamble, both for Romney politically and for Massachusetts, where health care has long been a dominant industry. The nation was watching. No state had tackled health reform in such sweeping fashion. No leader had ever driven a bargain quite like this. And it all started in a conversation with an old friend.

———

They had been close ever since Bain Capital had helped finance the early growth of Staples, the office supply superstore. The company's founder, Thomas Stemberg, had worked with Romney for about sixteen years as Romney sat on the Staples board. Stemberg liked to call him a "doubting Thomas," on account of his famous caution. This had long been Romney's habit—pushing, prodding, and firing skeptical questions until the risks of a decision were, if not eliminated, clearly understood. In the Staples boardroom, he was always asking, Does this add up? Are we sure this will work?

Now Romney had been elected governor of Massachusetts, and it was Stemberg's turn to provide the guidance, as a member of his friend's transition team. During one informal meeting about Romney's goals for his new term, Stemberg decided to challenge his old comrade to think big. "We were talking about a bunch of stuff," Stemberg recalled. "I said, 'Mitt, if you really want to do a service to the people of Massachusetts, you should find a way of getting health care coverage to them.'" The health care status quo made no sense, Stemberg went on. There was a "huge number of people going to the emergency room for care at a cost of three, four, five times what it costs to go to a doctor's office." That's wrong, he told Romney, because they were just passing along the expense of that care to others.

Romney was dubious about the feasibility of universal health care. "It struck me, when Tom suggested it, as being, frankly, impossible," Romney said. "Getting everybody insured would mean a huge tax increase. . . . I think I dismissed Tom out of hand." Romney also had other things to worry about, chiefly a looming state budget crisis. He had made some ninety-three promises during the governor's race, and universal health care was not one of them. "You can find a way," Stemberg told him.

It has been a familiar pattern in Romney's life. Presented with a monumental task—running for Senate, taking over the Olympics, launching a campaign for governor—he professes skepticism or outright opposition at first. Then, after a period of reflection—after he convinces himself the numbers just might work—he comes around. So it was with health care. As the months went by, Romney began to think: Maybe Stemberg was right. Maybe this is the time. "I must admit, the 'Aha' moment was in our conversations saying, . . . 'We are already paying for people who don't have health insurance,'" he later said. "Those people are getting treated; they're getting health care. The state is paying for those people—someone's paying for those people. If we could get our hands around that resource to help people buy insurance instead, it would be less expensive."

Over the first two years of his term, Romney's advisers worked quietly to understand the many complexities of health care, sifting through data, evaluating input from experts, and testing theories, hoping to craft a plan that would expand coverage to nearly everyone in the state without breaking the bank. Some of the concepts they settled on had never been tried before. For a long time, Stemberg did not hear from his friend in the corner office. "The next thing I know, he's hard at work, actually trying to do it," Stemberg said. "He called to tell me I gave him the idea."

Romney first thrust health care to the fore at one of his lowest moments as governor. Just weeks after his handpicked slate of Republican legislative candidates got walloped in the 2004 elections, he issued an unexpected plea for bipartisanship. In an op-ed piece in *The Boston Globe*, he pointed to previous achievements won through cooperation between his office and the Democratic legislature. "In the same bipartisan fashion," he wrote, "let's tackle the issue of healthcare." The initial response fell somewhere between skepticism and incredulity. "This administration hasn't been willing to work with anyone," Christine E. Canavan, a state representative, said at the time. "I just came out of a campaign where the man was trying to make sure I wasn't here anymore."

Within two days, though, Romney got a bracing response from an unlikely quarter. Ted Kennedy, the liberal lion who had thwarted Romney's freshman venture into politics a decade earlier, offered emphatic encouragement. One of the great champions of universal coverage, he saw promise in Romney's gambit. "We're basically stalemated [in Washington], so the states are going to have to try to come up with a response," he said. Romney had provided him an advance copy of the op-ed.

Romney didn't have a plan fully mapped out yet, but he knew what he didn't want to do. His first secretary of health and human services, Ronald Preston, had circulated a twenty-nine-page study concluding that near-universal coverage was attainable but would cost more than $2 billion a year, require expanding the Medicaid program, and mean imposing mandates not only on individuals to get coverage but on employers to offer it. To Romney, the employer mandate and Medicaid expansion were nonstarters. So was the huge bottom line. But there was increasing consensus on the Romney team that something had to be done. The U.S. Department of Health and Human Services began to signal ominously that a $385 million annual Medicaid grant the state received to extend coverage to the working poor was in serious jeopardy. That represented more than 5 percent of the state's Medicaid program, and losing the money would mean covering fewer people, not more. There was a great sense of urgency. "You were talking about causing real harm," said Christine Ferguson, who was state public health commissioner at the time. "The whole calculus changed."

So they turned to one of Romney's trusted sources of inspiration: data. It turned out that the state had a lot of it, and the picture that emerged was surprising. Of the estimated 460,000 Massachusetts residents without insurance, more than a third could afford to pay for basic coverage. Many others were eligible but not currently enrolled in the existing state Medicaid program, whose costs are split with the federal government. That left 36,000 short-term unemployed and about 150,000 who were working but caught in a health care vise—making too much for Medicaid but too little to afford private coverage. Those were the groups that would need subsidies to acquire insurance; universalizing health care boiled down to finding a way to cover them. Viewed this way, the numbers weren't small, but they weren't outlandish, either. Suddenly the problem seemed manageable.

The first task was finding a way to get uninsured people who could afford insurance to buy in. Romney's answer came in part from Jonathan Gruber, an economics professor at MIT and a leading adviser, usually to Democrats, on health care. Gruber had a computer model designed to predict how individuals and businesses would respond to changes in the health care system. What it showed, Gruber said, was that requiring individuals to carry insurance "made enormous sense." Romney was intrigued, because he believed in the principle of personal responsibility. And if younger, healthier people were coaxed into the system, the cost of premiums would moderate for a larger population. "Romney was like an engineer," Gruber recalled, "saying, 'We can actually get to this goal in a financially feasible way. Isn't that neat?' . . . He wanted to do what's right."

Romney's political advisers, however, were not keen on his taking on such a vast challenge, so uncertain of success. He was venturing into policy turf long dominated by Democrats. Even though Romney's approach was fundamentally different in most respects, one close aide privately called the idea Dukakis II, a reference to the failed 1988 effort by then-Governor Michael S. Dukakis to phase in near-universal coverage. "They were making the realpolitik argument, something along the lines that 'Republicans don't do that,'" Gruber said. But Romney challenged the dissenters. "Romney was arguing with them, saying, 'No, this is the right thing to do,'" Gruber added.

In the late spring of 2005, not long after his meeting with Gruber, Romney gathered with a dozen top policy and political advisers in a conference room near the governor's suite on the third floor of the statehouse. Romney's team was finalizing his health care bill, which he would soon file with the legislature. This was a pivotal meeting, and Romney was in CEO mode, presiding over a freewheeling debate to identify the strengths and weaknesses of their bold initiative. On the table was a critical question that would, in many ways, define the plan and also Romney's political ambitions, wherever they would lead him: should adults with sufficient income be required to buy basic health insurance or pay a penalty? "We discussed the need for people to take responsibility for their health care," said Amy Lischko, who was the state's health care policy director. Romney was drawn to that basic moral principle. "Everybody in the room was very aware it was a novel approach," recalled Timothy R. Murphy, who was Romney's point man on health care. "I don't think it was lost on anyone that Mitt had an emerging national profile, but there weren't many of us thinking there would be a national debate on this." Instead they were focused on what would both work in Massachusetts and fit the demands and philosophy of the man at the head of the table. "We felt that personal responsibility was consistent with what the governor stood for," Murphy said.

If Romney's policy team was on board, his political brain trust remained skeptical, believing the hazards spoke louder than the policy-making opportunity. By mid-2005, it was already expected within Romney's administration that he would serve only one term and seek the 2008 Republican presidential nomination. Moving hard on health care would put Romney way out front in a cause generally identified with the Left—even though the individual mandate was a concept with long conservative and Republican roots. In the early 1990s, the Heritage Foundation, a conservative think tank, endorsed the idea and some Republicans offered it as an alternative during debate on the doomed national health care push by then–first lady Hillary Rodham Clinton. One of the early GOP advocates for an individual mandate was then-senator John Chafee of Rhode Island, whose plan Romney had voiced support for during his race against Kennedy.

Despite some aides' skittishness, no one in Romney's inner circle was expressly opposed to a mandate when they gathered for the late-spring meeting. Romney himself was disposed to favor it but not ready to commit to it. Among other things, he was looking for a suitably spotlit setting in which to make his case, and one was coming up. He was to deliver a major speech on June 21, 2005, to a health care summit at the John F. Kennedy Presidential Library. As with most of his decisions, his final verdict on whether to include an individual mandate remained a closely held piece of information until the time was ripe.

Twelve days before the event, Andrew Dreyfus, who was then president of the Blue Cross Blue Shield of Massachusetts Foundation, was busy taking the temperature of those whose support any reform plan would need. He visited Murphy's office and told him he was assembling a chart, putting a mark next to each component of health care reform supported by the various players, including Romney and the legislature. Murphy balked when the individual mandate came up. The day before the summit, Dreyfus telephoned Murphy, again looking to fill in the blank. "What about the individual requirement?" Dreyfus asked. "There was some good silence on the phone," Murphy said. "I mean, it was supposed to be embargoed until the governor spoke." Murphy told him to put it in.

Romney delivered his speech, publicly backing an individual mandate for the first time. He told reporters afterward, "No more free riding, if you will, where an individual says, 'I'm not going to pay, even though I can afford it.'" He called it "the ultimate conservative idea," saying that people "don't look to government to take care of them if they can afford to take care of themselves." The reaction, in a harbinger of what would unfold over the coming years, was mixed, with many cheering the governor's chutzpah while voicing skepticism about requiring individuals to buy health care insurance. It was a huge step, even bigger than Romney understood at the time. And it was a bold one, the first in a furious debate that would consume state government and industry for months. Doctors, hospitals, insurers, researchers, business groups, advocates—just about every player in the vast medical-industrial complex in Massachusetts would be affected. Romney had stirred the pot, and stirred it with force. Everybody, from family doctors in rural towns to Republican opinion leaders in Washington, D.C., was anxious to see how the experiment would turn out.

The individual mandate was the biggest headline grabber, but there were other novel elements Romney and his team baked into their plan. They proposed using the state's threatened Medicaid money and much of the funding allocated to "free care" at hospitals to create a new subsidized insurance program for the working poor. Tim Murphy had a "eureka moment" after a meeting with officials of the Heritage Foundation, embracing their concept of a so-called exchange to provide one-stop shopping for small businesses and individuals seeking health coverage from commercial insurers. The goal was to create a health care marketplace driven by consumer choice, which had long been a bedrock principle of conservative thought on health care. That, too, would prove to be a forerunner to Obama's national plan.

If any state was going to figure out universal health care, Massachusetts had perhaps the best chance. As Romney's own review of the data had made clear, the state's percentage of uninsured residents was already among the lowest in the nation, because a higher percentage of employers offered coverage to workers than the national average. Massachusetts, too, was one of the few states in the country that had a pool of funds to pay for treatment of the uninsured at hospitals and health care centers. It was also one of only a small number of states that barred insurers from denying coverage or jacking up premiums based on people's health histories. And there was a long and deep institutional memory. Many local health care players had been down a similar road before and understood the possibilities and the pitfalls. The executive ranks of insurers, major health care providers, and business associations were marbled with veterans of past efforts to expand coverage, efforts that had fizzled in a hostile environment. But by the time Romney took office in 2003, the climate had changed, thanks in part to the foundation laid by a prior Republican administration, which had pushed hard to rein in Medicaid costs.

As Romney's team was assembling its plan, business leaders were mobilizing on another track. Leading the way were top officials of the state's largest insurer, Blue Cross Blue Shield of Massachusetts, and Partners HealthCare, a powerhouse provider network with nine hospitals—including two of the world's best regarded—and about 5,500 physicians, or roughly one-fifth of all doctors practicing in the state. The chief executive officer of Blue Cross, William C. Van Faasen, joined forces with John "Jack" Connors, Jr., the chairman of Partners. They supported expanded coverage, but their motives were not entirely altruistic. In exchange for their support, they wanted the state to increase substantially the long-lagging rate at which it reimbursed providers for treating Medicaid recipients, a major source of hospital revenue.

Mindful of the errors of the Dukakis reform drive of the 1980s, the Partners–Blue Cross alliance was determined to create a united front within the state's business community. Van Faasen and Connors, an advertising executive and one of Boston's business and civic leaders, approached John Sasso, a longtime Democratic political operative, to quarterback that effort at the statehouse. Sasso, who had been Dukakis's chief political strategist, had deep Democratic connections and a track record of getting big, complicated things done. He came with a national reputation, trusted by many Democratic heavyweights. He was also unfailingly discreet. But he faced an obstacle: he didn't know Romney and had no lines into his inner circle. Connors suggested Joseph J. O'Donnell, the region's undisputed king of local sports and entertainment venue concessions, who, like Connors, knew just about everyone but was also a great admirer of Romney, a friend and neighbor in Belmont.

O'Donnell warned Sasso that Romney harbored a grievance against him: Sasso had played a role in the negative attacks on Romney in the 1994 Senate campaign against Ted Kennedy. In a *Globe* op-ed piece after the 1994 election, Romney had written, "I've got to admit it was first-class work. Ted's top campaign strategist, John Sasso, had learned his lesson well from his 1988 Dukakis campaign drubbing by Lee Atwater." O'Donnell set up the meeting. When the three sat down, he told Romney he trusted Sasso. It was awkward and a little tense at first, but they soon put past unpleasantness aside. "This was talking about policy, not politics," Romney said. "[Sasso] said, 'Look, my clients . . . think we can work together on this.'"

If Sasso was one unlikely Romney ally, Kennedy was another. The two former rivals had fought over health care in their 1994 Senate campaign, but they had since proven they could work together, with Kennedy supporting the Mormons' bid to build a temple spire in Belmont and the two leaders working together to stave off military base closures in the state. Now it fell to Romney and Kennedy to win federal approval to save the endangered $385 million in annual Medicaid funding, a linchpin of any broader plan for health care reform. Romney and Kennedy went to Tommy G. Thompson, then the U.S. secretary of health and human services, seeking the federal government's blessing. On January 14, 2005, one of Thompson's last days in George W. Bush's administration, Romney, Kennedy, and their aides met with him and his staff for about two hours at the Hubert H. Humphrey Building in Washington, finally reaching an agreement. "I was never absolutely sure whether at the very end [Thompson] was going to sign it," Kennedy said later. As they worked out the details, they could hear the sound from Thompson's retirement party a floor above. "People kept coming down, saying, 'Everyone's at your party, waiting for you,'" said Stacey B. Sachs, Kennedy's top health care aide. "The next thing I knew, we were all heading up to the party." Romney and Kennedy were "kind of a riot on the stage, going on about being the odd couple," Sachs said. It was the first time they had shared a stage since debating in their 1994 campaign.

Kennedy's support of the Massachusetts health care law would later provide fodder for Romney's conservative political opponents, and Romney, under the glare of such scrutiny, would wrestle with how to discuss Kennedy's involvement. But most of Romney's critics exaggerate Kennedy's role. Some have even suggested that the legislation's success was the product of "a deal" between the pair. There was no deal. Neither Kennedy nor his staff was directly involved in writing the bill. Kennedy did, though, play a consequential role in providing cover for Democrats wary of cutting a deal with a Republican governor. He also helped move Robert Travaglini, the Senate president, and Salvatore DiMasi, the House speaker, to resolve a stalemate over competing versions of the bill. Kennedy made a "strategic decision" to support Romney once he became convinced that the governor was serious about extending coverage, Sachs said. "Everyone expected Kennedy to come out screaming, but he said, 'This looks pretty good, and if he's willing to work for this, let's work with him.' People wanted to be against it because it was Romney's, but because Kennedy came out in favor, they had to bite their tongues."

DiMasi and Travaglini were both barrier breakers, becoming the first Italian Americans to rise to the leadership of their respective chambers. They had both spent their lives in tight-knit Boston neighborhoods, separated by the city's inner harbor. And they were both savvy Democratic lawmakers who had devoted years to learning the mysteries, rites, and mores of the statehouse. But despite their close relationship, the speaker and the Senate president had developed colliding visions of what health care reform should look like. And as their dispute rumbled on, it threatened to derail the whole thing. Romney, fearing that his big push would amount to nothing, grew nervous.

For even though Romney was driving the health care debate, he knew well that he could do nothing on his own. He needed the House and Senate. And he knew that each would want to put its own stamp on such a watershed bill. The House and Senate versions shared some elements, but there were fundamental disagreements about how far the state should go toward universal coverage and how it would get there. DiMasi, for example, was adamant that any plan include a payroll tax on employers that did not offer insurance, a position shared by many liberal activists but not by Travaglini. And Travaglini initially favored a bill more narrow in scope. "It didn't go as smooth as silk," Romney said. "There were some real tough times."

Romney knew he wasn't the guy to be walking the hallways twisting the arms of lawmakers or trying to broker deals. Unlike his Republican predecessors, he had never enjoyed chummy relations with the Democratic legislature. So with the stakes so high, he wisely left most of the grubby horse trading to others. "Romney knew his best chance to get it passed was to be not too visible in the process," said Alan G. Macdonald, one of the business leaders active in the negotiations. But as negotiations wore on and the legislature remained deadlocked, Romney grew more fearful the effort would collapse. In early December 2005, Kennedy weighed in publicly, endorsing none of the competing versions but making clear that he supported the individual mandate Romney had proposed if it was part of a broad agreement. "I've never been one for individual mandates in the past, but I do think that the way this has been proposed, in that everybody will do their part, that's a compromise," Kennedy said. "I can buy into that." Still, the Travaglini-DiMasi standoff continued. The state missed a January 15, 2006, deadline to get a plan to Washington. Twelve days later, a Friday, talks broke down. The next day, Tim Murphy was out shopping when his BlackBerry buzzed with an e-mail from Romney's personal account. The governor had a personal letter he wanted to deliver to DiMasi and Travaglini, to get the bill back on track. "The plane is circling," Murphy quoted Romney as saying. "And we have to land it."

On Sunday, Romney, who had never pushed this hard for a bill, tried an unusual tack: he showed up at the legislative leaders' homes. DiMasi wasn't there, so Romney taped the letter on his door. Travaglini, in a sweatsuit and slippers, was. He invited Romney in. Romney made his pitch, warning of the impending loss of the precious federal funds if they did not resolve the matter soon. They talked for about five minutes. "How often does the governor ring your bell on a Sunday morning?" Travaglini recalled with a laugh. About two weeks later, Romney invited DiMasi to his office for a rare one-on-one meeting, according to DiMasi. Romney, always unflappable, was agitated, DiMasi said. "He did everything he could to put pressure on me to change my position on the business assessment. I tried to walk out, and he wouldn't let me go out the door. I actually opened the door, and he said, 'No, no, no. Get back in here. I have to talk to you.'" This was a side of Romney that was new to DiMasi. "He was very animated," DiMasi said. "It was the first time I'd seen it." When he left, the matter remained unresolved.

The group of business leaders, meanwhile, was ready to yield on a penalty for employers that did not offer coverage. On March 1, 2006, at his office in the John Hancock Tower above Boston's Copley Square, Jack Connors convened a high-powered group of executives from the four big business associations, Partners HealthCare, and Blue Cross Blue Shield. They discussed a proposed compromise, which called for an annual "fair share contribution" of $295 per employee from businesses with eleven or more employees that did not offer coverage.

The next night, DiMasi and Travaglini made peace, dining with their wives in Boston's Italian neighborhood, DiMasi's own North End. DiMasi's wife, Debbie, and Travaglini's wife, Kelly, also friends, had arranged the meeting to clear the air. "When communication broke down between the two of us, our wives may have overheard us describing each other in vulgar ways," Travaglini said. "Maybe around the third bottle of wine," he added with a chuckle, "things got a little clearer." On March 3, the day after their makeup dinner, DiMasi and Travaglini met with the business leaders at a closed-door session at the State House. They emerged with a resolution on the final sticking point: The bill would impose the $295-per-worker assessment. With that, the legislation was ready.

More than three years had elapsed since Tom Stemberg nudged Mitt Romney to tackle universal health care. Many more years had passed since the state's last attempt at it. But now Massachusetts was finally on the precipice. And Romney, who had invested whatever political capital he had left, was a big reason why. He was jubilant. "Today, Massachusetts has set itself apart from every other state in the country," he said, beaming, after lawmakers publicly unveiled the bill. "An achievement like this comes once in a generation." Romney stopped by DiMasi's and Travaglini's offices before his press conference to thank and congratulate them.

Given Romney's reticence about penalizing employers, his rhetoric on that issue caught the press corps by surprise. Asked if there were any parts of the bill he would veto, he said he still needed to review it all but, "We are where we'd hoped we'd be." Didn't he consider the penalty on employers a tax, as antitax activists did? And hadn't he pledged to veto any taxes? "It's not a tax hike," Romney responded. "It's a fee. It's an assessment." Businesses and workers who purchased health insurance already paid an assessment to help fund the "free care" pool, he noted, and "it makes sense to expand this assessment." As the press conference wore on, Romney reserved his right to review closely the employer assessment and other elements that had been added by the Democratic legislature. But his tone throughout signaled that he had no major objections. Toward the end, he was asked again: was he really okay with the new employer penalty? Romney said he was relieved that what he had feared most—a new, broad-based payroll tax on employers—was not in the plan. That was something, he said, that he "definitely would have been unable to sign." "This," he continued, "is of a different nature."

The next day, April 4, the bill sailed through the House and Senate with virtually no opposition. They called it "An Act Providing Access to Affordable, Quality, Accountable Health Care." A signing ceremony was set for April 12, 2006, at Boston's historic Faneuil Hall. Romney and state lawmakers knew they had done something big. The bipartisan bonhomie did not last long, however. The day before the ceremony, Romney published an op-ed in *The Wall Street Journal* whose cheery title—"Health Care for Everyone? We Found a Way"—masked a last-minute shift. His "Democratic counterparts," he wrote, had added a $295-per-person fee on employers that refused to help cover their workers. The same fee he had seemed comfortable with just a week earlier he was now labeling "unnecessary and probably counterproductive." He promised to "take corrective action," which meant a veto. Democrats felt sandbagged. This was a compromise that had been carefully worked out among many parties, including the business community. It would not raise a huge amount of revenue—$45 million was the estimate, and even that has proven to be overblown. But the principle was important to many involved in crafting the bill: employers had to put some skin into the game, too.

Right before the April 12 signing ceremony began, Romney aides announced in a press release that he was vetoing 8 of the 147 sections, including the employer mandate. There was little risk to Romney, who knew that the legislature would easily override him, as it did three weeks later. He could thus claim a large share of credit for the new law while washing his hands of something resembling a tax, which was problematic with the national Republican audience he was now courting heavily. But his eleventh-hour rejection of the employer mandate tainted the celebratory mood. DiMasi was outraged. Travaglini, however, said he wasn't surprised and didn't particularly care, because he knew the legislature would override Romney anyway. "It didn't bother me," he said, "because I had the votes."

Like most of Romney's momentous public events, the bill-signing extravaganza was a masterwork of political stagecraft, in a setting that evoked many important events in the nation's history. Guests received programs printed on faux parchment with commemorative lapel pins to match. The stage of the majestic hall was augmented with a pedestal for the podium and an extended platform with a desk on a circular oriental carpet. On either side were banners to mark the occasion: "Making History in Healthcare." The procession to the stage was led by a fife-and-drum corps clad in tricorn hats, breeches, and stockings, playing Colonial tunes. An audience of several hundred VIPs greeted Romney warmly with a thirty-second ovation.

As Romney spoke, the top of his head reached the frame of the majestic room's focal point, George P. A. Healy's enormous painting *Webster Replying to Hayne*, which depicts one of the most famous speeches ever delivered in the U.S. Senate during the debate over states' rights. "This is a politician's dream, you've got to admit," Romney began. He thanked Cecil B. DeMille, the late Hollywood producer of film spectacles, for organizing the event. "This does classify as being over the top." Privately, Romney had been uneasy about all the pomp, unsure how he would want to frame his association with the law politically down the road. Referring to the last time he and Kennedy had been at Faneuil Hall, for the pivotal 1994 debate, Romney quipped, "This for me feels a bit like the *Titanic* returning to visit the iceberg." He said to hearty laughs, introducing Kennedy as "my colleague and friend," "My son said that having Senator Kennedy and me together like this on the stage, behind the same piece of landmark legislation, will help slow global warming. That's because Hell has frozen over."

Kennedy was also greeted enthusiastically and after thanking Romney, he said, "My son said something, too, and that is when Kennedy and Romney support a piece of legislation, usually one of them hasn't read it." When the laughter subsided, he turned to Romney and said, "That's not true today, is it governor?" Then he turned serious. "This is an achievement for all the people of our commonwealth and perhaps for the rest of America, too," he proclaimed. "And we intend to make the most of it." Romney moved to the desk and used each of fourteen commemorative pens during the signing. "It's law," he said upon finishing. "Congratulations." The ceremony was over in forty minutes.

On that spring day, as he inked his name to the law, Romney knew he was going to launch a national campaign in the ensuing months. What he didn't know was how his health care push would play for him politically. "I have to admit that I'm very, very proud of having been part of this process," he said after the bill signing. "But I have no way of guessing whether it's going to be a help or a hindrance down the road. Time will tell."

————————

More than five years later, it remains an open question. Passage of the law remains his greatest political achievement, but it has also become linked in infamy, in the eyes of his conservative critics, with the national health overhaul pushed through by President Obama and his Democratic allies in Congress. Romney has said repeatedly that the Massachusetts law could serve as a model for other states, but he rejects comparisons with the national plan. "I think that there is a recognition that what we did with my leadership and that of others was to follow the constitutional principle of states' rights—that we were a laboratory of democracy," Romney said. "We carried out an experiment, and that's a right and proper thing to do under the Constitution. . . . What the president did was to impose a one-size-fits-all plan on the nation."

Michael Leavitt, who, as U.S. secretary of health and human services, gave final approval to the Massachusetts plan in July 2006, later said that about half the states had inquired about developing some aspect of what Massachusetts had done. "I don't know if what Mitt Romney did is a conservative idea or a liberal idea," Leavitt said. "But it is clearly an innovative idea."

But has it worked? A detailed examination by *The Boston Globe* in 2011 of voluminous health care and financial data, and interviews conducted with key figures in every sector of the health care system, found that although there have been some stumbles—and some elements merit a grade of "incomplete"—the overhaul has worked as well as or better than expected, especially in accomplishing its principal goal of expanding coverage to almost every citizen. The percentage of residents without insurance is down dramatically, according to one survey, to less than 2 percent; for children, the figure is a tiny fraction of 1 percent. Those are the lowest rates in the nation. Recent U.S. Census data, however, put the percentage of uninsured slightly higher, indicating an uptick from 2009 to 2010.

Many more businesses are offering insurance to employees than were before the law; the fear going in was that the opposite would happen. The plan remains exceptionally popular among state residents; indeed, its popularity has only grown with time. There are some unhappy sectors—notably small-business owners, who had hoped to see moderating premiums and chafe, in some cases, at the state's heavy-handed enforcement of the rules. And support for the requirement that individuals obtain insurance is down to a slender majority, a June 2011 poll showed.

Health care costs continue to grow at alarming rates, as they have nationally, but the consensus of industry leaders and health care economists is that the trend cannot be fairly traced to the makeover but rather to cost pressures baked into the existing health care payment system. Massachusetts does have among the highest health care costs in the nation, but it owned this dubious distinction long before Romney launched his push for universal coverage. The state's share of costs, however, has been rising, and hospitals are bearing an increasing share of the load.

By any reasonable assessment, failure—the blunt summary offered by Romney's foes—doesn't describe his push for universal health care. But neither is the law an unalloyed success. It remains a work in progress, an ongoing experiment, especially when it comes to bringing costs down.

As Romney left the statehouse for the campaign trail, he was clearly wary of how the health care law would look. He made sure to describe it as "conservative" and "market-based" before Republican audiences and suggested that its success would be in doubt once the Democrats "get their hands [on] it." He made light of Kennedy's appearance at the bill signing, saying later, "I was a little concerned at the signing ceremony when Ted Kennedy showed up." (Romney later contended that the remark had been merely a setup for a joke.) He later blamed his Democratic successor for fumbling the law's implementation.

But he has never fully disavowed it, as some suggested he do. "A lot of pundits around the nation are saying that I should just stand up and say this whole thing was a mistake, that it was a boneheaded idea, and I should just admit it," Romney said. "But there's only one problem with that: It wouldn't be honest. I, in fact, did what I believed was right for the people of my state." He recounted an instance when a man had stopped him as he came out of a supermarket near his house and said, "Your health plan saved my life." That, he said, "obviously warms my heart."

Romney's accomplishment on health care stood out in Massachusetts both for its merits and because it served as a reminder of what had made him such an attractive candidate for governor in the first place. He'd come in promising to shake up the system. And on the issue he had invested himself in more than any other, he had done just that. But that success raised a question, too: what else could he have achieved had he committed himself more to the job? "Significant successes," such as health care reform "showed what Romney might have accomplished as governor had he focused his efforts more steadily on state policy leadership," said Brian R. Gilmore, an executive vice president of Associated Industries of Massachusetts, which represents thousands of businesses. Others believe that Romney's record will hold up better over time. "I'm not naive enough to think that people obviously [don't] have some disappointment," Bradley H. Jones, Jr., the Republican leader in the Massachusetts House and frequent Romney ally, said at the end of Romney's term. "My hope is that, like many things in life, when there's a little bit of distance, people will take a broader view."

Romney himself said he was pleased at how much he had gotten done, despite a legislature so dominated by Democrats. "The truth is, if you look at the record, it's a heck of a lot more than I expected I'd get done in four years," he said. "I'll put it up against any other governor's in America, not because I'm brilliant, but because the legislature and we did pretty well together."

On the evening of January 3, 2007, Romney took the customary final walk out of the statehouse. And as usual, the stagecraft was off the charts. With cameras recording, he left the third-floor office and, with Ann at his side, descended the thirty-one steps of the statehouse. Along the way, he made a series of planned stops designed to highlight his record. He greeted the family of Melanie Powell, a thirteen-year-old killed by a repeat drunk driver and memorialized by Melanie's Law, the tough drunken driving bill he had signed. He met students attending state colleges under the John and Abigail Adams Scholarships he had created. And he welcomed two families able to afford health insurance because of the new law. Then the Romneys left Boston's Beacon Hill and returned home to Belmont for a quiet evening, a chapter of their life closed. The opening words of the next chapter had already been written. An hour before Romney departed the statehouse, the Federal Election Commission docketed a four-page form establishing a new political organization: the Romney for President Exploratory Committee.

ELEVEN

A RIGHT TURN ON THE
PRESIDENTIAL TRAIL

I know something about tailspins and it's
pretty clear Mitt Romney is in one.

—SENATOR JOHN MCCAIN ON HIS CAMPAIGN FOE

I t was just hours after passage of his health care bill—the single greatest achievement of his political life—and Governor Mitt Romney's thoughts had turned to Iowa. Settled into a top-floor suite at the Ritz-Carlton hotel, high above Boston Common, he was expecting a relaxed, chatty evening with a handful of Republican leaders from the first-caucus state. But the meeting quickly went awry. Romney had hoped that one of his guests, Doug Gross, would become chairman of his Iowa campaign. But Gross, an intense man who had grown up in a town called Defiance and was now a high-powered lawyer in Des Moines, wasn't convinced. He grilled the Massachusetts governor, beginning with a subject he knew might upset Romney: how would Romney handle questions about his Mormon faith in a state where the GOP caucuses are dominated by Christian conservatives, many of whom don't believe that Mormons are Christians?

"I'm not changing my religion," Romney said, growing testy, according to another participant.

"I'm not asking you to," Gross responded.

He moved on to his second question, and the atmosphere continued to cool. "We are sitting up here, up top of the Ritz-Carlton, you are fabulously successful, hopelessly wealthy compared to most people," Gross continued, noting Romney's starched shirt, fastidiously coiffed hair, and privileged upbringing. "Can you really relate to an average voter?" At that, Romney's wife, Ann, stormed out of the room. Romney, in turn, became so angry and insulted that "he didn't talk to me the rest of the night," Gross said—even when the two men later sat side by side at a women's collegiate basketball championship game. "It was the coldest shoulder I've ever experienced."

Mitt Romney felt ready to run for president, ready to take the one career step that would fulfill his dream and succeed where his father, George, had fallen short. But he wasn't ready for the most obvious questions that would come his way. He may not have felt he had to be, for solving problems and answering hard questions were skills he prided himself on, and with reason—his talents had shown through at Bain, at the Utah Olympics, and in the just-completed battle over health care. He had made millions of dollars and made his name. Still, there was a brittleness to his self-certainty, which came through loud and clear to Gross. And for potential allies like him, that was worrisome.

But it wasn't a deal killer; Gross decided to sign on. Romney, he concluded, had a unique set of attributes that could enable him to win the election and be a superb president. Romney had excelled in many realms, including as a Republican governor in a Democratic state, and—this was key—had boundless financial means. For his part, Romney, despite his irritation at the questioning, knew he needed Gross's connections. So Gross agreed to chair Romney's Iowa campaign, showering the governor with public praise. It was the start of a relationship that began in a bad place and was destined to grow worse. The things that troubled Gross at the outset—Romney's defensiveness when challenged, his resistance to advice from outside his immediate circle, his failure to face just how little he knew about running for president—would ultimately drive his campaign down a very bumpy road. It would leave behind bruises, frayed alliances, and a lingering question: what would he learn?

———————

The PowerPoint slides rolled across the screen, each equally brutal in its description of the candidate:

Perception—phony.
Slick—not human (hair?)
You do not know where WMR [Willard Mitt Romney] comes
   from . . .
No story beyond cold business, Olympic turnaround, CEO
   governor.

This was not an attack by one of Romney's opponents. It was a production of his own media team, war-gaming the likely lines of attack against him. It was a few months after the Ritz-Carlton meeting, and Romney's advisers had gathered to try to address a series of questions: Who is Mitt Romney? What is the perception of him among voters? How could the campaign shape or reshape that image? As the slides rolled on, a recommendation appeared on one. What was needed, it said, was the creation of a "Primal Code for Brand Romney"—a core message that could be embedded in the minds of voters.

The media adviser behind the presentation, Alex Castellanos, knew all about political war games. He was a Cuban native whose parents had had $11 in their pockets when they'd fled Fidel Castro and brought him to the United States. Castellanos had become a favorite of Republican presidential candidates for his deeply conservative principles, his unsparing assessments of the political landscape, and his tough tactics. Romney had hired Castellanos and directed him to pull no punches, so Castellanos compiled a scathing catalog of his new client's perceived liabilities. The PowerPoint presentation went on to contrast Romney with two leading likely rivals, Senator John McCain and former New

York City mayor Rudy Giuliani. McCain was a former prisoner of war and a hero, and Giuliani was known as "America's mayor" because of the way he had responded to the 9/11 attacks. Both had credible credentials for taking on the presidency during the war on terrorism. Romney had nothing similar.

One of the campaign's chief concerns was that Romney would be tagged, as one slide put it, as "Flip-flop Mitt," given his changes on issues such as abortion. The media team urged Romney to counter that with a forward-looking brand. One of the slides suggested that Romney use this as his catchphrase: "Yes, we can." But Barack Obama would take it before Romney could. Whatever the phrase, Romney had to be sold as an "optimistic, conservative leader who is calling upon the strength of the American people [to] lead us into the future, to a better place."

The Romney campaign team also zeroed in on what they called the problem of the three M's: Mormon, millionaire, Massachusetts. "There is a perception out there that there is this rich guy from a liberal state who's got a funny religion," as Romney's campaign manager, Beth Myers, later put it in a Harvard University Institute of Politics seminar. A poll by the *Los Angeles Times* and Bloomberg found that 37 percent of those surveyed would not vote for a Mormon for president, and the percentage was even higher among those likely to show up in Republican caucuses and primaries. A Gallup Poll taken around the same time found that 66 percent of those surveyed did not believe the country was ready for a Mormon president.

Yet winning in key early states such as Iowa and South Carolina would require support from religious conservatives, who were deemed least likely to back a Mormon. It was a topic that Romney would variously try to tackle head-on or try to dismiss as irrelevant. In the end, however, the campaign could not ignore the polls. Many people simply didn't understand Mormonism. The campaign's conclusion was that Romney needed a proxy, preferably an evangelical leader with national credentials, who could vouch for him.

So it seemed almost providential that, just as Romney was trying to decide how to deal with the issue, one of his aides received an unsolicited phone call from one of the most influential evangelicals in America—not a pastor but a savvy Atlanta public relations agent. Mark DeMoss had built a successful business that promoted conservative Christian organizations, giving him connections to nearly every important evangelical leader in the nation. DeMoss had never met Romney, but he believed that the governor was being unfairly tarred. Though DeMoss knew that some Christian leaders didn't consider Mormons to be Christians, he felt it made no sense to disqualify a Mormon from the presidency. Moreover, DeMoss was sold on Romney, impressed by his experience in government and business. The governor cleared his schedule to meet with DeMoss.

"You can't pay me anything, ever," DeMoss told Romney when the two met on September 11, 2006. If anyone thought DeMoss was profiting from the relationship, it would backfire. DeMoss urged Romney to meet evangelical leaders and face questions about Mormonism head-on. Romney agreed. And six weeks later, a who's who of evangelical leaders arrived at the governor's Belmont home, including the Reverend Jerry Falwell and the Reverend Franklin Graham, the son of Billy Graham. A plate of sandwiches was laid out in the kitchen, and then the guests joined Romney and his wife in the den. They took their seats in a circle of chairs as Romney said: Ask me anything you want.

Dr. Richard Land, one of the nation's most influential evangelical leaders, awoke at his hotel at 5 a.m. and headed to the airport in Lubbock, Texas, anxious to fulfill his agreement to join other evangelicals at the meeting with Romney. In his role as head of the Southern Baptist Convention's Ethics and Liberty Commission, Land was often cited as the most recognizable and respected voice in his denomination, which counts 16 million members in 42,000 churches in the United States. With his deep basso voice, ample girth, and blunt-spoken manner, he loomed large on any stage where presidential politics was discussed. Like many in his faith, Land questioned whether Mormons were Christian. He had said Mormonism might be considered a "fourth Abrahamic religion," the others being Christianity, Islam, and Judaism. But Land also believed that there should be—in Thomas Jefferson's famous words—a "wall of separation" between church and state. It was the persecution of Virginia Baptists that had helped convince Jefferson that government should not interfere with religion. Mormons, too, believed they had been persecuted by government—Romney's own great-grandfather had been pursued by armed U.S. forces seeking to arrest him on polygamy charges. Romney might never convince Land and other Southern Baptists of the virtues of Mormonism. But it was reasonable to believe that he and Land could agree about the need to keep the institutions of government and the church separate and that the concerns about Romney's religion would then begin to fade away.

As Land listened to his fellow evangelical leaders question Romney, one of them put the matter directly: "You do understand, Governor, that most evangelicals don't accept Mormonism as an orthodox Trinitarian faith?" Romney replied that he was well aware of that and assured Land and the others that he would keep his Mormon faith out of political decisions. Land then urged Romney to give a speech assuring Americans that his Mormonism would not influence him in the White House—and deliver it soon, preferably in Iowa. To underscore his point, Land showed Romney a copy of John F. Kennedy's famous 1960 address to Houston ministers, which Land considered a well-worded assurance by Kennedy that his decisions would not be influenced by the Catholic Church. Make a speech like that, Land said. Romney promised to think about it. As the meeting came to a close, the men and women bowed their heads. A prayer was said for Romney and his prospective presidential campaign. Shortly afterward, Colonial-style chairs were shipped to all of those who had attended. The chairs had an engraved brass plate on the back that said, "There is always a place for you at our table—Mitt Romney."

Confident now that they could swing Christian conservatives their way, the Romney team settled on their strategy. Instead of avoiding social issues and keeping a distance from evangelicals, Romney would "dive right" and bet on winning in a place where evangelicals held extraordinary power: the first-caucus state of Iowa. If Romney could win there, the Mormon issue would be off the table, and Romney might be on the path toward the nomination.

———————

They called it "Romney World," a campaign headquarters that filled the former Roche Bobois furniture store at 585 Commercial Street, a three-story gray-and-tan North End waterfront building. The windows in Romney's top-floor corner office provided sweeping views of Boston Harbor, the Leonard P. Zakim Bridge, and the Bunker Hill Monument. In prominent positions, pictures of George Romney were always the exemplar, in success and failure, for his son. Romney World was more than a physical location; it described a state of loyalty to Romney among his closest advisers that would be matched by few campaigns. To those even slightly outside this inner circle, getting through to the candidate seemed nearly impossible. At the head of this group, with an office next door to Romney's, sat Beth Myers. She had served as the governor's chief of staff for four years and was now charged with managing a national campaign, something she had never done.

It had long been presumed that Mike Murphy, Romney's political guru in his 2002 campaign, would reprise that role in 2008. He was renowned for a wisecracking and sometimes outrageous style that seemed at odds with the buttoned-down Romney. He relished playing the role of an inverse Romney, with his casual appearance, but was also like the candidate in his self-certitude. Romney aides variously called Murphy the campaign "Svengali," "mastermind," "alter ego," and a number of other double-edged superlatives. Just as Karl Rove was considered "Bush's brain" during George W. Bush's two presidential campaigns, Murphy was the muscle behind Mitt.

Murphy's success came partly from his ability to convince recalcitrant candidates to run negative attack ads against opponents, sometimes by leavening the commercials with humor. It was Murphy who, in his role as Romney's consultant in the 2002 gubernatorial campaign, had told Romney that Murphy's own soft-and-fuzzy family ads weren't working and convinced Romney to attack rival Shannon O'Brien. Romney, in his book *Turnaround*, credited Murphy with "sheer brilliance" and called him "hilarious." Murphy, meanwhile, was ensconced in his modern home atop Laurel Canyon, with its vista of the Los Angeles Basin, where he split his time between screenwriting and plotting Romney's presidential bid. Even when Murphy tried to give a convoluted explanation for Romney's turnabout on abortion— telling the conservative *National Review* that Romney had been "a pro-life Mormon faking it as a pro-choice friendly"—Romney stood by Murphy and accepted the explanation that the comment had been taken out of context.

In the early stages of Romney's presidential bid, it was widely understood that Murphy would be the lead strategist and bless the choice of campaign manager. Indeed, Murphy had already assured Romney that he was on board, and a number of Murphy associates had joined the team. Only one thing would upset the plan: Murphy had long worked for McCain. If McCain once again sought the presidency and ran against Romney, Murphy said he would not choose between them and thus would work for neither man. And so, when McCain announced he was in, Murphy called Romney and said: I'm out.

Murphy's departure would be widely lamented among top Romney aides outside the Boston inner circle. Doug Gross, the Iowa campaign chairman, had employed Murphy in his own failed run for the governorship of Iowa and knew how important the adviser had been to Romney. "Mike Murphy grew up in the streets of Detroit and understands what people think and what motivates them. Mitt Romney doesn't because he's not an average person in any respect." Without Murphy, "Romney had to rely on his own instincts" and heard conflicting advice from an array of advisers, none of whom had Murphy's influence, Gross said.

Romney authorized a search for someone who could run his campaign, and his top aides interviewed more than a half dozen of the nation's most experienced campaign managers. One by one, however, they said no—most because they could not meet the requirement to drop everything for a year or so and move to Boston. Attention turned to Myers. "Beth said, 'I don't want to be the campaign manager, and I certainly don't want to be the strategist.' She said that a hundred times," according to a Romney aide involved in the search. "She didn't seek the job; the job backed into her because there was no one else."

Throughout the campaign, Myers would be the subject of much criticism by Romney's state-level advisers. But her defenders stressed that she had accepted a difficult job when others would not and that she was always carrying out Romney's mandate. Myers directed a strategy that relied heavily on doing well in the early-voting states of Iowa, New Hampshire, Michigan, Nevada, South Carolina, and Florida. The mantra was "Win early and win often." Raising huge amounts of money, meanwhile, was expected to be no problem. The campaign set out to demonstrate Romney's credibility by hosting a one-day fundraising marathon on January 8, 2007, five days after Romney filed presidential exploratory committee papers. Romney ended the day with pledges for a stunning $6.5 million.

Speaking to reporters that day, Romney was asked about the possibility that he would have to use his own money to help finance his campaign. That, Romney said, would be "akin to a nightmare." Romney and his staff did not disclose that the candidate had already written checks totaling about $2.4 million. Later, when it was revealed that he had drawn heavily on his personal funds, an aide was quoted in *The Boston Globe* as saying that he was giving the money only temporarily. "A loan's a loan," the aide said. "That was just to start up." In fact, Romney eventually loaned his campaign $45 million, and, after the race was over, his staff admitted that the so-called loans would never be repaid.

———————

The campaign team under Myers was now set. The same could not be said for Romney's message. Nowhere was this problem more evident than in a state Romney was absolutely counting on winning—New Hampshire—and with a man he had personally recruited to chair his campaign there: Bruce Keough.

Keough was a real estate developer who had been elected to the state Senate from Exeter. He had run unsuccessfully for governor of New Hampshire the same year Romney had been elected governor of

Massachusetts. When Romney had first approached him to head up his Granite State effort, Keough had been intrigued but had a gnawing concern: what is the Romney message? He had urged Romney to focus on his reputation as an economic "Mr. Fix-it," someone who could transfer his talent for streamlining businesses into reshaping Washington. Romney had responded that it would be premature to settle on that message, saying, as Keough recalled it, "It is very early; who knows what the major issues of the day will be by the time we get to primaries?"

But Keough believed it wasn't too early; nothing was more important for Romney than to establish himself as the candidate who would turn the economy around. Romney kept on courting Keough, capping his effort at dinner at an Italian restaurant in Boston's North End. Keough eventually agreed to become the chairman of Romney's New Hampshire campaign and went to Romney's Boston headquarters for a meeting of top campaign officials from the early-voting states, including Iowa, New Hampshire, and South Carolina. Almost immediately, he said, there were "red flags." He reiterated his concerns about the candidate's message. He worried that the Romney team was focused more on building a campaign infrastructure than a rationale for the candidacy. Keough also thought that Romney was too focused on courting social conservatives. A strong pushback about this strategy came not only from top staff in New Hampshire (the nation's second-least-churchgoing state after Vermont, according to the Gallup Poll) but also from the more socially conservative bastions of Iowa and South Carolina. In all three cases, top campaign officials urged the Romney team at Boston headquarters to focus more on an economic message.

Iowa chairman Doug Gross and senior adviser Richard Schwarm—both of whom had been at the initial Ritz-Carlton dinner with Romney at which Gross had grilled the candidate—tried to convince the campaign of the pitfalls of running to the right on social issues, noting that many conservative Iowans, not to mention radio talk-show hosts, were not willing to accept that a former supporter of abortion rights was now "prolife." Schwarm said, "That was an argument that Doug and I had with the campaign over and over again." But Romney's Boston team said it understood Iowa and was sticking to its strategy.

The attack on the campaign strategy was soon joined by advisers in other key states. Romney's South Carolina team wrote an extraordinary four-page memo to Myers in which they essentially pleaded with the campaign to stop focusing on social issues and push the economic message. The memo noted that a number of politicians in South Carolina had been successful by focusing on that theme, that there was a "void" for such a candidate, and that Romney fit this role perfectly. "Every time Governor Romney talks about social issues, the flip-flopper accusations have been and will continue to be mentioned," the memo said. All that Romney had to do was "be acceptable to the pro-life crowd," not be its "champion." The concerns of the South Carolina team became so great that, several months later, they took the unusual step of insisting upon a private meeting with Romney to relay their fears. Meeting at a hotel during one of Romney's trips to the state capital, Columbia, two top South Carolina advisers told Romney that his continued effort to cast himself as a true social conservative was backfiring. Romney responded that he appreciated their concerns. But one of the South Carolina campaign officials at the meeting still didn't see much change and eventually became convinced that Romney "didn't want to deal with it."

———————

On February 13, 2007, a beaming Mitt Romney formally announced his candidacy in a setting designed to evoke the ties between him and his father. Striding onto the stage at the Henry Ford Museum in Dearborn, Michigan, he seemed a world away from the state he had governed. Instead, he was in the place where he had grown up and where his father had served as governor. An antique Nash Rambler, the original compact car that George Romney had staked his reputation on while serving as chairman of American Motors, was parked on one side of the candidate. A new Ford hybrid—a symbol of innovation—was parked on the other.

Mitt then began to talk about about his relationship with his father and about himself. "Dad and I loved cars," he said. "Most kids read the sports box scores. Dad and I read *Automotive News*. We came here together, him teaching me about cars that were built way before my time."

Throughout his campaign, one of the most asked questions would be: Who is the "real Romney?" Is he the moderate who once supported abortion rights or the self-described prolife Reaganite running for the presidency? Perhaps a clue was somewhere in that tableau on announcement day. His father had studied the automobile market and determined that there was demand for a compact car in an era better known for wide-bodied behemoths. Now here was the younger Romney, applying a business model to politics, shifting and adapting as the market of public opinion required. In business, changing positions in an evolving market can be the secret of survival. In politics, it can brand you a "flip-flopper." Romney's challenge was to show that his shifts were not expedient but reasoned and heartfelt. He argued that shifting gears represented an ability to innovate, not an uncertain core. And Washington needed innovation. In his speech he emphasized that point, using some variation of the word "transform" thirteen times.

Publicly, the Romney team expressed delight with the kickoff. They were just about alone in that feeling. A couple of weeks later, as Romney and his staff were traveling from an event to a local airport, a press aide pulled up an e-mail on his BlackBerry. A new ABC-TV/ *Washington Post* poll had just been released, the aide blurted out. Romney was at 4 percent, down from 9 percent a month earlier. The leader was Rudy Giuliani at 44 percent; McCain was at 21 percent; and former speaker of the House Newt Gingrich, who did not end up becoming a candidate, was at 15 percent. The press aide later recalled that as soon as he uttered the phrase "four percent," he knew it had been a mistake to speak out. He felt as if everyone in the van were looking straight through him. The only good news was that four out of ten people surveyed had no opinion about Romney. The campaign team needed to fill in the blanks. And stop the bleeding.

Romney's campaign faced a conundrum. The candidate's most obvious qualification for the presidency, beyond his business background, was that he was perceived to be a successful governor. But that success had been costly. He had won office by taking relatively moderate positions, and he had had to work with a host of liberal legislators in Boston and Washington to get things accomplished. To many conservative Republicans, Romney's success in working with liberal Democrats wasn't anything to be celebrated. It made him suspect.

Running against Kennedy in 1994 and for governor in 2002, Romney had sought to win over voters who favored abortion rights, and he had acknowledged during his presidential campaign that he once had been "effectively prochoice." He had also been supportive of gay rights, though not for gay marriage. And his signature achievement, the passage of a state health care bill that required nearly all residents to have insurance, sounded to some conservatives too similar to the proposals that Democratic presidential candidates wanted to be passed by Congress. The Romney team tried to address concerns about his years as governor by touting his record as evidence of his turnaround skills. They came up with an ad that portrayed Romney as "the Republican governor who turned around a Democratic state." It was an adroit pivot, but it was far from convincing to the conservative audience he was aiming at.

In the end, the Massachusetts message was something of a muddle, with Romney constantly trying to explain—or explain away—his positions. The real Romney, increasingly, seemed to be someone who didn't know where he stood or, worse, someone who would shift on core issues with the campaign winds.

Romney's team was also concerned that his wealth would turn voters off. Indeed, the third "M" in their trio of worries—millionaire—understated things. Romney's 2007 financial disclosure form pegged his wealth at between $190 million and $250 million. On paper, he seemed more at home on Wall Street than Main Street, more at ease with hedge fund managers than discussing the everyday concerns of voters. All that money, of course, had an upside: he could afford to quickly begin airing television ads. The campaign decided to start running ads unusually early—nearly a year before the first caucus or primary—seeking to define him as a job creator, business leader, family man, and above-the-fray politician.

An analysis of advertising during the Republican primaries, conducted by the Nielsen ratings firm, shows the extraordinary extent to which Romney relied on this strategy. From February to September 2007, Romney ran 10,866 television advertising spots, mostly in Iowa. During the same seven-month period, McCain ran five, and Huckabee and Giuliani ran none. The ad blitz boosted Romney's name recognition and poll numbers. Romney was moving forward, step by step. Even if the polls demonstrated more about name identification than deep support, progress seemed real. It gave the campaign confidence to push ahead with its all-out effort to win Iowa.

Despite the flood of ads and months of campaigning, Romney was still an enigma to many Iowa voters. Some eliminated him as a candidate based on his abortion flip-flop, and the campaign eventually realized that such voters were probably lost. But those who saw Romney as just another calculating politician might still be won over, if they could be convinced his rethinking of his views was sincere and evidence of his analytical cast of mind. Spokesman Eric Fehrnstrom explained Romney's mind-set in a particularly revealing comment to *The Des Moines Register*: "He's not a very notional leader. He is more interested in data, and what the data mean." The use of the word "notional" was telling. Fehrnstrom's comment apparently sought to highlight Romney's belief in facts over theories, but it may have left some voters questioning whether Romney saw himself as a man of big ideas, typically an ingredient in a successful campaign.

There was no doubt that Romney was data-driven, and his aides hyped that mind-set at every turn. Indeed, he loved numbers, and was convinced that a winning political strategy could be divined from them. That led him to become fascinated with the potential of "microtargeting." The idea is to come up with reams of data about each voter, ranging from magazine subscriptions to club memberships. The information is entered into a database and cross-referenced with the names of potential voters, enabling a campaign to assemble a detailed portrait of nearly everyone who might vote in the caucuses. The data are then used by a campaign to send specially tailored messages to key voters.

But, some within the campaign wondered, to what end? Doug Gross, the Iowa chairman, lamented that "we had a lot of data but no information." He saw little evidence that the strategy was helping build the coalitions needed to win the caucuses. Romney couldn't see it, but his problem was "macro," as one campaign adviser put it. He wasn't catching fire with voters on the ground, not like Mike Huckabee, the former governor of Arkansas, a man who shouldn't have had a chance against the well-financed, staffed-to-the-rafters Romney campaign, but who had something Romney didn't. And Huckabee didn't plan to spend a penny on microtargeting.

Still, in these early days Romney was more worried about McCain. On that front, the news seemed good: McCain's campaign seemed on the verge of collapse. By July, McCain's campaign estimated that it had blown through most of his $25 million campaign war chest. Most Romney advisers believed that their chief rival was finished. But one aide told Romney that the man who had survived five years as a prisoner of war in Vietnam was still a threat.

"Wait, he's not dead yet, let's stomp on him some more,'" the aide urged Romney.

"No, no, we're not going to spend money or time or effort," Romney said, according to Myers.

That decision helped allow McCain to survive and fight another day. But McCain also held his fire. His media team of Stuart Stevens and Russ Schriefer had produced a potentially devastating ad that slammed Romney for flip-flopping, using the candidate's own words against him. It showed Romney saying, "Abortion should be safe and legal in this country." And then saying, "We do have tough gun laws in Massachusetts. I support them." It also showed Romney equivocating on his own hunting background, at one point saying that he had hunted "varmints" a couple of times. Finally, it reran Romney's comment that he had been an independent during the time of the Reagan-Bush administration. The ad concluded by showing presidential candidate Romney saying, "This isn't the time for us to shrink from conservative principles."

"He's right," says the text in the ad, which concludes with a voice saying, "I'm John McCain, and I approved this message." But the voice wasn't McCain's. It was that of a producer, awaiting McCain's approval to complete the ad. McCain never authorized the ad and it did not run until months later in a leaked version on the Internet. The media team that had produced the ad, Stuart Stevens and Russ Schriefer, wound up jumping to Romney's campaign and would serve as Romney's chief advisers on strategy and media.

Meanwhile, a fierce argument was developing within the Romney campaign over whether to promote Romney's antiabortion position in a new round of television ads. Castellanos had proposed running a strong "prolife" ad. But Romney's South Carolina advisers dissented. It was several months after the South Carolina team had written its memo urging Romney to run as an economic Mr. Fix-it and not as a social conservative. Not only was the advice being ignored, but now an ad was being proposed that would underscore the social conservative message—and certainly draw new scrutiny to Romney's flip-flop on abortion. In one of the most heated conference calls of the campaign, a South Carolina adviser argued that the ad and the underlying strategy would hurt Romney. Castellanos said he saw no choice at the time but to acknowledge that "we were bleeding on our right and bleeding on our left, the worst of both worlds. We had run [ads] about Mitt the economic guy. . . . We weren't making Mitt some crazy right-wing figure."

The campaign was splitting dangerously into factions, further heightening the states-versus-Boston tension that had been boiling for months. The new media team was on board—Stevens and Schriefer— and Castellanos suddenly felt his authority in question. He protested to Romney and members of the inner circle, according to several people with knowledge of the conversations, and was told that this was not a demotion but rather was an implementation of "the Bain way," a reference to Romney's management style at Bain Capital. Romney said he wanted as many smart minds as possible in the room, with ideas fought over and the best rising to the top. Castellanos considered resigning but, out of loyalty to Romney, agreed to stay. Thus began what a number of those involved in the campaign said was an internal war in which strategists from the competing teams fought constantly and debates were often left to fester. A stream of complaints came from state-level campaign officials, who said they were never sure who was in charge. Romney, the man at the center, seemed disinclined to intervene. He liked to hear all the arguments, utterly confident in his ability to razor through to the right outcome.

Mandy Fletcher, the director of Romney's Florida campaign, said she had originally been attracted to Romney because "he was the turn-around guy and the business guy." But she also said that the delays and conflicts in the national campaign's decision making demonstrated that "running the campaign is a very different kind of business. In the business world you have a lot of time, weeks if not months and, on some projects, years" to make and implement critical decisions. "In the campaign it may be an hour or minutes."

Warren Tompkins, Romney's senior adviser in South Carolina, came to the same conclusion: "The glaring deficiency in the whole operation was the lack of an overall strategist, no single person that at the end of the day raised his hand and said, 'This is what we are going to do.' Somebody has to run the railroad. The irony of it is all is here's a man who sets up apparatus to make decisions, look at the bottom line, cut to the chase, and the campaign was everything but that." But Myers insisted it was all part of the plan. "There is definitely a creative dynamic," she said. Romney "wouldn't have it any other way."

Romney's strategy relied heavily on the idea that McCain and Giuliani would fight for the more moderate voters within the GOP prima-ries and caucuses, leaving Romney room to court the Right. The risk, however, was that he would be outflanked on his right by someone who came across as a more authentic conservative. The idea of a such a challenge from Mike Huckabee was barely on Romney's radar screen at this point, so when McCain and Giuliani both decided to make only a minimal effort in Iowa, the Romney campaign believed that the state was Romney's for the taking. That would turn out to be a huge miscalculation.

Iowa's caucuses have long been dominated by evangelicals who backed long-shot candidates—candidates who shared their religious beliefs but had little chance of getting elected. In 1988, the conservative Christian televangelist Pat Robertson came in second place in the caucuses, ahead of George H. W. Bush. That prompted the Bush campaign to come up with a new slogan: "Iowa picks corn, New Hampshire picks presidents." Bush did indeed win New Hampshire and went on to become the eventual nominee. That experience led some of Romney's advisers to wonder whether it was worth trying to win Iowa. But Romney was committed and invested about $2 million in his effort to win the Iowa Republican Party's "straw poll" in August 2007. It was, once again, a bid to make the religion issue go away. There was nothing scientific about the nonbinding poll; it measured organizational skill, not actual votes. Candidates were allowed to pay expenses for participants, so whoever had the most money and best campaign network was expected to win. For years, critics have called the event little more than an auction to the highest bidder—and no one was better positioned to be the high bidder than Romney. McCain and Giuliani stayed away.

Romney, however, went into the straw poll under fire from a pair of critics who were not on any ballot: conservative talk-radio hosts on the Des Moines station WHO. The 50,000-watt powerhouse, which had once employed a young announcer named Ronald Reagan and maintains a shrine to the late president, is one of the state's most influential media outlets. The station employed a fiery host named Steve Deace, who was to talk radio in Iowa what Rush Limbaugh is to the national audience. With a velvet voice, a passionate belief in Christian conservative values, and strong opposition to abortion, the radio host used his *Deace in the Afternoon* show to advocate for Huckabee and pulverize

Romney. "Given the nature of his ever-evolving positions on abso-lutely every issue," Deace said on one program, " . . . Governor Rom-ney either ranks first or second behind [Democratic candidate] John Edwards on my 'most despicable liar running for president' power rat-ings. No Republican in my mind is phonier than Governor Romney."

Addressing critics who accused him of being anti-Mormon, Deace responded, "I know some of you believe that I'm so anti-Romney on the basis of religious bigotry. Don't get me wrong, I am, you know, pretty much a religious bigot. I don't believe that traditional Mormon teaching and two thousand years of Christian doctrine and tradition are compatible. But I also don't believe in a religious test for office. . . . The problem with Romney is not that he's a Mormon, that's not what bothers me." What bothered him was that Romney's "flip-flopping makes him the most dangerous candidate in the GOP field." Deace's repeated attacks led one of Romney's top Iowa advisers, Brian Ken-nedy, to appear on the program just before the straw poll. After de-fending Romney's change on abortion as heartfelt, Kennedy could put up with Deace's barbs no longer. "You are waging a campaign both on and off the air to bring down Mitt Romney," he said, reflecting a view widely held in the Romney campaign.

Deace could hardly deny it. And although he was little known out-side Iowa, his power to sway voters was considered within the state to be one of the most important factors in Republican presidential politics. Huckabee's 2008 Iowa campaign manager, Eric Woolson, said that Deace had played a pivotal role in Huckabee's surge. "If you look at concentric circles of who listens to him most in this central Iowa area, where most folks listen to him, the numbers where you would expect Governor Romney to do well were suppressed and Governor Huckabee was up," Woolson said. "It was basically three hours of ad-vertisement against Governor Romney and for Governor Huckabee." Deace said later, "I don't believe that I am the reason Mike Huckabee won . . . but I might be a big reason why Mitt Romney lost."

Romney did not appear on Deace's show, but he did go on another WHO program, *Mickelson in the Morning*, hoping for a friendlier venue. But host Jan Mickelson pounced on the governor as soon as Romney slid on his headphones and pulled the microphone close. Mickelson used most of the interview to accuse Romney of breaking with Mormon doctrine by backing abortion rights when he ran for governor. During a commercial break in the show, Mickelson said he wanted to pursue the matter off the air. Romney, apparently not realizing that the conversation was being videotaped, tore into Mickelson.

"I don't like coming on the air and having you go after my church and me—" he said.

"I'm not going after—I agree with your church!" Mickelson said, referring to the doctrine against abortion.

"That's right. I'm not running as a Mormon, and I get a little tired of coming on a show like yours and having it all about Mormon," the angry Romney said.

As Mickelson persisted, Romney interjected, "So what should I do? Tell me what I should do! I should not have been prochoice. And therefore I'm just finished right there. 'Well, you're prochoice, therefore you've distanced yourself from your faith, therefore you're finished.' Well, that's not what my church says. There are leaders of my church that are prochoice! You're wrong! That's your problem."

Shortly afterward, someone posted a video of the supposedly off-air comments in an online forum. The video became an Iowa sensation. Deace had made the case against Romney, and now Mickelson had delivered what some considered a closing argument. The encounter was ballyhooed by some evangelicals as proof of Romney's lack of commitment on the abortion issue. Mickelson later said, "I thought this guy was the next great thing. Nobody was more surprised than I was." He said he called a Romney aide and said, "That was terrible, Romney didn't understand where I was coming from. . . . Let's fix it." Romney, however, did not appear on Mickelson's program again.

In Mickelson's view, Romney's performance on his show, combined with the daily pounding by his fellow host Deace, "cost Romney central Iowa." But some Iowans had been impressed: the supposedly cold, robotic candidate had showed passion, defended his faith, and taken on an Iowa icon. The private Romney, perhaps a more real one than most people had seen, could hold his own.

Romney won the straw poll with 31 percent, trumpeting it as "really the big kickoff for my campaign" and reinforcing the idea that he expected to win the caucus. He paid little heed to the man who came in second with 18 percent and who'd spent about one-tenth what Romney had. In the best tradition of political spin, Mike Huckabee called his runner-up finish "an amazing victory." In a dig at Romney, he told his supporters, "I can't buy you."

Despite Romney's bravado, some of his top advisers were alarmed. Several gathered in a hallway and shared their concern about the heavy bet on Iowa. It was the beginning of what became a months-long debate about whether the candidate should pull out most of his resources from Iowa and signal that he wasn't trying to win the caucuses. The debate, described by a number of participants, came as a result of concern about the rise of Huckabee and the collapse of the campaign of another Christian conservative, Senator Sam Brownback of Kansas. With Brownback pulling out of the race, Huckabee would have a fairly clear field in his bid for evangelical support. So although some of Romney's aides were celebrating the victory, others could see dark days ahead. Why not declare victory now, make a more modest effort in Iowa, and put aside millions of dollars to devote to other states such as Florida?

But Romney was committed to his all-in strategy for Iowa. His internal polling showed that he remained strong in New Hampshire. If he could win in Iowa, New Hampshire would surely fall into place and the nomination would be his, according to his advisers. So Romney went on Fox News to chastise McCain and Giuliani for failing to compete in the straw poll, further raising expectations that he would win the caucuses. "If you can't compete in the heartland, if you can't compete in Iowa in August, how are you going to compete in January when the caucuses are held?" he asked. "And then how are you going to compete in November of '08?"

He had set himself up for a fall. The "metrics" told him that he had won, that spending millions of dollars in Iowa to identify supporters was paying off. *The Des Moines Register*, however, told its readers in a front-page story that Romney's straw-poll win was "a bit hollow." Meanwhile, one of the nation's best-known Christian conservatives, Bob Jones III, the chancellor of the fundamentalist school named for his family in South Carolina, endorsed Romney's candidacy—but that, too, seemed a bit hollow. "As a Christian I am completely opposed to the doctrines of Mormonism," Jones said when he announced the endorsement in October. "But I'm not voting for a preacher. I'm voting for a president."

Romney could put off the questions no longer. From the time his candidacy had begun, he had hoped there would be no need to deliver a major speech about his Mormon faith. He thought he had dealt with the matter in countless interviews, and he worried that a big speech would draw more attention to the issue than ever. At the same time, he'd been collecting ideas and begun writing a draft. Finally, Huckabee's rise left Romney with no option; the speech had to be delivered. The question was where. Months earlier, Mitt had met privately with former President George H. W. Bush and discussed everything from the rigors of running a campaign to the impact of family to the role of religion. Now Romney accepted an invitation to deliver one of the most important speeches of his life at Bush's presidential library in Texas.

But what would he say? Some urged him to explain why he believes in Mormonism and to address directly charges by some evangelicals that it is a cult. Romney flatly rejected the idea. Instead, he followed the advice of Richard Land, the Baptist leader. The year before, Land had urged him to follow John F. Kennedy's example—to simply defend the right of any American, of any faith, to seek the presidency. Land privately thought it was "a mistake" that Romney had waited a year to deliver the speech. But now that Romney was going ahead, he agreed to take a prominent seat at the Bush library.

On December 6, 2007, about a month before the Iowa caucuses, Romney walked to the podium and took his place in front of a line of American flags. Several television news networks cut into their broadcasts to provide live coverage, with anchors predicting that Romney's words could make or break his campaign. But if listeners were hoping to hear about the Mormon doctrines he lived by, they were disappointed. He mentioned the word "Mormon" once and made a passing reference to the faith's former president, Brigham Young. He said he believes that Jesus Christ is the son of God and the savior of mankind. Referring only obliquely to Mormonism's belief that other religions are wrong and that Christ came to America, he said, "My church's beliefs about Christ may not all be the same as those of other faiths. Each religion has its own unique doctrines and history. These are not basis for criticism but rather a test of our tolerance. Religious tolerance would be a shallow principle indeed if it were reserved only for faiths with which we agree."

Romney assured voters that he would "serve no one religion" and would "serve only the common cause of the people of the United States." He emphasized repeatedly that his candidacy should be seen as evidence of the nation's belief in religious liberty. "There are some who would have a presidential candidate describe and explain his church's distinctive doctrines. To do so would enable the very religious test the founders prohibited in the Constitution. No candidate should become the spokesman for his faith. For if he becomes president he will need the prayers of the people of all faiths."

As he closed his speech, he concluded with the story of how Boston's Samuel Adams had been at a meeting of the Continental Congress in Philadelphia in 1774 when there was disagreement among members of different faiths about whether to say a prayer. As Romney told the story, Adams rose to say that "he would hear a prayer from anyone of piety and good character, as long as they were a patriot. And so together they prayed, and together they fought, and together, by the grace of God, they founded this great nation."

At that, the audience at the presidential library rose to its feet, applauding loudly, even though the speech was not finished. Romney looked pleased, and his aides afterward said they had never heard him speak with such passion. The reviews were good. Glenn Beck, who was then a host at CNN (and is himself a Mormon) told the nationwide audience that Romney had hit a "home run." But the glow did not last long, at least not in Iowa, where a candidate's views on religion are so important. A few days after Romney's speech, *The New York Times* published an article that described an upcoming magazine story on Huckabee. At one point in an interview with the *Times* writer, Huckabee seemed to go out of his way to stoke questions about Romney's religion. "Don't Mormons believe that Jesus and the devil are brothers?" he asked. The newspaper, in reporting Huckabee's rhetorical question, said that the authoritative *Encyclopedia of Mormonism* referred to Jesus as the son of God and Satan as a fallen angel, not "as brothers."

Romney was outraged by Huckabee's comment and sought to turn it to his advantage. "I think attacking someone's religion is really going too far. It's just not the American way, and I think people will reject that," Romney said. But Huckabee was the man of the moment. Though Romney's speech was generally well received, it had been delivered when Huckabee was rising in the polls. A withering dispute continued within Romney's campaign about how to respond to the Huckabee threat. Castellanos, the leader of Romney's original media team, urged a strong attack on Huckabee. A raft of opposition research—"oppo" in campaign lingo—had been prepared. One file detailed Huckabee's support for raising taxes. Another bulged with documents about Huckabee's support as Arkansas governor for paroling a convicted rapist. After a parole board had released the rapist, he had raped and murdered a woman and been convicted in 2003.

As Castellanos made his case to go on the attack, a member of Romney's second media team, Stuart Stevens, played down Huckabee's importance. "Why the sudden focus on Huckabee?" Stevens wrote to Castellanos and other campaign officials in an October 23, 2007, e-mail. "Is there any reason to believe everything has changed from a week ago or two weeks ago, when we got our data. We are reacting as if there was some new development in the race. . . . Let's don't suddenly get in the mindset that our Iowa mission is to kill Huckabee."

But that was the mind-set of Castellanos, who had been through many campaigns that had risen or fallen on such decisions. He ordered the production of an ad that attacked Huckabee's parole record. It was envisioned as one of the most powerful spots of the campaign, a more empathetic version of the infamous "Willie Horton" ad that had helped sink the 1988 presidential campaign of Democrat Michael Dukakis of Massachusetts. Instead of using a heavy-handed narrator, Castellanos and his team tracked down and filmed the mother of a woman murdered by a convicted rapist who had been released during the Huckabee administration. The ad showed the mother holding her daughter's locket and accusing Huckabee of having supported the rapist's release. The emotional words of the mother were accompanied by a frame that said, "Mike Huckabee granted 1,033 pardons and commutations."

Other Romney advisers feared that running the ad would backfire. It elevated Huckabee's importance in the race and might turn off some voters, they argued. The final decision was left to Romney. Concerned that the attack would seem desperate and create sympathy for Huckabee, he killed the ad. Instead, the Romney campaign aired what it considered a "soft" spot. Titled "Choice: The Record," it began by comparing Romney and Huckabee favorably, praising them as "two former governors, two good family men, both prolife, both support a constitutional amendment protecting traditional marriage." The difference, according to the ad, was that Romney had cracked down on illegal immigrants (a claim undercut by a report that he had hired a landscape firm that employed illegal immigrants) while Huckabee supported more lenient immigrant policies. The ad infuriated some Romney advisers, who considered it only a glancing blow at Huckabee and a questionable one at that.

By the time the Iowa campaign was over, Romney would spend nearly $10 million, much of it on ads and microtargeting. That is a stunning amount for Iowa and about ten times what Huckabee wound up spending. It greatly diminished the resources available for fights to come, especially in South Carolina and Florida. The financial advantage was so overwhelming that Huckabee was initially given little chance to win. Indeed, a Huckabee aide said the campaign was so short of cash that it couldn't afford the $30,000 cost of buying a state Republican Party list of people who had voted in the previous caucuses, the bare minimum information needed by most campaigns. But Huckabee had an unexpected ally. A Colorado-based political consultant named Patrick Davis, who would not set foot in Iowa during the campaign, orchestrated an independent $1 million effort designed to boost Huckabee and hurt Romney. Davis, who had been a political director of the National Republican Senatorial Committee, believed that Romney's flip-flop on abortion disqualified him. He set up a telephone operation to conduct what was called a "political survey." The operation was massive, with calls to 850,000 Iowa households—twice—in the days before the caucuses. The survey was known in the political business as a "push poll," in which questions are designed to influence opinion, not just gauge it.

One question went like this: "Does the fact that of the leading five candidates for president, only Governor Huckabee has always been pro-life and made protecting the lives of the unborn and the vulnerable in our society a top priority throughout his public life make you want to learn more about Governor Huckabee?" After the "survey" was finished, the listener was told that more information was available at "TrustHuckabee.com." The Romney campaign was outraged but could do nothing to stop the calls. Inside the Romney campaign, meanwhile, there was frustration that no similar independent group was stepping up to promote Romney. "We were waiting for some kind of help from somebody, and it never seemed to arrive," said Brian Kennedy, Iowa's former Republican Party chairman, who oversaw Romney's campaign in the eastern part of the state.

Romney's troubles in Iowa were having potentially disastrous repercussions in New Hampshire. Romney raced back to the Granite State, where his campaign's overconfidence was looking increasingly ill founded. Romney's latest strategy was to blast McCain for supporting an immigration bill that Romney said would let "everybody who came here illegally . . . stay forever." But Romney was under attack by two of the state's most important newspapers. The *Concord Monitor*, which had a more liberal bent, ran an editorial calling Romney a "phony" and urging voters to back someone else. Though Romney's campaign dismissed the *Monitor*'s editorial policy as liberal, it was unprepared for the endorsement of McCain by the influential and reliably conservative *Union Leader*, which blasted what it called Romney's untrue attacks on the Arizonan's record.

McCain promptly authorized a devastating ad that cited the *Monitor*'s characterization of Romney as a "phony" and quoted from a *Union Leader* editorial that said, "Granite Staters want a candidate who will look them in the eye and tell them the truth. John McCain has done that. . . . Mitt Romney has not." Romney seemed rattled. The former governor of neighboring Massachusetts, the owner of a New Hampshire lakefront home, was in danger of losing everything on what should have been home turf. McCain could not resist sending a zinger Romney's way. "I know something about tailspins," the former Navy pilot said, "and it's pretty clear Mitt Romney is in one."

Romney could stay in New Hampshire no longer. His campaign in Iowa seemed to be imploding, even as his staff remained publicly optimistic, convinced that the "metrics" predicted a victory. They were openly dismissive of Huckabee and his organization. A few days before the Iowa caucuses, a writer for the conservative *National Review*—which had given Romney its valuable endorsement—asked Romney spokesman Eric Fehrnstrom to assess the contest.

"We're going up against a loose confederation of fair taxers, and home schoolers, and Bible study members, and so this will be a test to see who can generate the most bodies on caucus day," Fehrnstrom responded.

"Not that there's anything wrong with any of those groups?" the *National Review* writer asked.

"Not that there's anything wrong, but that's just a fact," Fehrnstrom replied. "That's just where he has found his support. I have a theory about why Mike Huckabee holds public events in Iowa like getting a haircut or going jogging, or actually leaving Iowa and going to California to appear on the Jay Leno show. It's because he doesn't have the infrastructure to plan events for him. And when he does do events in Iowa, he goes to the Pizza Ranch, where you have a built-in crowd, so you don't have to make calls to turn people out. We're very proud of the organization we have built in Iowa." Huckabee's national campaign manager, Chip Saltsman, was as bemused as he was angered. "Eric's quote just shows the disrespect they had for us," Saltsman said. But Saltsman was happy to have his candidate underestimated.

On January 3, 2008, the day of the Iowa caucuses, Romney's campaign aides had a target that they believed would bring them victory: 25,000 votes. They were initially overjoyed as it became clear that they would exceed the goal, eventually garnering 30,021. But as the evening wore on, the Romney team's confidence—some would call it hubris—proved to be misplaced. The overall turnout far exceeded their estimates, and Huckabee wound up with 40,954 votes, vastly outdistancing Romney.

Richard Schwarm, the former Iowa Republican Party chairman and senior adviser to the Romney campaign, was convinced that anti-Mormonism had played a major role in Romney's defeat. "There are a lot of Iowans who view themselves as not bigoted but just don't believe that Mormon is a Christian religion," Schwarm said. "Or worse, I heard lots of times, 'I'm not a bigot, but my aunt and sister are.'" Myers said anti-Mormonism in Iowa "was a major issue" and that—despite warnings from the campaign's Iowa team—she hadn't initially realized what a big role evangelicals would play. Had she understood the extent of their influence, she said, Romney would have deemphasized the state. "If we had known that there would have been 110,000 caucus goers, with a majority of those being evangelical Christians, I would have thought that would have been a tough situation for Mitt to win." (The actual Republican turnout was 119,188.)

It was a remarkable statement. Evangelical power had long been the story of the Iowa caucuses. How could Romney, the man with a passion for data and details, have so underestimated it? More amazing, how could his Boston political team? In short, Romney's Iowa strategy had been a disaster. Instead of coming into New Hampshire with unstoppable momentum, he was struggling to keep his candidacy alive.

Still, some of the fundamentals seemed good: he was well known throughout the state, had a home there, and faced few questions about his religion. But the campaign had failed to heed the advice of some of its staff in New Hampshire, and now the state seemed to be slipping from Romney's grasp.

———————

Weeks before the New Hampshire primary, U.S. Senator Judd Gregg, Romney's national cochairman, had traveled to Boston for a briefing at campaign headquarters. Settling into a seat in a conference room, Gregg listened to Romney's advisers describe how easily the candidate was going to win the New Hampshire primary and the nomination. Gregg was shocked. From his years of experience, he knew all too well how a candidate could gain or lose twenty polling points in New Hampshire in a matter of days, earthquake-type shifts that depended on momentum and emotion. Yet here were Romney's aides, with their quantitative analysis and polling charts—the "quant," as Gregg called it—insisting that the nomination was all but sealed. "The pollsters were absolutely sure Romney would be the next president of the United States. I thought it was an absurdly 'quantish' view of the campaign," Gregg said. "I can just remember walking out of the room shaking my head." He felt out of sync with the campaign of which he was national cochairman. He appeared at some events and sometimes introduced Romney, but he was otherwise mostly shut out of the decision making. The Romney campaign "didn't want me or my organization to do any-thing," and, he said, he didn't try to push his way into the inner circle. Asked why a campaign wouldn't take advantage of one of the most experienced political operations in New Hampshire, Gregg responded with a single word: "Ego." He explained, "In these campaigns, people tend to be very resistant to outsiders. So my decision was not to get involved in how to run their campaign because they didn't appear to want to know."

Bruce Keough, the chairman of Romney's New Hampshire campaign who had been wooed so strongly by the candidate at a North End restaurant, decided on a far more aggressive approach.

For months, he had been frustrated that the Boston team wasn't listening to concerns raised by him and others who knew best what was happening on the ground. The campaign held conference calls with staff in the key states, but they were mostly one-sided, he said, with Romney aides providing a briefing on their latest plans. Moreover, after having been courted strongly by Romney, Keough had had little chance for interaction with the candidate. Throughout the entire New Hampshire campaign, he said, he'd had only two brief opportunities to ride with Romney in his car as he traveled across New Hampshire. He said Romney preferred to travel with only an aide, press secretary Eric Fehrnstrom. "His preference in traveling between events in New Hampshire was to be alone with Eric in the car and not use those gaps in his schedule as opportunity" to learn what was happening in the state, "and I thought it was curious," Keough said. He was frustrated that Romney didn't apply "the old adage about management by walking around. If you want to know how things are going on the factory floor, go talk to the factory workers." Keough fired off a memo to Myers and Bob White, Romney's close friend and associate from Bain Capital. He recapped the problems with the campaign and argued that Romney had to hammer away at a message of fiscal discipline. He never learned if Romney had seen the memo.

Romney was careening in the wrong direction and time was running short. A campaign memo had warned the candidate that the early primaries would be a "rocket sled process, 2 1/2 weeks, little time for adjustments." Usually, there had been at least a week between voting in Iowa and New Hampshire. This time there were only five days.

Huckabee now posed a real challenge on Romney's right. Not only had the Romney campaign failed to anticipate Huckabee's strength in Iowa, but it also hadn't thought he would go all-out in New Hampshire. Myers later said she had thought Huckabee would skip New Hampshire—where social conservatives are not nearly as important as in many other states—and go directly to South Carolina, which held the South's first primary. "I was surprised [Huckabee] went to New Hampshire." But Huckabee's ascent prompted his campaign chairman, Ed Rollins, to fly into New Hampshire, three weeks before Huckabee won Iowa, making it clear that his candidate would campaign across the Granite State. "We are going to be full bore here," he said in Concord, the state capital.

Worse, Romney faced new trouble on the left. The Romney campaign plan had counted on McCain and Giuliani splitting the moderate vote. But Giuliani was waging a surprisingly poor campaign. Never popular with the party's conservative wing, which disapproved of his stance in favor of abortion rights, Giuliani was having trouble gaining support when a series of reports hurt his image among voters. Stories surfaced that the former New York City mayor had billed the city for a security detail at times when he was visiting a woman with whom he was having an extramarital affair. Around the same time, his former police commissioner Bernard Kerik, whom Giuliani had recommended as the secretary of homeland security, was indicted on tax-related federal charges.

Wayne Semprini, the chairman of Giuliani's New Hampshire campaign, had believed the state was tailor-made for his candidate. Voters weren't fazed by Giuliani's support of abortion rights and liked his position as a fiscal conservative and strong leader. At the beginning of the campaign, Semprini, the former chairman of the state's Republican Party, had mapped out an ambitious advertising budget for Giuliani, convinced that the former New York City mayor could win the state and use it as a launching pad to victory. Now Semprini wanted to make his move. As he drove in a campaign car with Giuliani, he told the candidate that the time had come to make a major television advertising push in New Hampshire. Giuliani picked up his cell phone and, as Semprini recalled it, ordered an aide at New York headquarters to start running the ads. Semprini was thrilled. But the expected ad blitz never came. Semprini was later told by a Giuliani aide that the campaign didn't have the money; that was why Giuliani was spending so little time in New Hampshire and so much time running around the country raising funds.

The Romney campaign was aghast that Giuliani was pulling back in New Hampshire. It had believed all along that it needed Giuliani to do reasonably well in order to draw support away from McCain. Now the McCain campaign seized on Giuliani's collapse, telling voters that a vote for the former mayor was effectively a vote for Romney. "When Rudy stopped advertising in New Hampshire, that was one of the worst days in our campaign," Myers said later. McCain, deemed all but dead a few months earlier, was gaining momentum in New Hampshire. He was once again at his most comfortable, spending most of his time at town meetings and talking to voters.

Romney struggled as he tried to match McCain's success in connecting with voters, and his every utterance came under scrutiny. Asked about his lack of foreign policy experience in the wake of the assassination of former Pakistani leader Benazir Bhutto, Romney responded, "The President is not an expert. The president is a leader who guides America in making the important decisions which must be made to keep us safe." After he said he had seen his father march with the civil rights leader Martin Luther King, Jr., the assertion was questioned and he backed away from it.

Romney knew he had to change his message. Five days before the New Hampshire primary, he sat in his Portsmouth hotel room, took out a yellow legal pad, and sketched out a theme: "Washington is broken." The subtext was that only an outsider could bring real change. It would be Romney's mantra for the rest of the campaign and would be revived for the 2012 effort. But Romney's opponents, sensing that his carefully constructed campaign was coming unglued, pounded him the following night at a January 5 debate at Saint Anselm College in Manchester. McCain mocked Romney's effort to describe himself as the candidate of change. "I agree, you are the candidate of change," McCain said, alluding to Romney's change of positions on various issues. When Romney charged that McCain had supported a plan that granted amnesty to illegal immigrants, McCain said there were penalties involved and quoted Romney as having previously said the plan was "reasonable and was not amnesty." McCain then used the exchange as a vehicle to portray Romney as a rich liar: "You can spend your whole fortune on these attack ads, but it . . . won't be true."

On the morning of the January 8 New Hampshire primary, Romney's top campaign staff gathered at headquarters in Manchester. Even as the vote came in, the consultants and pollsters talked at length about Romney's message and his "brand." Keough, the New Hampshire campaign chairman, could not believe what he was hearing. A year earlier, he had attended a similar meeting in Boston at which all of Romney's staff had discussed the need to settle on a message. Now, as the same debate rattled through the headquarters on Elm Street, the frustrated Keough could take no more.

"What have you people been doing for the last year?" he demanded, bristling.

As he looked around, the problem seemed obvious. It was, as one aide put it, the "Noah's Ark campaign." There were two of everything, it seemed, including the competing media teams. The Romney campaign spin had been that the candidate loved the creative tension, but Keough had noticed a change in the candidate's attitude since the Iowa loss. "Mitt was a little less certain that he had the best campaign that money could buy," he said. That night, election returns showed that McCain had beaten Romney in New Hampshire by a margin of 37 to 32 percent.

Romney had one last chance, his advisers believed, to return to the original idea of campaigning as Mr. Fix-it. Echoing the Obama campaign, he would promise to bring change to Washington. The emphasis on social issues would be lowered, if not dropped. As the New Hampshire results came in, Romney sent an e-mail to Castellanos, saying he had at last latched onto the message the adviser had long been pressing on him. "Alex. Well, change was it—just like you said from the beginning," Romney wrote. "Never found a better word for it. Change it is. And change we will have—soon. Hope for the better . . . Mitt." But now money was an unexpected concern. After giving millions of dollars of his own money, Romney was nearing the limit of how much he was willing to contribute out of his fortune. The financial need, however, was great. There were upcoming primaries in Michigan, Nevada, and South Carolina. Romney's campaign believed it could win with an economic message in Romney's birth state of Michigan and easily win in Nevada, which has a large Mormon population. He did go on to win both of those states.

The question was whether it was worth following through on his initial vow to go all-out in South Carolina. As Romney struggled over such questions of strategy, his frustrations with the direction of his campaign surfaced in one of his rare public displays of anger. With his plane grounded by a snowstorm, he held an impromptu news conference at a Staples store in South Carolina. As he took his position in front of a rack of ballpoint pens, reporters scrambled for position. Associated Press reporter Glen Johnson (who subsequently joined *The Boston Globe*) nabbed a seat on the floor in order to plug his laptop into a nearby outlet. Johnson listened as Romney sought to contrast himself to his opponents, saying, "I don't have lobbyists running my campaign, I don't have lobbyists that are tied to my—"

"That is not true, Governor, that is not true," Johnson interjected from his ground-level position. He had been traveling on the Romney campaign plane and knew that one of the best-known lobbyists in Washington, Ron Kaufman, was a valued Romney strategist. "Ron Kaufman's a lobbyist."

Romney angrily denied that Kaufman was running his campaign. "Did you hear what I said, Glen, did you hear what I said? I said I don't have lobbyists running my campaign, and he's not running my campaign." He did acknowledge, however, that Kaufman was "an adviser." Fuming, he kept up the argument, denying that Kaufman had participated in "senior strategy sessions," but under further questioning from Johnson, he seemed to contradict himself by acknowledging that Kaufman had helped him in debate preparation sessions. Still, the incident might have passed without further notice had Romney, piqued at his loss of message control, not sought out Johnson after the press conference. A videographer caught the confrontation as the candidate got close to Johnson's face. "Listen to my words," Romney said. Then it was the turn of Fehrnstrom, Romney's ever-protective press secretary, who chastised Johnson for "being argumentative with the candidate. It's out of line. You're out of line."

For many reporters covering the campaign, the moment crystallized their frustration with Romney's inaccessibility, as well as with the off-putting and combative style of his inner circle. Later, Romney himself seemed to acknowledge as much when, after the Staples encounter, everyone settled into the campaign plane. Playing the role of flight attendant, Romney strolled into the press section with a tray of hot hors d'oeuvres, offering some to Johnson while an aide made a remark about making peace. Speaking shortly afterward on *The Tonight Show*, Romney went public with his damage control, telling Jay Leno that reporters "have a tough job to do. I respect the fact they've got to ask me tough questions and get in my face, but if I don't agree, I'm going to come back hard as well."

Some saw the episode as evidence of Romney's thin skin. But others saw something unintentionally authentic, and it doubtless made for great television. The bottom line, however, was that Romney was off message and, when it came to competing in South Carolina, lacked the commitment.

Looking ahead in the calendar, Romney's campaign wanted to save money for upcoming contests in Florida and a slew of "Super Tuesday" states. At Boston headquarters, South Carolina began to look like a financial drain with little prospect of success. So, shortly after the New Hampshire loss, Romney's South Carolina advisers made a difficult call to Boston headquarters. Unless Romney spent a lot of time in South Carolina leading up to the primary, they said, he had no chance of winning. Romney decided to pull out most of his resources. Beth Myers would later insist that the "the truth of the matter is we hadn't been in South Carolina that much. We did not have that much of an

investment going in . . . we never really pulled the trigger on South
Carolina."

––––––––––––

Nearly everything was bet on Florida. But if the fight between the
states and Boston headquarters had been bad in earlier states, it was
even worse in Florida. The state bore the burden of trying to turn
around a dying campaign with ever-diminishing resources. It hadn't
started that way. When Romney first met with his Florida team, he
had won them over with assurances that he would defer decisions to
them and that he would tap his personal fortune to make sure they
had the needed resources. Then, on October 19, 2007, a crucial meeting took place at a hotel in Orlando. Romney and his Boston team,
including Myers and the competing teams of strategists, met with the
Florida campaign staff to discuss how they would win the state and
then the nomination.

The state staff prepared a PowerPoint presentation, Romney's favorite form of communication, titled "All Roads Lead to Florida." The
presentation laid out the challenge of winning in a state with about
a dozen major media markets and an expected Republican turnout of
1.5 million voters—about thirteen times the number who would turn
out for the Iowa caucuses. Yet although the campaign would spend
$10 million on its gamble to win Iowa, it had far less to reach many
more voters in Florida. Though an exact number is not available, about
$5.5 million would be spent in Florida on television ads, the bulk of
the campaign's spending in the state.

A key to winning Florida, according to the PowerPoint presentation, was to send millions of pieces of mail, utilizing the costly "microtargeting" data that Romney valued so highly. The mailings would
be tailored to win over three groups: 450,000 social conservatives,
450,000 progun voters, and 250,000 households where someone was
expected to vote by absentee ballot. In a state dominated by older
people, who are likely to be most receptive to mailings, the strategy
was considered particularly crucial.

The Florida team wanted to hear reassurances from Romney that he would follow through on his commitment to deliver the needed money to the state. If Romney lost Iowa and New Hampshire, the Florida team said, everything would rely on winning the Sunshine State. But Romney's national strategists shot down even the discussion about the possibility of losing both Iowa and New Hampshire. "This was a major debate that took place in that room of 'Let's not put all our eggs in one basket, let's not be shortsighted,'" said Mandy Fletcher, Romney's Florida director. "But the top decision makers in the campaign were very confident that the strategy in either Iowa or New Hampshire would bring us a win and that would carry us. From the Florida standpoint, the concern was 'What's Plan B?' and there was none."

As the money dried up, Boston changed its mind so often about strategy that the Florida team was required to write at least seven different plans, the last version coming just a couple of weeks before the primary. The Florida staff would recommend where to spend money on television, and Boston sometimes overruled them. Florida argued against spending money on television in heavily Democratic areas, but Boston disagreed. Florida urged that money be invested in the Panhandle region, where McCain was strong, and Boston resisted.

Sally Bradshaw, Romney's senior adviser in Florida, exemplified the frustration. "I really cared about Governor Romney," she said. "I really believed he was the right guy. I was willing to fight for these things. I was a squeaky wheel, I pushed so hard. That probably left me feeling more disappointed than most. We laid it out on the line."

Most remarkably, after spending so heavily to produce extensive data about voters, Boston rejected pleas from the Florida team to follow through on the plan to harvest that information into millions of pieces of carefully targeted mail. In fact, the Florida strategists said, not a single piece of that mail was sent, a decision that was kept secret by the campaign and upset a number of Romney's Florida strategists. "If you had to boil [Romney's failure in the state] down to one thing, not sending mail in Florida was a really big thing and I truly believe it could have made a difference," said Fletcher, the Florida director. "If he had come out of Florida as a winner, he could have been the nominee."

The McCain team now saw it as a race against Romney, with Florida the place to finish off the former Massachusetts governor. The McCain plan was to draw Romney into a debate about whether there should be a "timetable" to withdraw troops from Iraq, and it worked. "Governor Romney wanted to set a date for withdrawal, similar to what the Democrats are seeking, which would have led to the victory by Al Qaeda in my view," McCain said. Romney, who had never suggested such a date, took the bait. "I don't know why he's being dishonest," Romney responded. "But that's dishonest."

The exchanges went on for several days, dominating news coverage. A top McCain adviser later said, "We played Romney like a fiddle," keeping him away from his strength of an economic message and on McCain's turf of national security. McCain beat Romney by a 36-to-31-percent margin in the January 29 Florida primary.

Romney's last hope was for a comeback in a debate to be held in advance of the February 5 slate of Super Tuesday primaries. The exhausted candidates flew from Florida to California for what would be Romney's last chance, a debate at the Ronald Reagan Presidential Library in California. Romney pushed back against McCain's accusation that he backed a timetable for withdrawing U.S. troops from Iraq. In one of the bitterest exchanges of the campaign, he accused McCain of pulling "the kind of dirty tricks that I think Ronald Reagan would have found to be reprehensible." McCain fired back, accusing Romney of lacking "the experience and the judgment" to be president in a time of terrorism.

As time passed, McCain regretted his performance. He said he'd been worn out from flying cross-country and offered an apology to Romney. "I was really tired," McCain said. "I was testy in that debate and clearly lost it. I just overreacted. I kind of took after Mitt. That just was a bad performance on my part, the moral being to not be exhausted when you are doing these debates."

But the attack had been effective. On February 5, McCain won a number of key states, including Illinois and New York, all but assuring him the Republican nomination. Romney, whose wins included Massachusetts and Colorado, vowed to fight on, saying, "We're going to keep on battling, we're going to go all the way to the convention, we're going to win this thing and go to the White House." At the time, it seemed that the animosity between the two men was insurmountable. But Romney's vow to continue proved to be bluster. Two days afterward, Romney pulled out, and a week later he endorsed McCain. It was time to make peace and look to the future. The vice presidency could be on the line.

TWELVE

BACK INTO THE FIRE

My power alley is the economy.

—MITT ROMNEY

They stream into the Elks Club in sweatshirts, ties, windbreakers, Boston Red Sox T-shirts, cowboy hats, baseball caps, plaid Oxfords, and homely sweaters. And still they come, every name dutifully logged by a campaign aide with a laptop, a growing catalog of New Hampshire voters eager for fresh leadership in the White House. Each one is asked: Are you backing Mitt Romney? Many say yes and take a sticker. Some are still shopping. The seats are soon full. Latecomers have to pile in behind the TV cameras, back in the cheap seats with the press. The low-ceilinged room, festooned with Romney placards and red, white, and blue bunting, is starting to feel claustrophobic. The cigarette smoke wafting up from the basement doesn't help.

This partisan crowd of a few hundred people in Salem, New Hampshire, on a brisk fall evening in 2011, doesn't mind, though. They're fed up—with President Obama, with the lack of jobs, with the fact that everywhere they go they see signs and instructions in tongues other than English. "There's like eight to ten different languages on the ATM machine!" a woman complains. Mitt Romney, his watch glistening under the stage lights, is talking hard truths, tough choices, and economic realities. But he's offering an appealing antidote—himself—promising to dive headlong into the job and ask the nation to follow him. "I will demand more of the American people," he says, "in terms of work, energy, and passion, and commitment to the country, attention to the challenges we face, harder work from our kids in school, demanding higher standards of our teachers and our young people, and parents working with their kids. We're going to have to do better."

He is now sixty-four, with more gray at his temples and more lines on his face, but Romney's energy shows no signs of abating, his skills of earnest persuasion as sharp as ever. You can still imagine him, if you try hard enough, at the home of a Frenchman years ago, asserting the merits of his faith until the door slammed. You can imagine him putting the hard sell on Ann, imploring her to wait for him, not to fall for the hunk at Brigham Young University who was just filling her loneliness. You can imagine him, armed with reams of data, showing a company how to operate more profitably and then taking over a company of his own. You can imagine him giving the charge to a demoralized Olympic community, igniting its spark anew. And you can imagine him peering into the Byzantine world of state government, rubbing his hands together—eager to take it apart and rebuild.

His confidence, on this day, veers close to arrogance at times: "It's so hard for some people who haven't spent their life—or haven't spent a day!—in the private sector to know how it works," he says. But he leavens those words with humility, a sign of a more experienced candidate. Not that he's figured everything out. In a riff on retail politicking, he briefly laments the ubiquity of camera lenses. "It's the nemesis of a campaign, by the way," he says. "Everybody you meet has a camera and wants your picture—with them! It just takes a lot of time." Then the gears turn in his head. He realizes he's talking to a bunch of people who will probably want a photo of him afterward—with them. And he pivots quickly. "It's kind of fun, my face is all over Facebook," he says cheerfully. "This is like free advertising. Keep it up, guys. Keep it up."

It had all started with him marveling at American innovation, still in many ways a Detroit boy who hasn't lost his sense of wonder at the latest invention, gadget, and late-model car, still drawn to the creative alchemy that is technological advancement. "I can take a blooming picture with my phone!" he says, sounding just like the young Mitt might have years before, at his father's side, surveying the latest automotive prototype with wide eyes.

Today, he's a long way from Detroit, more than six decades on the road and still traveling. And he is still finding his way. He got close to his ultimate goal four years ago, before he was forced to fall in line behind John McCain, putting a bitter rivalry aside. Indeed, their makeup press conference at Romney's campaign headquarters in Boston was as awkward as they come. But to Romney failure became just another hurdle to mount. It didn't take him long to start trying again.

Mitt Romney hadn't been anyone's number two since the earliest days of his business career—and then not for long. But after bowing out of the 2008 race, that was the job he was gunning for. It was also the only one open. His hope was to salvage his first national campaign by becoming McCain's chosen running mate and, if that didn't work out, put himself in position to make another run. In his trademark analytical way, he embarked on the courtship as if it were one more primary, this time with the goal of gaining not voters' approval but McCain's. He shook off the pain of defeat, or at least put it out of mind, and set off to campaign hard for the man who had vanquished him.

Romney barnstormed from state to state as a McCain surrogate and headlined fund-raisers for the chronically cash-poor candidate. He earned praise as the ultimate good scout, a man who put helping his party above nursing his wounded pride. And McCain took notice. "I had every confidence of his loyalty," McCain said. "Anytime anybody asked him to do something anywhere for our campaign, Mitt did it."

As McCain pondered his running-mate options, Romney was in the mix and at times leading the short list. He was, to many McCain aides, clearly the most accomplished and qualified of the names in contention, but "accomplished" and "qualified" were not necessarily what the Arizona senator was looking for. McCain, who was running way behind in the polls, felt pressed to consider an unorthodox choice, a surprise, someone who could shake things up and wrest the mantle of change from Barack Obama. McCain was also determined to run with someone with whom he felt personally comfortable. The fact was that he and Romney, though they had been campaign rivals for months, barely knew each other. The question of whether to go with a bold choice instead would require further pondering on McCain's part. But closing the personal gap between the two men could be dealt with straightaway.

And so, in May 2008, a few months after Romney pulled out of the presidential race, he and Ann climbed into a white Ford Mustang and drove toward the canyons around Sedona, Arizona. He was still frustrated at having squandered his chance with a flawed campaign strategy, a fractured staff, and his own uneven performance as a candidate. But now was not the time for self-criticism; it was time to sell himself anew. Along with other prospective candidates, Romney had been invited to spend a weekend at McCain's ranch. There McCain could take his measure of Romney on friendly territory, at his own pace, as they barbecued, walked to the creek, and hiked around the area's breathtaking Red Rocks.

To break whatever tension still lingered, McCain invited the Romneys to join his family at one of his favorite hangouts, a restaurant in Jerome, an old, spooky mining town thirty miles away in the high desert. It was a historic Spanish Mission–style venue, once a hospital and psychiatric ward for miners. Now it was an upscale eatery called The Asylum. The McCains and the Romneys settled into their seats, taking in the sweeping views of the Verde Valley. McCain soon found himself seeing the man across the table in a way he never had. "I'd always gotten the impression during the campaign that he was a little stiff," McCain said. "As I got to know him and his family," he said, he found "that's just not the case. In informal settings, he's a very talkative, entertaining guy with lots of experiences."

They were opposites in many ways: the disciplined, straitlaced Romney, who didn't swear or drink, and the fiery, tempestuous McCain, a former navy man who'd seen a bit more of life. But there in

Arizona, they found something in common. There was history, for one. Both had family ties to the state. And they shared something else: both had struggled to step out of, and beyond, a paternal shadow. McCain, the son and grandson of admirals, and Romney, the son of a governor, had both grown up with the highest of expectations. Gradually, to the surprise of aides in both camps, "Mitt and I became friends," McCain said. "I think Mitt is one of those guys, certainly was in my case, the better I got to know him, the better I liked him."

After the trip to Arizona, Romney remained a top prospect to be McCain's running mate, acknowledging that he would be "honored" if McCain picked him. McCain aides were asked to list pluses and minuses of various candidates, and Romney inevitably came up with a passel of pluses: competency, conservatism, the ability to raise money. But some McCain aides emphasized the minuses. Romney had been a success in business, yes, but sometimes at the expense of workers. Many of his deals had left "blood on the floor," as the aides put it, and that might be hard to explain to voters. And then there was his image as something of a weather vane on social issues. One of the McCain aides, who had never gotten over Romney's gibes at McCain during the primaries, made an argument that a fellow adviser summed up this way: "This is a Massachusetts flip-flopper potentially being coupled with the unflinching man of honor, and the two brands don't add up."

Romney's supporters in McCain's inner circle continued to push for their man, and Romney was thoroughly vetted by the team of lawyers McCain hired to evaluate potential picks. But McCain began to focus harder on making an unconventional pick. His pollster Bill McInturff told him that more than two-thirds of the public thought the country was on the "wrong track" and that the majority of such people would vote for the Democratic nominee. Given those numbers, McInturff told McCain, "There is no precedent in American political history, post–World War II, for the Republican nominee to win." McCain's campaign manager, Rick Davis, was equally grim, telling McCain, "If we don't do anything that significantly mixes it up, you're going to lose."

McCain took the advice, looking past Romney's attributes to abruptly choose a little-known but promising Alaska governor named Sarah Palin. McCain wanted a game changer who could rally the Republican base, siphon support from women voters who had backed Hillary Clinton for the Democratic nomination, and draw attention away from Obama, whose appealing backstory as a black man of Kenyan-Kansan heritage bathed his campaign in historical significance. The moment McCain settled on the rambunctious, youthful Palin, Mitt Romney's campaign year was over. It was time to go home.

———————

It was early February 2011, and Romney had just arrived at an upscale Washington restaurant for a private meeting with one of the nation's most influential evangelicals. Across the table sat Richard Land, the head of the Southern Baptist Convention's public policy arm, the Ethics and Religious Liberty Commission. It felt, in many ways, like a reprise of four years earlier, when Land had flown to Boston and met Romney at his home to discuss campaign strategy. Land was among many who had counseled Romney to court social conservatives, and now he prepared to make the pitch again, but this time in a different economic and political climate.

Romney and Land, along with a few others, settled into a booth at the Acadiana restaurant and ordered plates of Louisiana cuisine. Land offered his take on the political landscape. He saw advantages for Romney in a U.S. economy still staggering out of a deep recession, and he saw risk for him in the controversial health plan he had pushed into law in Massachusetts. But Land's key message was that Romney should not be swayed by those who were now advising him to downplay his positions on abortion, same-sex marriage, and other social issues. That would be a fatal mistake, he said, because the Tea Party movement, still ascendant within the GOP, had zeroed in on downscaling government but included many who cared deeply about social issues.

But this wasn't 2008, and Romney's response to Land's counsel showed how much the candidate had changed. He thanked Land for his advice without saying how much of it he would take. Some months later, the answer seemed clear: Romney kicked off his campaign by focusing heavily on fixing the economy. He did not mention abortion in his announcement speech. Even when speaking to a faith-based group the day after his kickoff, he mentioned abortion in a sentence and quickly moved to more comfortable terrain: kickstarting the U.S. economy. It was a striking change and intended as one.

In the aftermath of 2008, Romney closely analyzed his campaign, talking through the failure with his closest advisers. True to form, he wallowed in the data, crunched the numbers, and evaluated the results thoroughly. Several things had gone wrong. His message had been muddled. He had spent far too much time and money in Iowa. He had miscalculated his popularity in New Hampshire. He'd relied too heavily on expectations about how his competitors would fare. Romney, admitting the limits of his own political instincts, also seemed particularly rueful, two advisers said, at not having had a campaign architect such as Mike Murphy, who had stayed out of the 2008 race because of his ties to McCain. "I never had a strategist," Romney told his friends. "I had all the pieces of the puzzle but didn't fit them together." He had needed a team he could trust implicitly, a key ingredient in his success at many points in his life and career. In 2008, his team had been divided. Some advisers insisted that Romney thrived in that environment, refereeing the collision of ideas and making the end call as the CEO. But the lesson of 2008 seemed to be that running a presidential campaign, with its compressed time frame and unpredictable currents, is nothing like leading a state, helming the Olympics, or buying and selling companies.

Looking ahead to 2012, Romney concluded that he needed a different kind of campaign. He looked again to his close circle of advisers in Boston, who had learned from their mistakes and grown and changed in the intervening years. One group had helped Republican Scott Brown achieve a stunning victory, winning the U.S. Senate seat formerly held by Ted Kennedy. Another group had played a key role in midterm election victories by Republicans across the country. In preparation for the second try, Stuart Stevens, who came with years of experience in presidential campaigns, moved to Boston and was empowered as chief strategist. The two bickering media teams of 2008 were reduced to one. After spending $2 million to win Iowa's straw poll in 2007, Romney would refuse to participate four years later. Instead of spending millions of dollars on early campaign ads, he would hoard his campaign cash. And rather than devoting countless hours to wooing evangelical leaders, he would say that the time for discussing his religion had come and gone. Read Article VI of the Constitution, he would say, quoting it: "No religious test."

In a frank admission, Romney acknowledged that his major mistake in 2008 had been quite simple: he had failed to get across what he was really all about, a problem he had also identified after his 1994 Senate race. Once again, he had lacked definition. "I think that one of the things that's very important in running a campaign is to make sure that you're known for the things that really motivate you," he said. "And I needed to do a better job to focus my campaign on the economy and getting the economy right and creating jobs. And whether through my ads or through my responses to debate questions or on the stump, my power alley is the economy."

This time, his aides said, Romney would play to his strengths every possible minute. It would be a calculated risk, though. He still had no strong foreign policy credentials. Social issues would remain important to many in the Republican base. Conservatives disliked his Massachusetts health care plan, notwithstanding his promise to repeal "Obamacare" if elected president. And his career as a leveraged-buyout specialist could backfire if voters still saw him as more in tune with Wall Street than with the squeezed middle class. But if the economy remained the dominant issue, then perhaps, just perhaps, he and his team thought, the pieces of the puzzle might fit.

Before making the leap in 2012, Romney first consulted with Ann, as he had always done. She told him that he should go for it, and without regret. He also wrote a book, *No Apology*, that revealed little about himself or his family but was filled with policy prescriptions that, like campaign white papers, provided a sketch of his conservative principles. With so many people asking where he really stood, Romney now had something concrete that he could show. It's in the book, he'd say. Just read it.

It's telling, though, what Romney didn't emphasize. *No Apology* concluded with a list of sixty-four "action steps" that formed what he called his "agenda for a free and strong America." There is no mention of abortion or same-sex marriage on the list, although he did make room for recommendations such as "Adopt dynamic regulations." In the text of the 309-page book, he only briefly referenced his "unapologetically pro-life" stance and his "opposition to same-sex marriage." Asked why he put so little emphasis on social issues, which had played such a significant role in his 2008 campaign, he responded blandly, "It's always a great interest on the part of those questioning a candidate to know where they stand on social issues, but I don't know it's a topic that's going to be resolved with rhetoric and analysis. It's rather a topic where one has one view or one has the other view and you're not going to persuade someone." In other words, that wasn't a battle he planned to wage this time around.

On April 12, 2011, Romney entered the Harvard Club in New York City, which advertises itself as "the city's most exclusive private club," and appeared before more than a hundred of the country's most powerful and wealthy Republicans. Most in the audience had pledged to raise a minimum of $25,000 from their friends, and many were expected to raise much more. They included Wall Street traders, executives, and others who liked Romney's business background and applauded his approach to economic issues. With their help, Romney told the gathering, he could raise tens of millions of dollars and be on the road to the White House. The gathering underscored Romney's decision not to distance himself from Wall Street and the business world. At the same time, he kept a distance from some elements of the Tea Party. He praised the Tea Party's concerns about big government and endorsed some of its candidates but also said that voters should be wary of the "temptations of populism." "The populism I'm referring to is, if you will, demonizing certain members of society: going after businesspeople, going after Wall Street, going after people who are highly educated, people who are CEOs," he said. "That kind of 'all of our problems are due to that group' is something that is unproductive."

There was no doubt that Romney was, to some degree, speaking in self-defense, given his long ties to Wall Street and the investment community and his two Harvard degrees. At the same time, he began making a better effort to connect with regular people. In the 2008 campaign, his staff had argued with a Florida adviser who wanted the candidate to take off his tie when meeting with retirees. In this campaign, as he prepared for an interview by Piers Morgan on CNN, Romney solicited his wife's advice. "I'm going to be on with Piers today, what should I wear? I think I should wear a tie, don't you?" Ann replied, "No, no, no. Just wear the shirt you've got on, a blue shirt, and a sports coat." As Romney said in recounting the exchange, "I do as I'm commanded." And in fact, Romney has often adopted the business casual look, tieless and in simple slacks and a shirt. Sometimes he even pulls out the jeans (alternating between the fashionable Gap 1969 variety and a pair of Levi's 514s). In 2008, his press spokesman criticized rival Mike Huckabee for inviting photographers to watch Huckabee get a trim at an Iowa barbershop. This time, Romney invited photographers to watch him get a $16 cut, posting on his Twitter account afterward, "Just got a Trim at Tommy's in Atlanta."

But the effort to recalibrate his image could come off awkwardly. In August, Romney climbed atop bales of hay at the Iowa State Fair and, per tradition, began hawking his message like a carnival barker. Facing a crowd that included a number of hecklers, he argued against raising taxes as a way of saving Social Security, Medicare, and Medicaid. Someone in the crowd yelled, "Corporations!" Romney took the bait. "Corporations are people, my friend," he said. "No, they're not!" a heckler shouted. "Of course they are," Romney responded. "Everything corporations earn ultimately goes to people." The comment, which drew catcalls from the Left, was fully in harmony with Romney's convictions about how capitalism works to benefit all. The phrasing seemed impolitic for a man trying to broaden his appeal to working-class voters, but he stood by the remark.

When he ran the 2002 Olympics, Romney came up with a vision for the Games: "Light the Fire Within." It was intended as a celebration of persistence and inner strength. In retrospect, it can be read as a guiding principle for Romney's life; he has never lacked for drive. Not as the teenager who stumbled to the finish line in the race at Cranbrook. Or the missionary whose trials pushed him nearer to his faith. Or the son who spent so many years trying to match and exceed the ambitions of his dad. Or the businessman who pushed and pushed until the deals made sense. Or the governor who followed his own compass to achieve a historic breakthrough on health care.

Romney, indeed, has always had persistence, always had ambition and exceptional stamina. What he has struggled with, in politics, is exactly who he is, with decoding his political DNA. For years, he could just operate in his father's shadow or avoid those hard questions in the private sector, getting by on brains and leadership alone. But if, as he said, his die was now cast—if he had settled on a true sense of self—he would have to prove it and stick to it under the hot glare of a presidential campaign. And sticking to one vision of what he is about has always been the hardest thing for Mitt Romney.

One autumn day in 2011, Romney headlined a fund-raiser in a ballroom of a Marriott hotel in Bethesda, Maryland. J. W. "Bill" Marriott, Jr., the chairman and CEO of Marriott International and a longtime friend, introduced him by celebrating his service on the company's board of directors. He told the roughly 250 people there that Romney had helped save his company when it was facing economic difficulties. Now, Marriott said, he could do the same for the country. Among those in the audience was Dane McBride. Few people have a better insight into Romney's development over the years. McBride had grown up as the only Mormon in his North Carolina high school. His childhood hero had been George Romney. Then, as a missionary, McBride had been paired with Mitt and believed from the start that this was a young man who would one day match his father's ambition. He had attended Romney's wedding, and they had both gone to BYU. Now he was supporting his friend in the biggest test of Romney's life. Watching him that day in Maryland, McBride felt his mind wandering back across the years. "He was the same passionate person that I had known since we were nineteen years old," he said.

Back in those missionary days, Romney had learned how to speak carefully, to avoid coming across as some kind of extremist from a fringe religion, McBride said. He'd learned how to work in an environment in which he was the minority, the one swimming against the tide, the one who did not have the luxury of being doctrinaire or absolute. Flexibility, adaptability, humility—those were the essential skills. But in politics, McBride lamented, finding common ground "can be interpreted as being inauthentic." That tag of inauthenticity remained a serious risk as Romney charged deeper into the 2012 race. And there were others. For one, could the multimillionaire persuade the everyman that he knows what it means to struggle? Romney and his campaign hoped that the moment—one of great economic anxiety—perfectly matched his turnaround message. If Romney couldn't sell that, he couldn't sell anything.

———

Back in Salem, New Hampshire, Romney is delivering his closing argument in characteristically grandiose terms. The United States, he warns, must remain a beacon of strength and liberty in an uncertain world. "If we don't," he says, "freedom itself is at risk."

He lingers on the wood floor afterward, his face framed by a red-and-blue sign that reads, NEW HAMPSHIRE FOR MITT. He signs a few baseballs, saying to no one in particular that he feels like a pro athlete, an aside that his sons would find especially hilarious. He crouches down to meet people at eye level. A woman in patriotic garb challenges him on tax cuts, but he dispatches her with a pointed retort. "If you want to raise taxes, that's easy to do," he says; just vote for a Democrat.

Slowly he makes his way toward a side door. He doesn't have time to shake every hand. Or sign every brochure. Or give a word to the press. He has one last deal to close, and there's a lot of work still to do. Time to go.

AFTERWORD

O n a humid night in mid-April 2012, Mitt Romney bounds onto a makeshift stage inside a grand rotunda at the Franklin Institute in Philadelphia. It is an auspicious setting, a tribute to one of the wisest and most acerbic of the founding fathers, Benjamin Franklin. Tonight, Romney is focused not on the past, but on the future—his future—looking ahead with sunny sanguinity to the new "American century" he has promised, with him leading the way.

He can, finally, savor the taste of victory. At least for now.

On this night, with his last serious rival, former U.S. Senator Rick Santorum of Pennsylvania, now vanquished, Romney at last has grounds to celebrate. The strange rhythms of the GOP primary race have subsided. He has swallowed the indignity of waiting for the party faithful to settle on him as plainly the best in an underwhelming field. He has endured. And he is now, after hard-fought wins and unsettling losses, the presumptive Republican nominee, his biggest political achievement to date. Having mastered the game of long ball, Romney outlasted his dogged competitors, each with a claim on a segment of the GOP electorate and each having, in turn, battered Romney as too moderate, too weak, or too much the greedy capitalist to defeat President Obama in November. The attacks were wounding, but he has survived, a measure of his stamina, the appeal of his economic message, and, not least, the efficacy of the negative ads he and well-funded allies unleashed on his Republican brethren.

In Philadelphia, Romney scans the rotunda, modeled after the Pantheon, its ornate interior lined with marble columns. Arcs of chairs encircle the stage, filled with several hundred enthusiastic supporters. The candidate, typically a dominating presence, seems overshadowed as he takes his position next to a giant white statue of Franklin. Few politicians standing here would miss the chance to muse about Franklin's meaning to America, but Romney pays him little mind. Instead, he asserts, questionably, that he never expected to run for president. "I've found myself having the time of my life," he says. "I've had the chance to meet everyday Americans living their lives in states across the country, and I come away with more confidence and more optimism about our future." As he speaks, some of Franklin's most famous epigrams from his *Poor Richard's Almanac* appear on the walls, beamed from a projector in large type. One particularly apt saying flashes repeatedly within the candidate's eyesight: "Diligence is the mother of good luck."

It could be Mitt Romney's motto.

He has been nothing if not diligent building his campaign apparatus, apprenticing within the Republican hierarchy, and positioning himself carefully in a fluid political landscape. He and his team exploited major changes in the rules of running for president, mastering new ways of raising and spending money and collecting party delegates. He has filled no arenas with adoring fans, sparked no wildfires of political enthusiasm. But what Romney has lacked in buzz he's made up for with persistence, methodically dispatching the threats to his candidacy. His perseverance has proven inexorable.

As has his luck. The most determined political foes on the right, the Tea Party forces who largely rejected Romney as insincere and misguided on key issues such as health care reform, never sufficiently coalesced to deny him his dream. Romney's efforts to win them over were often awkward and ultimately unavailing. As his friend and onetime political adviser Mike Murphy said, "watching Mitt Romney try to connect with the Tea Party is like watching the Queen of England try to eat a chili cheese hot dog." Romney has made it to this triumphant day because the Tea Party—and conservatives overall—could not settle on an alternative candidate. GOP voters hungry for Anyone But Romney flirted with nearly every competitor. Their power, however, was diffuse, leaving Romney always a step ahead of a splintered opposition. Santorum and Newt Gingrich, the polarizing former

House speaker from Georgia, battled for the conservative mantle, and neither ever got the clean shot he wanted at Mitt Romney. In the end, those who found Romney so difficult to stomach had to do just that.

Now, standing at Franklin's feet, Romney has transcended the success of his father, who came up short in 1968 amid wide divisions in the Republican Party. It is forty-four years later, and Mitt Romney faces the challenge of uniting a newly fractured GOP and inspiring many voters not thrilled to see him on their ticket. To anyone watching in Philadelphia, Romney's profile in victory remains a familiar one: the strategy geek who's good with data but less so with people; the businessman with a record of buying and selling that cuts both ways; the shape-shifter within the Republican Party, still searching for acceptance, and the ultimate prize.

Into the mid-summer of 2011, Mitt Romney was cruising. The other presidential prospects deemed most threatening—namely Indiana Governor Mitch Daniels and Chris Christie, the brash, beguiling governor of New Jersey—had decided not to run. It would be, so it seemed, Romney versus the long shots. He had played a classic front-runner's game, taking few risks, rarely giving interviews or taking press questions, all in the name of preserving his air of inevitability. He would run out the clock. His posture said, *Call me when it's over.* And then, one Saturday morning that August, 30,000 people streamed into Houston's Reliant Stadium for a prayer rally.

The rally's host was Rick Perry, Texas's long-time governor and a devout Christian, who for weeks had been making noises about jumping into the presidential race. Perry had leadership experience in Austin, conservative credentials, and homespun charm. His state had also created jobs at a good clip. On this morning, he sounded less like a politician than a revivalist. "Father, our heart breaks for America," he said, eyes closed and leaning into the microphone. "As a nation, we have forgotten who made us, who protects us, who blesses us." Worshippers stood or kneeled in the aisles and in front of the stage, many with tears in their eyes and shouting "Amen!" One Texas woman who drove two hundred miles to attend said, "I believe God has prepared Rick Perry for such a time as this." With many Republicans practically begging him to run, he began courting activists in early primary states, building his national profile, and lining up campaign donors. He entered the race with high expectations and a promise of smaller government, saying, "I'll work every day to try to make Washington, D.C., as inconsequential in your life as I can."

Perry's entrance, at least initially, unnerved the Romney team, and with reason. After barely two weeks, Perry leaped to the top of the polls. But his moment proved brief. Over the next few months, it was Perry who became inconsequential, felled by some goofy policy positions, disastrous debate performances, and a lackluster campaign. ("Oops," he would famously say in one debate after forgetting which federal agencies he had vowed to eliminate.) By October, his poll numbers were sinking into the single digits. The Texan's swagger was gone, and soon enough he was, too. But his improbably quick rise showed a tremendous desire, perhaps even desperation, among many Republican voters for someone to take on Romney. Romney himself seemed to have a hard ceiling in most national polls.

The next audition for a Romney alternative came from Herman Cain, a former pizza chain executive from Atlanta, whose folksy, magnetic style was winning over voters, despite some quirky ideas, including a much-derided tax reform proposal he called the 9-9-9 plan (nine percent income, corporate, and national sales tax rates). Cain matured from a curiosity into an improbable front-runner, overtaking Romney in national polls in October. Then his star dimmed, too, as his platform and his past received more scrutiny.

Then it was Newt Gingrich's turn. It seemed an unlikely comeback by one of the more divisive figures in modern American political history, but Gingrich's rise, like the others', was revealing. The closer the actual voting grew, the shakier Romney's presumed lock on the nomination became. His own missteps—including a foolishly combative performance in a Fox News interview in late November, in which he bristled at legitimate questions—further weakened his grip. It was plain that many GOP voters did not trust him to lead a winning fall campaign. As veteran Democratic strategist James Carville put it later, forcing Romney on the Republican electorate was like "trying to feed a dog a pill. The dog keeps spitting the pill up."

Nonetheless, the wide-open nature of the race, and the fickleness of Republican voters, gave Romney's team an opening it didn't anticipate. Burned in Iowa in 2008, his campaign had devoted relatively little attention to the January 3 caucuses, focusing instead on notching a big win in much friendlier territory—New Hampshire, whose primary was a week later. But as the Iowa caucuses neared, Romney's campaign started believing a win was within reach. He and his family began campaigning there aggressively. The rallies got bigger. "Eye of the Tiger" blared at campaign events. Romney grew more confident. He knew an unexpected win could be an accelerant, and so the once reluctant participant was now all-in. There was just one problem. So was Gingrich, who was surging. If Romney were going to pull off an Iowa surprise, he would need to stop Gingrich. Fast. And he would need some help from his friends.

It was a cold winter day, two years before the primaries began, and about a half dozen top Romney advisers gathered at the Madison hotel in Washington. Romney was all but certain to launch another bid for president. The advisers had worked for him in 2008. None wanted to repeat the mistakes of the previous election cycle. One of the biggest had, at the time, seemed like one of Romney's greatest strengths: his ability and willingness to spend his own money on the race—$45 million of it, by the time it was over. Access to his enormous personal wealth had made the campaign lax about controlling expenditures, which ran counter to Romney's thrifty nature. It had also sent a message to donors that their money was not crucial. This time, the advisers said, Romney was expected to keep his wallet closed. Besides, Romney wanted a leaner operation. As the meeting ended, the advisers agreed that the tentative decision not to tap Romney's wealth during the primaries must be kept secret. But they purposefully let the opposite impression filter out, suggesting that Romney was prepared to spend heavily again, thus helping scare away potential competitors with far smaller bank balances.

Around the same time, Romney and his team got a boost from a landmark Supreme Court decision known as Citizens United. The court ruled, 5–4, that certain kinds of corporate and union political contributions constituted free speech. Together with other court rulings and decisions by the Federal Election Commission, this enabled companies and wealthy individuals to give as much as they wanted to independent committees that overtly supported or opposed presidential candidates. In the rationale of the federal courts, such spending was proper because it was supposed to be done independently of the campaign. But this would prove a porous safeguard. The new rules promised to completely change the political landscape, and no candidate was in a better position to reap the benefits than Romney.

Three senior aides from Romney's previous campaign decided to take action. In the wake of his 2008 defeat, this troika had closely analyzed why their candidate had lost, concluding that one of Romney's biggest problems was the failure of any well-financed independent group to air negative ads about his opponents—the kind of spots Romney was reluctant to run himself. In Iowa, for example, his media specialists had prepared a commercial that attacked rival Mike Huckabee's record, as Arkansas governor, supporting the parole of prisoners who later committed horrible crimes. Romney refused to approve it, wrongly believing that he could win Iowa without it. Huckabee then surged past him to win the caucuses, crippling Romney's bid for the nomination. Larry McCarthy, one of Romney's 2008 ad men, believed that, to win in 2012, Romney needed allies who would mount such scathing attacks. When McCarthy learned that two other former Romney aides were thinking of forming just such an organization, it seemed the ideal way to fill the void. "We knew what needed to be done. We all came out of Romney World," McCarthy said. With little notice, McCarthy and the two other Romney veterans—Carl Forti, the campaign's national political director in 2008 and a close ally of Republican strategist Karl Rove, and Charles R. Spies, a former lawyer for the Federal Election Commission who had served as counsel to

Romney's 2008 campaign—established a so-called Super PAC they dubbed Restore Our Future.

While individual donors to the Romney campaign were legally limited to giving $2,500 for the primary and another $2,500 for the general election, they could give whatever they wanted to Restore Our Future. Fundraisers were held around the country, from New York City to Beverly Hills. Although Romney said during an appearance on MSNBC that he had no communication "in any way, shape, or form" with Restore Our Future because otherwise he would face jail time for violating election laws, he did go to at least one fundraiser for the group, in New York City. A public interest group, Democracy 21, asked the Justice Department in March 2012 to investigate, but the Super PAC and Romney campaign dismissed the accusation as baseless, saying a federal ruling allowed Romney to be a speaker at the PAC's fundraisers.

The fundraising effort was a phenomenal success, raising about $51 million through the first three months of 2012. No donor to Restore Our Future was more important than a reclusive, seventy-nine-year-old Texas homebuilder named Bob Perry, who had given $4 million by the end of March. Perry had been a major benefactor of Swift Boat Veterans for Truth, the 2004 group that attacked John Kerry's service record in Vietnam. After Perry, the most generous donors to Restore Our Future were some of Romney's former colleagues from Bain Capital and wealthy Wall Street investors. None of Romney's competitors could muster anything like that firepower. The Super PAC supporting Santorum raised $8 million. Gingrich's campaign was kept alive by the Super PAC supporting him, which raised $24 million, most of it from two individuals, casino mogul Sheldon Adelson and his wife, Miriam. Restore Our Future, however, with its narrow focus on running tough ads against Romney's competitors, was arguably the most powerful force in the primary campaign.

Iowa was the first laboratory. In an open-ended question, a pollster for Restore Our Future had been asking voters what words they would use to describe Gingrich. One negative word repeatedly came up: "baggage." Voters were fuzzy, though, on the specifics. What was Gingrich's record, anyway? McCarthy quickly came up with a concept for a TV ad: pieces of luggage on an airport carousel that would represent old Gingrich controversies. It was a devastating visual image, broadcast on TV stations all over Iowa. Gingrich, with little money in his campaign coffers, told reporters that Romney should instruct the Super PAC to pull the ads, which Gingrich said included falsehoods. But he did not respond in kind. Vowing to deliver a positive message, Gingrich's campaign produced a widely panned TV spot that did not directly respond to the charges. Instead, he wished Iowans a Merry Christmas.

"My goodness, what a squeaker!" It was the day after the Iowa caucuses, and Romney was marveling at what looked then to be his razor-thin victory over Rick Santorum. At 2:30 a.m., Iowa Republican Party officials declared Romney the winner by a mere eight votes. Romney redubbed his campaign bus the Landslide Lounge. For the moment, he could afford to joke: His unlikely Iowa victory was in hand, and Gingrich, knocked down by the attack ads, was relegated to a distant fourth. Santorum's strong showing in Iowa was a warning sign, but Romney looked ahead to New Hampshire, landing in Manchester for a rally with John McCain, a Granite State favorite whose primary win there over Romney four years earlier helped secure him the 2008 Republican nomination.

Romney had long enjoyed a comfortable lead in New Hampshire, where he was well-known and had spent years building a political operation. His campaign kept an eye, however, on former Utah Governor Jon Huntsman, a more moderate voice in the party who had recently been ambassador to China under President Obama. Huntsman, whose detailed platform and foreign policy experience set him apart from the field, had staked his presidential hopes on a strong showing in New Hampshire. U.S. Representative Ron Paul of Texas, with his libertarian worldview, was also hoping that voters in the "Live Free or Die" state would swarm to his side. Santorum, meanwhile, frittered away his momentum. He had no money to spend on television ads and lost traction by talking about social issues, which hold less currency in

New Hampshire than in many other states. In the end, none of the rivals proved a real threat. Romney won big, and seemed on a roll. Following the votes in Iowa and New Hampshire, he built a sizable lead in polls in South Carolina, which voted next. To many, the nomination fight seemed just about over.

Gingrich had other ideas. He had marched through New Hampshire with his trademark Cheshire Cat smile, insisting he was sitting precisely where he wanted to be. While Romney campaigned in a carefully controlled manner, refusing many interviews and erecting barriers around the stage when he addressed audiences, Gingrich acceded to countless media requests and was easily approachable to voters. Shortly before the New Hampshire results came in, Gingrich could be found lounging on a couch at a Residence Inn off Interstate 93 in Concord. He insisted he would do just fine in South Carolina and would stay in the race for months. "Think of this like the Super Bowl," he said, looking as comfortable as a channel-surfer on his favorite sofa. "The first few minutes have gone by and we learned how the other team plays. Now we are in the next possession." Much to Romney's surprise, Gingrich was about to be proven right.

When Romney arrived in Florence, South Carolina, on the morning of January 17, his aides had expected an enthusiastic celebration. A large hall was rented, and a stadium rocker, "We Are the Champions," blared from the speakers. But to the shock of the campaign, the audience was barely a hundred people strong. Romney's voice echoed through the mostly empty hall as he clasped a microphone. "Gosh," he said, sounding embarrassed and proffering an excuse. "This is a work day, isn't it?" It wasn't a good omen.

Two days later, in a January 19 debate in Charleston, CNN anchor John King asked the former governor if he would release multiple years' worth of income tax returns, just as Romney's father had done in running for the presidency. Romney waffled on the answer—he had hoped to keep his returns private, as he had in previous campaigns. His hesitation only fueled growing questions about what kind of businessman he had been in the years when he amassed an estimated $250 million fortune. His years at Bain Capital, which were supposed to give him credibility as the man who could fix the nation's economy, were suddenly under assault. Rick Perry called the Bain way of buying and selling firms "vulture capitalism." The Super PAC backing Gingrich released a video called "When Mitt Romney Came to Town," suggesting that Romney had earned a windfall for himself by shutting down factories and throwing hardworking Americans out of their jobs. The video had its share of exaggerations, and some Republicans warned that the intra-party squabbling over the merits of free enterprise was damaging the GOP's fall chances. Romney himself accused his rivals of engaging in "the bitter politics of envy." But the cumulative effect of all the attacks took a toll.

The tax return controversy compounded the problem. His taxes had been in the headlines already, following his acknowledgment two days before the debate that his effective tax rate was much lower than that paid by most Americans. It turned out to be 13.9 percent, based mostly on the lower rate levied on investment income, which accounted for most of Romney's earnings. By comparison, many Americans received their income from salaries and many paid a much higher rate. Romney's tax status was not a surprise to those who had followed his career closely, but because he had never released his tax returns, this had not been widely publicized. When King asked his question, Romney looked miffed and answered, "Maybe." He let out an awkward laugh. For a candidate facing serious questions about his core convictions, "maybe" was not the word that his advisers, who were watching from a nearby room, were hoping to hear. As one put it later, "It was a pained look around the table." Romney said he might release some returns in April, when they were ready—in other words, after the primaries were likely over.

Behind the scenes, the issue divided the campaign, with top advisers arguing with each other about whether the issue would blow over. Chris Christie, who earlier had boosted Romney with an endorsement, went on national television to make clear where he stood, and thus may have forced Romney's hand. "What I would say to Governor Romney is that if you have tax returns to put out, you know, you should put them out sooner rather than later, because it's always better in my view to have complete disclosure, especially when you're the front-runner," Christie said on NBC's *Today* show. Romney ultimately released his 2010 returns and an estimate of his 2011 earnings. But he insisted he would never release earlier returns—those which would detail what he made during his time leading Bain Capital.

Romney's dithering over the tax returns was one in a series of political missteps he would make throughout the primary campaign. And they weren't just harmless mistakes, easily forgotten. A series of ham-handed comments helped cement the impression that he was an out-of-touch multimillionaire. He told job-seekers in Florida, "I'm also unemployed." He casually bet Rick Perry $10,000 during a debate exchange. He told a New Hampshire crowd that he knew what it was like to fear a pink slip. In celebrating the freedom to choose a health plan, he said, "I like being able to fire people who provide services to me." He characterized nearly $375,000 he made in speaking fees one year as "not very much." He said, in explaining his focus on the middle class, that he was "not concerned about the very poor." Later in the campaign, in economically depressed Detroit, he would say that his wife drives "a couple of Cadillacs." And while at the Daytona 500, he would remark that while he doesn't follow NASCAR that closely, "I have some great friends who are NASCAR team owners." The comments left observers and voters shaking their heads.

As Romney's foot-in-mouth moments kept piling up, so did doubts about whether he could effectively carry the Republican message into November. A few days before South Carolinians went to the polls, things got worse. Iowa Republicans had revised their caucus results and declared Santorum the winner. He had edged Romney by thirty-four votes. The reversal did not matter in terms of delegates, but it was a blow to Romney's public standing. And it came just as Gingrich was overtaking Romney in South Carolina polls. The Romney team was reeling. Gingrich was surging. It was clear the nomination race was nowhere near over. Instead of carrying two wins to South Carolina and preparing for a third, Romney was poised to emerge from the first three contests with just one victory. "Everybody got knocked off our track in South Carolina," one Romney adviser said. "Our hand was being forced, and when your hand is being forced in a campaign, that's a bad time. It was the lowest three or four days." Gingrich would win South Carolina's January 21 primary by 12 points. Romney's march to the nomination slowed to a crawl.

Ten days. That was all the time Romney had to recover. A defeat in Florida's primary, on January 31, would severely threaten his chances. Polls in the wake of the South Carolina vote put Gingrich ahead. As Romney headed to the Sunshine State, his campaign and Restore Our Future pursued a common, if not overtly coordinated, strategy: go negative. Restore Our Future stuck to what worked, airing an echo of its "Baggage" commercial. A Romney ad hit the same themes, showing a picture of Gingrich and the words "resigned in disgrace," referring to his speakership. An analysis of campaign ads showed that out of the $15 million Romney and his supporters spent on media, every commercial but a single Spanish-language spot was negative. As ABC News later put it, "less than 0.1 percent of the ads in Florida were positive Romney ads," calling this "the statistic of the campaign." Romney possessed a superior organization and benefited from an early flood of absentee ballots, but the bottom line was that going negative won campaigns. Gingrich's numbers slipped, and Romney won Florida by 14 points, easing doubts about his viability. He had shown toughness and resilience.

He would continue to need both. Over the next two months, Romney and his rivals settled in for a protracted, state-by-state competition, with momentum seesawing back and forth. Romney posted wins in states such as Nevada and Arizona, only to see Santorum, who emerged as his last credible challenger, roar back to life with victories of his own, including a three-state sweep of Missouri, Minnesota, and Colorado on February 7. At each turn, Romney faced new queries about why he couldn't put the other guys away. It was an especially pointed question in Romney's native Michigan, which many assumed he would carry easily. Romney ran neck-and-neck with Santorum until the end and prevailed narrowly.

For all of his conspicuous political struggles, Romney and his team were quietly focused on the one thing that truly mattered: collecting the 1,144 delegates required to secure the Republican nomination. As the race stretched into February and March, he built a commanding delegate lead; indeed, because of the way delegates were apportioned, Romney collected plenty even in states he lost. It was a task suitable for a guy who loved strategy and numerical certainty: forget all the noise and concentrate on hitting the target. This was especially evident one Friday in early March. Amid campaign stops at a farmers' market in Jackson, Mississippi, and a tractor plant in Birmingham, Alabama, Romney took time out to call into a poolside luncheon in the Northern Mariana Islands, a U.S. territory in the Pacific with nine GOP delegates. "If I become president, I can assure you that you will have my attention and my interest, and I will update myself with the concerns of the Northern Marianas," Romney said on the call, according to the *Saipan Tribune*. The effort paid off; Romney collected all nine.

The tussles with Gingrich and Santorum, though, laid bare once again Romney's difficulty drawing support from conservatives. Indeed, through much of the voting, Romney had lost every state in which a majority of primary voters called themselves evangelicals. He tried his best to fit in but was still often treated as ideologically suspect, still the uninvited guest at the party. In early February, Romney sought to reassure the Conservative Political Action Conference in Washington, calling himself "severely conservative." Those words weren't in his prepared speech, and for good reason. Who had ever described himself that way? In his eagerness to show solidarity, he had slipped into self-parody.

While Romney was pilloried for his inability to relate to voters, his wife, Ann, was having much better luck. She had been instrumental in her husband's decision to run again—telling him that if he really believed he could fix the economy, he really had to—and on the trail, she was as natural as her husband could be awkward, effortlessly finding a connection with voters, particularly women. Behind the scenes, though, Ann sometimes struggled in her role. She proudly told her audiences how she managed her multiple sclerosis by riding horses and through other therapies. What she didn't say was that the campaign sometimes took a heavy toll. Shortly before March 6, or Super Tuesday, she experienced telltale signs that her disease was affecting her. She started to lose control of speaking. She couldn't get some of her words out and had a hard time composing her thoughts. "Uh-oh, big trouble," she thought to herself. Despite her fatigue, she decided, "I couldn't quit. I didn't tell anybody I was tired." Ann's enthusiasm for her husband's cause never flagged, even if she sometimes had an odd way of expressing it. During an interview with a Maryland radio station, the host asked about her husband's reputation for being stiff. She responded, "Well, you know, I guess we better unzip him and let the real Mitt Romney out, because he is not!" The remark prompted a flood of mocking tweets. But the episode also drew attention, once again, to the enduring question: Who was the real Mitt Romney?

In fact, there was no new Romney who would suddenly appear before voters. What voters saw of him on the campaign trail in 2012 had only brought into sharper relief the same qualities—and liabilities—of the man who had been running for one office or another for two decades. He was comfortable in his tight family and within his Mormon faith, considerably less so in other settings. He was among the wealthiest Americans and wasn't terribly at ease talking about it. He had, in fact, made some of his fortune by building companies and creating jobs, but also by firing people and closing factories. He was not a credible ideological warrior, not a political fighter who came from a tradition of rigid partisanship or right-wing purity. His parents had taught him tolerance and moderation, after all, and he'd begun his political career as an independent. To the extent that he'd displayed an overriding philosophy it was pragmatism. The lesson of the primaries was clear to his advisers; Romney was at his best when he focused on Obama's handling of the economy, not when he defended his own positions on social issues, or his record at Bain Capital, or his wealth. Four years earlier, he'd been an unconvincing social conservative. This time, he returned successfully to an economic message, the one he knew how to sell best.

After all the zigs and zags of the nomination race, after voters took a hard look at him and his fellow competitors in countless debates, TV ads, media interviews, and campaign appearances, it was hardly a surprise that they returned to Romney, even if tepidly. He had a persuasive case to make to a general election audience, and many Republican voters clearly put a higher premium on winning in November than anything else. As one adviser explained to *Politico*, Mitt Romney is not, and will never be, the thrilling, free-spirited guy who picks you up for a date on the back of his motorcycle. "He's the guy you take home to Mom."

In all the time he'd spent in New Hampshire during the campaign, Mitt Romney had delivered countless stump speeches, shaken untold hands, and toured more Elks clubs and mom-and-pop businesses than he'd ever remember. Back when it began, he had even served his wife's homemade chicken-and-bean chili from crockpots to supporters on a New Hampshire farm. But the night of April 24 wasn't merely another campaign appearance in this friendly neighboring state. This was special. This marked an end. And a beginning.

Romney ostensibly was celebrating wins in five more primaries. The race, though, was over by this point. This would be no traditional victory speech, no run-of-the-mill thank-you to Republican voters. He was looking ahead now. Presiding over a raucous rally in a ballroom at the Radisson in downtown Manchester, he turned to his true audience—those watching on national TV at home—and implicitly accepted the nomination he'd all but won. At this moment, he wasn't just another Republican candidate; he was the guy people would begin imagining as a potential president. And he had to own the part.

His speech hit grand themes—love of country, economic malaise, the deep hole he blamed President Obama for digging. He took the chance to introduce himself to any voters just tuning in, casting his business career in a positive light and celebrating the success story of his father—how he had risen from selling paint from the trunk of his car to being the governor of Michigan. And then, echoing Ronald Reagan's famous question from 1980—"Are you better off than you were four years ago?"—Romney asked what the country had gotten in more than three years of an Obama presidency.

"Is it easier to make ends meet?" he said.

"No!" the crowd yelled back.

"Is it easier to sell your home or buy a new one?"

"No!"

"Have you saved what you needed for retirement?"

"No!"

"Are you making more at your job?"

"No!"

"Do you have a better chance to get a better job?"

"No!"

"Are you paying less at the pump?"

"No!"

Obama's responsibility for all that is debatable, of course, but for political purposes it was the president's economy to defend. Romney went on to hint at some of the policy changes he would seek as president, and he closed with a stirring peroration. "Today, the hill before us is a little steep but we've always been a nation of big steppers," he said, approximating a Texas drawl. "Many Americans have given up on this president, but they haven't ever thought of giving up. Not on themselves. Not on each other. And certainly not on America."

One word Romney didn't use was "conservative," a striking departure from a major speech just a couple months earlier, during which he uttered "conservative" or "conservatism" some two dozen times. This was a new audience, a new market to conquer. Already, Romney had felt the tension inherent in pivoting to the general election, backing an extension of low student loan rates after embracing a budget plan from House Republicans that rejected exactly that. Longtime adviser Eric Fehrnstrom implied that there would be many tactical shifts ahead, describing the general election this way to CNN in March: "Everything changes. It's almost like an Etch A Sketch—you can kind of shake it up, and we start all over again."

At the end of his nearly fifteen-minute speech, flag-waving supporters erupted in applause. Romney stepped away from the podium, which bore the words "A better America begins tonight," and into a new chapter of his political life, one that would test him like he'd never been tested before. Winning the primary race had been a laudable accomplishment; he was in the history books now. But the biggest challenge lay ahead. As Romney looked toward November, all those familiar questions about his political skills, his ability to connect, and his record would only grow more urgent.

For all the echoes of Romney's past political adventures, though, the general election campaign would begin with fresh hurdles and opportunity. The extended primary contest forced him further to the right than he probably wanted to go, particularly on pressing issues for women and immigrants. He emerged from the rough-and-tumble GOP primary race with many voters holding negative impressions of him. And yet voters' doubts about President Obama's stewardship of the economy remained powerful. The enthusiasm he rode to office in 2008 had ebbed. Romney, if he could persuade people that he could succeed where Obama hadn't, promised to be a strong general election candidate. The choice would be stark. "The contrast could not be bigger," Romney had said in Philadelphia. "We have a very different vision for America."

ACKNOWLEDGMENTS

The names of two coauthors appear on the front of this book, but this is, in many ways, the work of numerous reporters and editors over the years who have helped tell the story of Mitt Romney in the pages of *The Boston Globe*. We are proud to count ourselves among them.

This book was edited by one of the best in the business, Mark S. Morrow, a *Globe* deputy managing editor. Mark's guidance, insight, and grace are imprinted on every page. We are extraordinarily grateful for his leadership.

Support for this project came from every level of the newspaper, starting with Martin Baron, the editor of the *Globe*, who saw the great value—and service—the paper could provide by writing an exhaustive and fair-minded book rooted in the *Globe* staff's unparalleled knowledge of Romney's life and career. We are thankful for his vision and devotion to the craft, and to Christopher M. Mayer, the *Globe*'s publisher, for his belief in, and support for, serious, in-depth journalism in all its forms. Other senior *Globe* editors, including Managing Editor Caleb Solomon, Deputy Managing Editor Christine S. Chinlund, and Metro Editor Jennifer Peter, have likewise stood behind us.

The Real Romney would simply not have been possible without a critical group of fellow *Globe* writers and editors. Beth Healy, whose knowledge of Boston's private equity world is without equal among reporters, was an indispensable partner, as was Brian C. Mooney, who devoted many weeks to capturing Romney's Massachusetts political record. The skillful hands of Neil Swidey and Peter S. Canellos, who has long been an advocate for this project, built the foundation on which much of the book rests. Michael Paulson was a constant resource for our understanding of Romney's faith and his missionary work in France. Bob Hohler's deep dive into the 2002 Winter Olympics was instrumental. Stephanie Ebbert's work on Romney as a youth, family man, and politician was invaluable. And Robert Gavin and Sacha Pfeiffer's reporting on Romney's business career helped immensely.

At HarperCollins, our editor, Tim Duggan, understood the potential in the project from the start and, with skill and enthusiasm, guided it to swift completion. His assistant editor, Emily Cunningham, provided invaluable help and support throughout. Tina Andreadis and Beth Harper of HarperCollins publicity have championed the book with spirit, and John Jusino, Shannon Ceci, and Lynn Anderson were key partners in bringing it to light.

Janice Page, the *Globe*'s book development guru, enthusiastically guided this project from the beginning, and the *Globe*'s literary agents, Lane Zachary and Todd Shuster of Zachary Shuster Harmsworth, expertly brought the book to reality. We are also indebted to our fact checkers, Stephanie Vallejo and Matt Mahoney, whose sharp eyes never ceased to amaze us. The *Globe*'s chief librarian, Lisa Tuite, and her staff, including Jeremiah Manion, Charlie Smiley, and Marleen Lee, fielded countless research requests with vigor. Others who contributed significantly to the project's success include Ann Silvio, Scott LaPierre, Thea Breite, and Julie Chazyn. Glen Johnson, the political editor of Boston.com, generously shared his wealth of Romney knowledge. Donald MacGillis, the *Globe*'s national political editor, and Michael J. Bailey, the deputy national political editor, were supportive at every stage. We have drawn on stories written by scores of *Globe* writers over the years, but a few names stand out: Mitchell Zuckoff, Ben Bradlee, Jr., Charles Stein, Frank Phillips, and Scot Lehigh. Carolyn Ryan and David Dahl also provided strong editorial leadership as Romney debuted on the national stage.

The *Globe*'s publicist, Mary Zanor, as always, provided energetic support. Assistance also was provided by Malgosia Myc, assistant reference archivist at Bentley Historical Library at the University of Michigan, and Anna Schuessler of the *Stanford Daily*. Archivists at the University of Utah and Brigham Young University also provided valuable help, as did Richard B. Anderson, Grant Bennett, and John Wright.

We conducted more than one hundred interviews in 2011 for this book, many with people who had never previously talked publicly. From Stanford classmates and Bain Capital partners to Mormon leaders and 2008 campaign aides, we are deeply grateful to everyone who made time for us and saw the value in a complete, independent biography, including many sources who requested anonymity and whose names do not appear in this volume. Romney's cousin Mike Romney graciously took Michael Kranish on a tour of the Mormon colony in Mexico established by the Romney family. Essdras M. Suarez, a *Globe* photographer, was a delightful companion in retracing the journey of Romney's ancestors in the American Southwest and Mexico.

From Michael Kranish:

I am grateful to the *Globe*'s Washington bureau chief, Christopher Rowland, for allowing me to step aside from many daily duties and work on this project. He and my Washington colleagues were a source of much wisdom. They include Matt Viser, who followed Romney on the campaign trail; Donovan Slack, who monitored Romney's campaign money; and Tracy Jan, Bryan Bender, and former *Globe* reporter Sasha Issenberg. I am also indebted to my family for allowing me to keep the "Author at Work" sign on the door and understanding the press of deadlines and passion for this type of project, as they have since I worked a similar project for the *Globe* about John Kerry and then for an independent book about Thomas Jefferson. My wife, Sylvia, and daughters, Jessica and Laura, are always on my front page. My mother, Allye, who owned a Scandinavian design store one block from the White House, and my late father, Arthur, the first Washington journalist in the family, fostered my interest in presidential politics and history from the earliest days. To my fellow reporters and editors at the *Globe*, you are an inspiration, daily.

From Scott Helman:

I would like to thank my editors at the *Globe* magazine for freeing me, without complaint, to work on this book and seeing its value, starting with Anne V. Nelson and Doug Most, but also Francis Storrs, Veronica Chao, and Melissa Schorr. All have been enthusiastically supportive. My wife, Jessica, was a voice of love and support from the beginning, abiding, with my sons, Jonas and Eli, many nights and weekend days with me holed up in the attic. I'm forever indebted to my parents, Kay and Larry, whose midwestern values—and memories of Michigan's George Romney, albeit from the superior state to the south—informed this book in more ways than they know. And to my *Globe* family—a heartfelt thank-you for all you've taught me.

NOTES

Interviews for this book were conducted by the authors and a team of *Boston Globe* reporters, including Neil Swidey, Beth Healy, Brian C. Mooney, Michael Paulson, Bob Hohler, Ann Silvio, Stephanie Ebbert, Rob Gavin, and Sacha Pfeiffer. Much of the 2007 interview material first appeared in a seven-part *Globe* series, "The Making of Mitt Romney."

Prologue

6   "There is no leader": A. F. Mahan, "Busy George Romney Can't Resist Challenge," Associated Press, February 8, 1962.

7   "Politics is like washing diapers": Niraj Warikoo, "Wife of Former Governor Had Opinions of Her Own," *Detroit Free Press*, July 8, 1998.

Chapter 1: Praying for a Miracle

11   Now, in 1946: Neil Swidey, "Lessons of the Father," *The Boston Globe Magazine*, August 13, 2006.

11   She needed a major operation: George Romney, letter to his family, March 13, 1947. Bentley Historical Library, University of Michigan, call number 852178 Aa 2.

12   "I remember my father's face": Neil Swidey, "Lessons of the Father," *The Boston Globe Magazine*, August 13, 2006.

12   "long and heated arguments": Clark Raymond Mollenhoff, *George Romney, Mormon in Politics* (New York: Meredith Press, 1968), 44.

12   "never marry him": Ibid., 45.

13   "Miss Lenore LaFount": "Inspiring Lives: George Romney," http://byutv .org/watch/31c61253-ffe4-42ad-808d-e3b3486c1c7e#!page=2&season= All-Seasons.

13   "the biggest sale": George W. Romney, *The Concerns of a Citizen* (New York: Putnam, 1968), 259.

13   "never had any regrets": "Fate Changed Mrs. Romney's Plan to Become a Film Star," *The Holland (Michigan) Evening Sentinel*, February 24, 1962, p. 4.

13   "This was the legend": Interview of Jane Romney by Neil Swidey, 2006.

13   He left his lobbyist job: Mollenhoff, *George Romney*, 44–48.

14   "We consider it a blessing": George Romney, letter to his family, March 13, 1947, Bentley Historical Library, University of Michigan, call number 852178 Aa 2.

15   In 1953: "Republicans: The Citizen's Candidate," *Time*, November 16, 1962, www.time.com/time/magazine/article/0,9171,829375,00.html.

15   "We're going to make": Mitt Romney, *Turnaround: Crisis, Leadership, and the Olympic Games* (Washington, D.C.: Regnery Publishing, 2007), 11.

15   One of the keys: Mollenhoff, *George Romney*, 118–119.

16   "If Ramblers are such great cars": Interview with Scott Romney, 2007.

16   thirty miles per gallon: Mollenhoff, *George Romney*, 103; "Autos: Gamble on the Rambler," *Time*, December 19, 1955; U.S. Department of Transportation, "Summary of Fuel Economy Performance," April 28, 2011, www.nhtsa.gov/staticfiles/rulemaking/pdf/cafe/2011_Summary_Report.pdf.

16   30-cents-per-gallon gas: The price of gas was about 30 cents a gallon in 1955. The inflation calculation comes from www.westegg.com/inflation/.

17   "It was more like": Romney, *Turnaround*, 11–12.

17   When Mitt was fourteen years old: Arthur O'Shea and Bert Emanuele, "Mormons: Each a Teacher, Each Taught," *Detroit Free Press*, April 16, 1961.

17   "Dream and dream big": Romney, *The Concerns of a Citizen*, 263–267.

17–18   "I grew up idolizing him": Romney, *Turnaround*, 11.

18   "Dad was more settled": Neil Swidey and Michael Paulson, "Touched by Tragedy, a Leader Emerges from a Life of Privilege," *The Boston Globe*, June 24, 2007.

18   "a real breakthrough": "Back at the Mansion," *Time*, January 11, 1963.

18   "the Bickersons": Interview with Tagg Romney, 2007.

18   "a loner": Brock Brower, "Puzzling Front Runner," *Life*, May 5, 1967.

19   Mitt wanted no repeat: Jack Thomas, "Ann Romney's Sweetheart Deal," *The Boston Globe*, October 20, 1994.

19   "Mitt's more like": Neil Swidey, "Lessons of the Father," *The Boston Globe Magazine*, August 13, 2006.

19   When George held: Swidey and Paulson, "Touched by Tragedy, a Leader Emerges from a Life of Privilege."

20   "one of the most enchanted": www.cranbrookart.edu/Pages/History.html.

20   "He was tall": Swidey and Paulson, "Touched by Tragedy, a Leader Emerges from a Life of Privilege."

20   "Mitt is doing well": Mitt Romney 1961 report card, Cranbrook School, www.boston.com/news/daily/24/romney_reportcard.pdf.

21   "He came up to the car": Interview with Graham McDonald, 2011.

21   "was genuinely distressed": Interview with Gregg Dearth, 2011.

22   "It was definitely looked upon": Interview with Sidney Barthwell, 2011. Barthwell, the lone black in Romney's class, went on to become a magistrate judge in Michigan as well as a successful runner.

22   had cramped up: Interview with Graham McDonald.

22   "It is something": Interview with Gregg Dearth.

22   running through the New Hampshire woods: "Ad Watch: Romney's 'Lead-
        ership' and Lakeside Jog," *The Washington Post*, August 8, 2007, http://blog
        .washingtonpost.com/channel–08/2007/08/ad_watch_romneys_leader
        ship_an.html.

22   George agreed to work with: Mollenhoff, *George Romney*, 163–167, 173; re-
        cession information at http://recession.org/history/early-1960s-recession.

23   "You know, I think": "Cover Story," New England Cable News, October 31,
        2002.

23   George decided: Mollenhoff, *George Romney*, 167.

23   "I would introduce myself": Romney, *Turnaround*, 12.

23   "You can't find": Tom Wicker, "Kennedy Assails G.O.P. over Trade; in
        Detroit Speech, He Says Republicans Impeded Bill," *The New York Times*,
        October 7, 1962.

23   But even President Kennedy: Romney would have many interactions later
        in life with the Kennedy family. He would lose a Senate race to President
        Kennedy's brother Edward but then, as governor, work with him on health
        care legislation.

24   a pamphlet: "Who Is the Real George Romney?" (Detroit: Michigan AFL-
        CIO, 1966).

24   "too much as a business party": "The Citizen's Candidate," *Time*, November
        16, 1962.

24   "Romney Son": Swidey and Paulson, "Touched by Tragedy, a Leader
        Emerges from a Life of Privilege."

24   "It was lots more fun": "Mitt Romney Keeps Vigil as Clock Is Running
        Out; Hears Dad Argue, Win Point with Solons," *The Benton Harbor (Michi-
        gan) News-Palladium*, December 19, 1963.

25   "They would hug": Interview with Dick Milliman, 2007.

25   "like living in a drama": Interview of Jane Romney by Neil Swidey, 2006.

25   "The rights of some": Mollenhoff, *George Romney*, 235.

25   "Dogmatic ideological parties": Ibid., 230.

26   "A liberal in his treatment": "BYUtv—Inspiring Lives: George Romney,"
        http://byutv.org/watch/31c61253-ffe4-42ad-808d-e3b3486c1c7e#!page=
        2&season=All-Seasons.

26   When Mitt was asked: Mitt Romney, appearance on *Meet the Press*, Decem-
        ber 16, 2007. During the same appearance, Romney incorrectly said that
        his father had marched with Martin Luther King, Jr.

27   "It was primarily": Interview with Sidney Barthwell, Jr.

27   During one of his father's: Interview with Scott Romney.

27   since his early teens: Swidey and Paulson, "Touched by Tragedy, a Leader
        Emerges from a Life of Privilege."

28   Mitt had first met Ann: "Mitt Talks About Ann," http://web.archive.org/
        web/20080215013646/http://www.mittromney.com/Learn-About-Mitt/
        Photo-Album/The-Romney-Family/Mittxs_Tribute_to_Ann.

28   "I caught his eye": Thomas, "Ann Romney's Sweetheart Deal."

28   "I fell in love": Mitt Romney campaign ad, 2002.

28   Mitt learned to keep up: Thomas, "Ann Romney's Sweetheart Deal."

28   "no matter where he was": Ibid.

29   Her father: "Mitt Romney Marries Ann Davies," *The New York Times*,
        March 22, 1969.

29   "creative genius": Thomas, "Ann Romney's Sweetheart Deal."

29   "Dad," said Ann's older brother: Interview with Roderick Davies, 2007.

30   "What," she asked: Interview with Mitt Romney, 2007.

Chapter 2: Following the Call

31   "I believe": Mitt Romney, speech on faith, Houston, Texas, December 6, 2007.

33   as many as five children: There are differing accounts on how many children
        came with the Romneys. The report of five children comes from the Mor-
        mon church's ancestry web site, www.familysearch.org, which names five
        children who were born in England and died in the United States, although
        it is possible that some arrived separately from their parents. The five were
        George, Elizabeth, Sarah, Joseph, and Ellen. See Miles Romney family tree
        at https://www.familysearch.org/search/treeDetails/show?uri=https%3A%2
        F%2Ffamilysearch.org%2Fpal%3A%2FMM9.2.1%2FMB23-ZR3.

33   "master mechanic": Thomas Cottam Romney, *Life Story of Miles P. Romney*
        (Independence, Missouri: Zion, 1948), 7.

33   "a pillar of light": "Joseph Smith's First Vision," www.lds.org/library/
        display/0,4945,104-1-3-4,00.html.

33   "divine pronouncement": "Truth Restored," www.lds.org/manual/truth
        -restored/chapter-13-years-of-endurance?lang=eng.

34   several dozen wives: Fawn McKay Brodie, *No Man Knows My History: The Life
        of Joseph Smith, the Mormon Prophet* (New York: Vintage, 1995), 457–488.

34   "despotism": Ibid., 381.

34   "We are earnestly seeking": *Nauvoo (Illinois) Expositor,* June 7, 1844, www
        .solomonspalding.com/docs/exposit1.htm.

34   The Nauvoo City Council: The order to destroy the Nauvoo *Expositor* is at
        http://law2.umkc.edu/faculty/projects/ftrials/carthage/expositororder.html.

34   "point of a bayonet": "William Farrington Cahoon, 1813–1897," www.boap
        .org/LDS/Early-Saints/WFCahoon.html. Just as the Romneys fled, Miles's
        sister, five-year-old Ellen, died of an illness on February 24, 1846; see www
        .familysearch.org/search/treeDetails/show?uri=https%3A%2F%2Ffamilyse
        arch.org%2Fpal%3A%2FMM9.2.1%2FMB23-ZR3.

34   twenty thousand: "Nauvoo, Illinois: 1839–1846," http://lds.org/gospellibrary
        /pioneer/02_Nauvoo.html.

35   Instead, the Romneys fled: Romney, *Life Story of Miles P. Romney*, 18–19. Ellen Romney died May 12, 1846, according to http://awt.ancestrylibrary.com/cgi-bin/igm.cgi?op=GET&db=flakey&id=I555148915&ti=5542.

35   And the elder Romney: Romney, *Life Story of Miles P. Romney*, 8.

35   "twin relics of barbarism": www.ushistory.org/gop/convention_1856republicanplatform.htm.

35   desertion: One of those who deserted was a Prussian émigré, Carl Heinrich Wilcken. After he converted to Mormonism, his daughter married the son of a Mormon leader. That couple, Dora and Helaman Pratt, would be one set of Mitt Romney's great-grandparents.

36   "They were trying": Mitt Romney interview, *60 Minutes*, CBS-TV, May 13, 2007.

36   "Miles had grown into": Romney, *Life Story of Miles P. Romney*, 23–25. The reference to playing Hamlet is from an unpublished family biography of Miles's son Gaskell.

37   "attractive Scotch lass": Romney, *Life Story of Miles P. Romney*, 25.

37   The couple had a month together: Ibid., 27.

37   "all day from sunup": Hannah Hood Hill Romney, "Autobiography of Hannah Hood Hill Romney," *Our Pioneer Heritage*, vol. 5 (Salt Lake City: Daughters of Utah Pioneers, 1962), 264–265.

37   "You go back": Michael S. Durham, *Desert Between the Mountains: Mormons, Miners, Padres, Mountain Men, and the Opening of the Great Basin, 1772–1869* (Norman, Oklahoma: University of Oklahoma Press, 1999), 164.

37   "Many, now, wonder": Miles P. Romney, "Persecution," *Millennial Star*, October 1864, 629.

38   "never milked a cow": Romney, *Life Story of Miles P. Romney*, 294–297.

38   "We were happy": Romney, "Autobiography of Hannah Hood Hill Romney," 266.

38   "Brother Miles": Ibid., 266.

38   "Nothing short of a firm belief": Romney, *Life Story of Miles P. Romney*, 61.

38   "I felt that was more": Romney, "Autobiography of Hannah Hood Hill Romney," 266.

39   "There will yet be built": Brigham Young, quoted at www.stgeorgetemplevisitorscenter.info/by/byowner-baker2.html.

39   "in a little shanty": Romney, "Autobiography of Hannah Hood Hill Romney," 266–267.

39   one of the era's most lavish residences: Visit of the author to Brigham Young house in St. George, Utah. Miles A. Romney built the main part of the house, and Miles P. Romney built an addition. The restored home is visited today by Mormons from around the world, who are told the story of Miles P. Romney's role in building the house.

39   "was very jealous": Romney, "Autobiography of Hannah Hood Hill Romney," 267.

40   "trials met with": Romney, *Life Story of Miles P. Romney*, 74.

40   "the Anti-polygamy bill": Ibid., 71–73. Romney and the others said the legislation violated the Declaration of Independence's guarantee that all men had the rights of "life, liberty, and the pursuit of happiness" and the Bill of Rights' guarantee of freedom of religion.

40   "prettiest girl in St. George": Jennifer Mouton Hansen, ed., *Letters of Catharine Cottam Romney, Plural Wife* (Urbana and Chicago: University of Illinois Press, 1992), 1, 10, 19.

40   "weakness for wine": Miles's descendants said that he later gave up alcohol.

40   "one of the noblest": Romney, *Life Story of Miles P. Romney*, 13–14.

41   "I said I would take it": Romney, "Autobiography of Hannah Hood Hill Romney," 269.

42   "Here you can see": Hansen, *Letters of Catharine Cottam Romney*, 26.

42   "With the noise and confusion": Romney, "Autobiography of Hannah Hood Hill Romney," 269.

42   The Romneys eventually found: Meryl Romney Ward, "Biography of Gaskell Romney" (unpublished family document, 1984), 5.

43   "Hang a few": David King Udall, *Arizona Pioneer Mormon: His Story and His Family, 1851–1938* (Tucson: Arizona Silhouettes, 1959), 116. The quote was included in the May 30, 1884, edition of the *Apache Chief*. Udall's progeny would include two famous grandchildren: Stewart Udall, who served as secretary of the interior in the Kennedy and Johnson administrations; and Morris Udall, an Arizona congressman who ran unsuccessfully for president in 1976.

43   "a mass of putrid pus": Carol Sletten and Eric Kramer, *Story of the American West* (Pinetop, Ariz.: Wolf Water Press, 2010), Kindle edition, location no. 3709.

43   Still, although Miles had promised: Hansen, *Letters of Catharine Cottam Romney*, 13.

43   "hate upon Romney": Udall, *Arizona Pioneer Mormon*, 116.

43   "American Siberia": Edward William Tullidge, *History of Salt Lake City* (Salt Lake City: Star Printing Company, 1886), 149.

44   "due to lack of evidence": Romney, *Life Story of Miles P. Romney*, 160–166.

44   "I told him Mr. Romney": Romney, *Autobiography of Hannah Hood Hill Romney*, 271–272.

44   "disguising himself so completely": Hansen, *Letters of Catharine Cottam Romney*, 111.

44   "Eventually Miles was called upon": Mitt Romney, *Turnaround: Crisis, Leadership, and the Olympic Games* (Washington, D.C.: Regnery Publishing, 2007), 8.

44   colony of Juárez: Visit by coauthor Kranish to Colonia Juárez.

45   "I sometimes think": Hansen, *Letters of Catharine Cottam Romney*, 113.

45   Hannah, meanwhile, was still trying: Romney, "Autobiography of Hannah Hood Hill Romney," 271–275.

45   "When it rained": Ibid., 276–277.

45   "21 of us all together": Romney, *Letters of Catharine Cottam Romney*, 118–119.

45   Gaskell: Meryl Romney Ward, "Biography of Gaskell Romney" (unpublished family document, 1984), 9–10. The Cliff Ranch is no longer in use by the Romneys and the original buildings no longer exist, according to Mitt Romney's cousin Mike Romney, who lives in a nearby town.

46   "I now publicly declare": "Official Declaration," http://lds.org/scriptures/dc-testament/od/1?lang=eng.

46   "There was a great closeness": Ward, "Biography of Gaskell Romney," 10.

46   "Yes, and go to hell": Ibid.

46   In 1895, after completing: Romney, *Life Story of Miles P. Romney*, 317–318. After Gaskell's wife Anna died, he married her sister, Amy.

46   "accumulated a great deal": Romney, "Autobiography of Hannah Hill Hood Romney," 281.

46   The children went: Tom Mahoney, *The Story of George Romney: Builder, Salesman, Crusader*, 1st ed. (New York: Harper, 1960), 53–54.

47   One day in early March 1904: Romney, *Life Story of Miles P. Romney*, 289.

47   "breathed his last": Romney, "Autobiography of Hannah Hill Hood Romney," 281.

47   "very prosperous": Ward, "Biography of Gaskell Romney."

47   His wealth enabled him: Clark Raymond Mollenhoff, *George Romney: Mormon in Politics* (New York: Meredith Press, 1968), 26.

47   George Wilcken Romney: His middle name was in memory of the Prussian deserter Carl Heinrich Wilcken, who had left U.S. forces and joined the Mormon side and was the great-grandfather of George.

47   At one point: Mollenhoff, *George Romney*, 24–26.

47   "would die": "Refugees Camp in Lumberyard; Mormons Are Still Fleeing to City," *El Paso Herald*, July 30, 1912.

48   Vastly outnumbered: Mollenhoff, *George Romney*, 26.

48   "the first displaced persons": Mahoney, *The Story of George Romney*, 60.

48   "I was kicked out": George Romney, *The Concerns of a Citizen* (New York: G. P. Putnam and Sons, 1968), 263–267.

48   "until it is safe": "Salazar Reported at Kilometer 52; Gives Food and Water to Mormons," *El Paso Herald*, October 25, 1912.

48   But Gaskell's family: Hannah made a trip back to the Mexican colonies and died there in 1928; see Romney, "Autobiography of Hannah Hill Hood Romney," 284. Gaskell and his family, including George, returned to visit the Mexican colony in 1941, according to an unpublished family manuscript titled "Mormon Towns and Trails."

49 "Even though father": Ward, "Biography of Gaskell Romney," 16–19.

49 Gaskell requested $26,753: *Gaskell Romney v. the United States of Mexico*, Salt Lake City, Utah, March 15, 1938.

50 "Miles Park was a pioneer": Interview with Mike Romney, 2007.

Chapter 3: Outside the Fray

52 Hoover Tower: "Books: The Hoover Library," *Time*, June 30, 1941.

52 "The campus was quite isolated": Interview with Wayne Brazil, 2011.

53 Mark Marquess: Interview with Mark Marquess, 2011. Marquess has served as Stanford's baseball coach for more than three decades. As of late 2011, he had not talked with Romney since their college days.

53–54 Ditching his coat and tie: Neil Swidey and Michael Paulson, "Touched by Tragedy, a Leader Emerges from a Life of Privilege," *The Boston Globe*, June 24, 2007.

54 Harris was protesting a war: Interview with David Harris, 2011.

54 "It sounds silly now": Interview with Mike Roake, 2007.

54 "I don't think": Interview with Mark Marquess.

54 "He was conservative": Interview with James Baxter, 2011.

54 "He didn't put on any airs": Interview with Mark Marquess.

55 Also on Mitt's floor: Robert Mardian, Jr.'s, father was Robert C. Mardian, who later served as assistant attorney general in the Nixon administration and worked in President Richard Nixon's reelection campaign. He was convicted during the Watergate scandal for conspiracy to obstruct justice, but the conviction was overturned on appeal. See Dennis McClellan, "Robert Mardian, Watergate Scandal Figure," *Los Angeles Times*, July 21, 2006.

55 "I would say publicly": Interview with Robert Mardian, Jr., 2011.

56 Ken Kesey: Christopher Lehmann-Haupt, "Ken Kesey, Author of 'Cuckoo's Nest,' Who Defined the Psychedelic Era, Dies at 66," *The New York Times*, November 11, 2001. Kesey's "acid test" exploits were famously chronicled in Tom Wolfe's *The Electric Kool-Aid Acid Test*, first published in 1968.

56 "There was this cultural current": Interview with David Harris.

56 PEACE: http://a3m2009.org/archive/photos/1965–1966_photos/stanford_com mittee/scpv/index.html

57 Joan Baez: David Harris, *Dreams Die Hard* (San Francisco: Mercury House, 1993), 123–124.

57 "the same that brought": Andrew L. Johns, "Achilles' Heel: The Vietnam War and George Romney's Bid for the Presidency, 1967 to 1968," *Michigan Historical Review* 26, no. 1 (April 1, 2000): 10.

57 Returning from Vietnam: George Romney's gubernatorial papers include a schedule that shows he stopped at the San Francisco Hilton on his way back from Vietnam. One of Mitt's classmates said that George stopped at Stanford on his way from Vietnam to Detroit. George may have made one

other visit because he is remembered to have been on campus with his wife and Ann Davies. The dinner described here appears to have been during the San Francisco stopover. Romney stayed at the Hilton on the evening of November 12, 1965. George Romney Papers, Bentley Historical Library, University of Michigan.

57   "He spoke about": Interview with Peter Davenport, 2011.

57   "He didn't want his parents": Jack Thomas, "Ann Romney's Sweetheart Deal," *The Boston Globe*, October 20, 1994.

58   Mitt's older brother, Scott: Interview with Scott Romney, 2007.

58   "cut back": Interview with Alan Abbott, 2011.

58   "It was especially": Interview with Mike Roake.

58   A campus rally was headlined: Harris, *Dreams Die Hard*, 125.

59   "You don't stand a chance": Ibid., 135.

59   Harris . . . won the election: Ibid., 133–135.

60   850 students: "Draft Tests at Stanford Under Way," *Oakland Tribune*, May 21, 1966.

60   Harris not among them: Harris had marched to the sit-in with other protesters but said he did not join them in staying overnight because he had concerns about the legality of it.

60   SPEAK OUT: Jay Thorwaldson, "Governor's Son Pickets the Pickets," *Palo Alto Times*, May 20, 1966, George Romney collection, box 3-V, Bentley Historical Library, University of Michigan. The circumstances of that day were described in an interview with Romney's classmate William Black in 2011.

60   "Down with mob rule": Thorwaldson, "Governor's Son Pickets the Pickets."

60   "Come out": Interview with Mitt Romney, 2007.

60   "We had animated": Ibid.

60   "a zero": Interview with David Harris. Harris said he knew George Romney's son was on campus but didn't remember interacting with him.

61   "There were some wards": Interview with Barry Mayo, 2007.

62   Some non-Mormons in Utah: Wallace Turner, "Suit over Draft in Utah Attacks Mormon Missionary Deferments," *The New York Times*, February 7, 1970.

62   "the substantial number": Interview with Richard Leedy, 2007.

62   "I was supportive": Interview with Mitt Romney.

62   "I was not planning": Joe Battenfeld, "GOP Senate Hopeful Romney Got Draft Deferment for Vietnam," *Boston Herald*, May 2, 1994.

63   the number 300: Mitt Romney's Selective Service record, No. 20-67-47-252.

63   "based on thin tissue": David Kirkpatrick, "Romney's Life Took a Turn in France," *The New York Times*, November 17, 2007.

63   Though Romney's parents: Interview with Dane McBride, 2011.

63   If he didn't, she told him: Ben Bradlee, Jr., "Romney Seeks New Chapter in Success," *The Boston Globe*, August 7, 1994.

64   In the 1950s: "Country Information: France," January 29, 2010, www.ldsch urchnews.com/articles/58571/Country-information-France.html.

64   LeHavre: "LeHavre, the City Rebuilt by Auguste Perret," http://whc.unesco .org/en/list/1181.

65   Romney shared a one-bedroom apartment: Interview with Donald K. Miller, then Romney's senior companion, 2007.

66   "Bang!" Interview with Mitt Romney.

66   Romney would take a long, hot bath: Interview with Donald K. Miller.

66   "There were about 20 guys": Swidey and Paulson, "Touched by Tragedy, a Leader Emerges from a Life of Privilege."

67   "had a personality": Interview with Marie-Blanche Caussé, 2007.

67   *Think and Grow Rich*: Napoleon Hill, *Think and Grow Rich*, ebook, www .archive.org/details/Think_and_Grow_Rich, 1, 160–175.

68   "We were red-blooded": Interview with Dane McBride.

68   "great message": Interviews with Dane McBride, 2007 and 2011.

68   "singing, basketball exhibitions": Mitt Romney to his parents, 16 July 1968, Box 3—Early Series, George Romney Papers, Bentley Historical Library, University of Michigan.

69   Romney later said: Bradlee, "Romney Seeks New Chapter in Success."

69   "As you can imagine": Lawrence Wright, "Lives of the Saints," *The New Yorker*, January 21, 2002.

69   "On a mission": Kirkpatrick, "Romney's Life Took a Turn in France."

69   "more of a teaching experience": Interview with Mitt Romney.

70   "History was changing": Interview with Paul Richardson, 2011.

70   "There were plenty of people": Interviews with David Harris, 2007 and 2011.

71   "inherit the kingdom of God": "Conversion," http://lds.org/study/topics/con version?lang=eng&query=conversion.

71   So Jim stood outside: Swidey and Paulson, "Touched by Tragedy, a Leader Emerges from a Life of Privilege."

71   "one of the greatest missionaries": Brock Brower, "Puzzling Front Runner," *Life*, May 5, 1967.

72   "a social outcast": Swidey and Paulson, "Touched by Tragedy, a Leader Emerges from a Life of Privilege."

72   "I was thrilled to stand in": George Romney to Mitt Romney, February 16, 1967, and March 6, 1967, George Romney papers, Bentley Historical Library, University of Michigan.

73   "as far as I am concerned": Clark Raymond Mollenhoff, *George Romney, Mormon in Politics* (New York: Meredith Press, 1968), 256.

73   "It is unthinkable": Ibid., 337.

73   "been unable to bring": B. J. Widick, *Detroit: City of Race and Class Violence* (Detroit: Wayne State University Press, 1989), 170–172.

73   Romney took the riots to heart: Mollenhoff, *George Romney*, 287–288, 292.

74   the greatest brainwashing: Neil Swidey, "Lessons of the Father," *The Boston Globe Magazine*, August 13, 2006.

74   She was listening: Swidey, "Lessons of the Father."

74   Chuck Harmon, Romney's press secretary: Swidey, "Lessons of the Father."

75   "blurt and retreat habits": Mollenhoff, *George Romney*, 301.

75   "I would be VERY happy": Mitt Romney to his parents, 16 July 1968, Box 3, Early Series, George Romney Papers, Bentley Historical Library, University of Michigan.

75   The news got worse: Mollenhoff, *George Romney*, 253; Associated Press, "Nixon Favored by Poll," November 20, 1967.

75   "Mitt was very passionate": Interviews with Byron Hansen, 2007 and 2011.

75   "When my dad said": Interview with Mitt Romney.

76   Romney had initially believed: Battenfeld, "GOP Senate Hopeful Romney Got Draft Deferment for Vietnam."

76   "I think we were brainwashed": Newsweek Feature Services, "Cabinet Kids Against the War," *The Boston Globe*, May 31, 1970.

76   "The brainwash thing": Swidey, "Lessons of the Father"; interview with Jane Romney. Neil Swidey, a reporter for *The Boston Globe*, showed the footage to Mitt Romney.

76   Shortly afterward, George headed: Clayton Knowles, "Romney, Back in U.S., Calls His Trip 'Beneficial,'" *The New York Times*, January 4, 1968.

76   "It is clear to me": "Romney Quits Race for President," *The Boston Globe*, February 29, 1968.

77   "We were the only Americans": Interviews with Dane McBride.

77   demands: "The French Worker Wants to Join the Affluent Society, Not to Wreck It," *The New York Times*, June 16, 1968.

77   Communication was difficult: Kirkpatrick, "Romney's Life Took a Turn in France." The article quoted a fellow missionary, Byron Hansen, as saying the lack of respect for authority in France "affected Mitt."

77–78   "The feeling that we had": Interview with Dane McBride.

78   "It's a horrid climax": Leola Anderson journal, May 28 to June 5, 1968, Anderson family document.

Chapter 4: A Brush with Tragedy

79   "We'll be right there": Interviews with Richard Anderson, 2007 and 2011, and http://freepages.history.rootsweb.ancestry.com/~timbaloo/SuchALife/pages/Ch3/8Miss/8Miss05.htm.

79 On their way south: http://freepages.history.rootsweb.ancestry.com/~timbaloo/SuchALife/pages/Ch3/8Miss/8Miss24b1.htm.

80 climbed into the front seat: Interviews with occupants of the car, 2007.

80 A thirty-four-year-old man: Article in *Sud-Ouest*, a French regional newspaper, June 17, 1968.

80 Romney stopped: Interview with Mitt Romney, 2007.

80 "We were all talking": Ibid.

80 A Catholic priest: *Sud-Ouest* article, June 17, 1968; interviews with people in the car.

80 The collision collapsed: Interview with Bruce Robinson, 2007.

80 "It happened so quickly": Interview with Mitt Romney.

80 They were driving slowly: Interview with Suzanne Farel, 2007.

80 *"Il est mort"*: Interview with Bruce Robinson.

80 Rescuers had to pry him: Interview with Mitt Romney.

81 "I remember the call": Interview with Jim Davies, 2007.

81 Crushed by the impact: Richard Anderson, writing in Leola Anderson's journal, September 19, 1968.

81 "a blur of pain": Ibid.

81 "Tragedy struck": Interview with Byron Hansen, 2007, and Hansen's journal.

81 "When we initially arrived": Interview with Byron Hansen.

81 The doctors had declined: Interviews with Joel McKinnon, 2007 and 2011; and Richard Anderson, writing in Leola Anderson's journal.

81 "I was making rounds": Interview with Bruce Robinson.

81 The family was also worried: Interviews with André Salarnier, 2007; and Bruce Robinson, 2007 and 2011.

81 "Mitt was just coming out": Interview with Bruce Robinson.

82 "He probably came": Ibid.

82 The driver of the Mercedes: *Sud-Ouest* article, June 17, 1968; and interviews with French Mormons who responded to the accident. A priest at the parish in Sireuil confirmed in 2007 that the church's former pastor, now deceased, was Albert Marie.

82 Romney said the truck driver: Interview with Mitt Romney.

82 "Duane Anderson refused": Interview with André Salarnier.

82 David Wood said he remembered: Interview with David Wood, 2007.

82 Romney has no such recollection: Interview with Mitt Romney.

82 "We were conservative": Interview with David Wood.

82 "Oh, yeah, I was": Interview with Mitt Romney.

82 "Mitt was not": Interview with Richard Anderson.

83 Duane Anderson: Interviews with Bruce Robinson; and Richard Anderson, writing in Leola Anderson's journal.

83   For Mitt, the fatal accident: Interview with Mitt Romney.

83   advice on his love life: Interview with Richard Anderson.

83   "It was a very difficult": Interview with Mitt Romney.

83   Church leaders called: Interview with J. Fielding Nelson, 2011.

83   A mission president: Ibid.

84   But when Nelson arrived: Ibid.

84   "You had this shock experience": Interview with Dane McBride, 2007.

84   He resisted suggestions: Interview with Bruce Robinson.

84   "His resilience": Interviews with Joel McKinnon.

85   But he wanted to keep: Interview with Mitt Romney.

85   Duane Anderson returned: Richard Anderson, writing in Leola Anderson's journal.

85   he did lose one small battle: http://freepages.history.rootsweb.ancestry.com/~timbaloo/SuchALife/pages/Ch3/8Miss/8Miss25c.htm.

85   He personally brought few new members: Interview with Mitt Romney.

86   "I wasn't there for": Henry B. Eyring, "Waiting upon the Lord," speech at Brigham Young University, September 30, 1990.

86   "He came to respond": Interview with Mitt Romney.

86   "This made me very painfully aware": Ibid.

86   "I was frightened": Ibid.

86   Romney apparently had: Interviews with Mitt Romney and Bill Ryan, 2007.

86   On his way home: Interview with Mitt Romney.

87   "Is this what's in store": Ibid.

87   "That was terrifying": Ibid.

87   "He became really, really distraught": Interviews with Dane McBride, 2007 and 2011.

88   Ann's roommate: Interviews with Mitt Romney, Cindy Davies, and Kim Cameron, 2007; interviews with Dane McBride.

88   "Gosh, this feels": Interview with Mitt Romney.

88   "I think Lenore had a hard time": Interview with Cindy Davies.

88   "accomplish things of significance": Interview with Mitt Romney.

89   In the spring of 1969: Laurena Pringle, "Ann Davies, Mitt Romney Married," *Detroit Free Press*, March 1969; Associated Press, "Couple Wed in Civil Ceremony," March 22, 1969; "Mitt Romney Marries Ann Davies," *The New York Times*, March 22, 1969.

89   named for a friend: Tagg Romney, "Campaigning for Dad," online chat with *The Washington Post*, December 20, 2007.

89   The new parents were thrilled: Interview with Dane McBride; Associated Press, "Couple Wed in Civil Ceremony"; "Mitt Romney Marries Ann Davies," *The New York Times*.

89   Romney, despite having soured: David D. Kirkpatrick, "Romney, Searching and Earnest, Set His Path in '60s," *The New York Times*, November 15, 2007.

90   He was invited: Interview with Dane McBride.

90   "highest honors": E-mail from Michael Smart, BYU communications, August 26, 2011; Mitt Romney, *Turnaround: Crisis, Leadership, and the Olympic Games* (Washington, D.C.: Regnery Publishing, 2004), 13.

90   "I pray": Ben Bradlee, Jr., "Romney Seeks New Chapter in Success," *The Boston Globe*, August 7, 1994.

91   "He looks up": Interview with Howard Serkin, 2007.

92   "When we all got there": Interview with Janice Stewart, 2007.

92   "We viewed ourselves": Interview with Howard Brownstein, 2007.

92   Stephen Breyer: Hugh Hewitt, *A Mormon in the White House? 10 Things Every American Should Know About Mitt Romney* (Washington, D.C.: Regnery Publishing, 2007), 46.

93   "There was nothing jaded": Interview with Garret Rasmussen, 2007.

93   "Most of us dressed": Interview with William Neff, 2007.

93   The snub: John Ehrlichman, *Witness to Power: The Nixon Years* (New York: Simon and Schuster, 1982), 104.

94   "We've got to": John Herbers, "Romney Making His Greatest Impact Outside Government by Challenging U.S. Institutions," *The New York Times*, May 15, 1969.

94   Romney declined: Christopher Bonastia, *Knocking on the Door: The Federal Government's Attempt to Desegregate the Suburbs* (Princeton, N.J.: Princeton University Press, 2006), 108.

94   "It was a stunning blow": Lenore Romney, letter to John Ehrlichman, www .boston.com/news/daily/27/lenore_letter.pdf.

94   Arriving in Wilkes-Barre: Louise Cook, "Rebuilding After Agnes' Floodwaters Continues," Associated Press, August 10, 1972.

95   Romney let loose: George Romney–Richard Nixon meeting, August 11, 1972, Nixon tapes.

95   Then, in a tart resignation letter: www.presidency.ucsb.edu/ws/index .php?pid=3709.

95   He briefly considered: Bob Bernick, Jr., "Bennett Makes Some Changes; Senator Plans to Seek Third Term Despite '92 Vow," *Deseret News*, August 29, 2001.

96   Mitt and Ann had: Bradlee, "Romney Seeks New Chapter in Success."

96   "He didn't mind": Interview with Howard Serkin.

96   One guest: Interview with Howard Brownstein.

96   "He mentioned it once": Interview with Mark Mazo, 2007.

96   "You got the feeling": Interview with Howard Brownstein.

97   "He was an outstanding recruit": Interview with Charles Faris, 2007.

97   was admitted to practice: Michigan Board of Law Examiners records.

97   "That's where my friends are": Jo-Ann Barnas, "Romney Gives His All to Games; Michigan Native Saves Scandalous Olympics," *Detroit Free Press*, February 14, 2011.

98   "At BCG, analysis was king": Interview with Lonnie Smith, 2007.

98   "He worked": Interview with Charles Faris.

98   "For me and everybody there": Interview with Lonnie Smith.

Chapter 5: Family Man, Church Man

99   It was shaping up: Interview with Mark and Sheryl Nixon, 2011.

100   Shortly after leaving: Tom Moroney, "Aftermath," *The Boston Globe Magazine*, December 8, 1996.

101   The Romneys have long cited: Ben Bradlee, Jr., "Romney Seeks New Chapter in Success," *The Boston Globe*, August 7, 1994.

101   Over the next decade: E-mail from Eric Fehrnstrom, August 25, 2011.

101   Like many Mormons: Bradlee, "Romney Seeks New Chapter in Success"; interview with Tagg Romney, 2007.

102   "For us": Interview with Mitt Romney, 2007.

102   Tuesday evenings: Bradlee, "Romney Seeks New Chapter in Success"; e-mail from Eric Fehrnstrom, September 22, 2011.

102   Before high school: Interview with Tagg Romney.

102   She had left: "Ann's Biography," www.mittromney.com/Ann-Romney/Biography, accessed September 2, 2011 (no longer online); e-mail from Eric Fehrnstrom, August 28, 2011.

102   She would become active: "Ann's Biography"; Bella English, "Where Mitt Leads, Ann Romney Follows," *The Boston Globe*, October 29, 2002.

102   Her husband's preferred term: Ann Romney, "Chief Family Officer," http://fivebrothers.mittromney.com, June 22, 2007 (no longer online).

102   "So far as the family": Interview with Douglas Anderson, 2011.

102   Eagle Scout badges: Hugh Hewitt, *A Mormon in the White House? 10 Things Every American Should Know About Mitt Romney* (Washington, D.C.: Regnery Publishing, 2007), 79.

102   "It's one which is challenging": Bradlee, "Romney Seeks New Chapter in Success."

102   Once, Ann forgot: Interview with John Wright, 2011.

103   a skill she had absorbed: Ann Romney, "Memories of the Salt Lake Olympics," http://fivebrothers.mittromney.com, September 20, 2007 (no longer online).

103   even ran a small cooking school: Ann Romney, "The Sunshine State," http://fivebrothers.mittromney.com, November 23, 2007 (no longer online).

103   One of the most popular: Ibid.

103   "I was willing": Meredith Bryan, "Free Fatherhood Advice from Mitt Romney," *GQ*, June 2007, 194.

103   "I've broken almost every bone": Hewitt, *A Mormon in the White House?*, 84.

103   One winter day: Interview with John Wright.

104   In time, each of the boys: Interview with Tagg Romney.

104   "They were very impressive young men": Interview with Philip Barlow, 2011.

104   "Mitt tried to teach": Interview with John Wright.

104   Indeed, all five sons: *This Week*, ABC, February 18, 2007.

104   Mitt Romney said: Bryan, "Free Fatherhood Advice from Mitt Romney," 194.

104   "Growing up": Interview with Mitt Romney.

104   Tagg said his father: Interview with Tagg Romney.

105   "We've tried to civilize": TV spot, Romney for Governor, 2002.

105   For fifteen years: Barnstable County, Massachusetts Registry of Deeds, http://barnstabledeeds.org; interview with Elaine Price, 2011.

105   Bennett was up: Interview with Grant Bennett, 2011.

105   learned to water-ski: Interview with John Wright.

105   In June 1981: Frank Phillips, "GOP Hopeful Arrested in 1981," *The Boston Globe*, May 5, 1994; interview with John Wright.

106   Tagg said some: Interview with Tagg Romney.

106   Matt, Josh, and Craig: E-mail from Eric Fehrnstrom, September 22, 2011.

106   "Overnight," Tagg said: Interview with Tagg Romney.

107   One night: Ibid.

107   To Mitt, the special one: Ibid.

107   On Mother's Day: Jose Antonio Vargas, "Romney Brothers Dish on Dad," *The Washington Post*, June 9, 2007.

107   Mitt is driven first by reason: Interview with Tagg Romney.

107   But she has remained: Interview with John Wright.

108   "Mitt's not going": Interview of Jane Romney by Neil Swidey, 2006.

108   "I know there are things": Interview with Tagg Romney.

108   Mitt imposed strict rules: Ibid.

108   The destination of this journey: The story of the Romneys' trip to Canada with their dog Seamus was first reported by Neil Swidey, based on interviews with Romney family friends and confirmed by members of the Romney family.

109   "He's very engaging": Interview with former aide, 2011.

109   "He wasn't overly interested": Interview with former aide, 2011.

110 "He has that invisible wall": Interview with Republican, 2011.

110 "A lot of it is": Interview with former aide, 2011.

110 they had moved: Registry of Deeds, Middlesex County, Massachusetts, http://masslandrecords.com/malr/controller.

111 "How can you afford": Interview with Tagg Romney.

111 The family's modest getaway: Real estate records, news clips, visit by author.

111 Romney acknowledged: Bradlee, "Romney Seeks New Chapter in Success."

111 The family had no cook: Neil Swidey and Stephanie Ebbert, "Journeys of a Shared Life," *The Boston Globe,* June 27, 2007.

111 And he was frugal to the core: Romney family videos, provided to *The Boston Globe* in 2007.

111 "I sometimes thought": Interview with Joseph O'Donnell, 2011.

111 "His strategy": Interview with John Wright.

112 "Mitt," said Kem Gardner: Interview with Kem Gardner, 2007.

112 "Compared to my dad": Interview with Tagg Romney.

112 "Mitt was always working": Interview with Helen Claire Sievers, 2011.

112 "He opened it": Interview with Grant Bennett.

112 John Wright remembered: Interview with John Wright.

112 There were stretches: Interview with Mitt Romney.

113 "To us, he was just Dad": Interview with Tagg Romney.

113 Mormon lineages: Interviews with Helen Claire Sievers and Grant Bennett.

113 Mormon congregations: Interviews with Helen Claire Sievers and Ken Hutchins, 2011.

113 In another departure: Peggy Fletcher Stack, "Mitt and His Faith: Remembering When Candidate Romney Was Bishop Romney," *The Salt Lake Tribune,* January 11, 2008.

113 Their selection is carefully vetted: Interview with Ken Hutchins.

114 "It really is quite a tremendous": Interview with Tony Kimball, 2011.

114 Mitt Romney first took on: Interview with Grant Bennett.

114 close to four thousand members: Interviews with Grant Bennett and Ken Hutchins.

114 His leadership in the church: Interview with Philip Barlow, 2011.

114 Up to then, church practice reflected: Peggy Fletcher Stack, " 'Black Curse' Is Problematic LDS Legacy," *The Salt Lake Tribune,* June 6, 1998.

114 "one of the most emotional": Scot Lehigh and Frank Phillips, "Romney Hits Kennedy on Faith Issue," *The Boston Globe*, September 28, 1994.

114 "I heard it": *Meet the Press*, NBC, December 16, 2007.

114 But though the church: Peggy Fletcher Stack, "Exiles in Zion," *The Salt Lake Tribune*, August 16, 2003.

115 But a dichotomy exists: Lisa Wangsness, "GOP Rivals Have Different Takes on Mormon Faith," *The Boston Globe*, August 15, 2011.

115 The Boston area: Interview with Tony Kimball.

116 In the early-morning darkness: Janet Peterson, "Belmont's Blessing in Disguise," *Ensign*, April 1987.

116 And the blaze: Interview with Philip Barlow.

116 "I don't know when": Peterson, "Belmont's Blessing in Disguise."

116 Some locals objected: Mark Miller, "Nearly Built Mormon Church Hit by a Suspicious Blaze in Belmont," *The Boston Globe*, August 2, 1984; Peterson, "Belmont's Blessing in Disguise."

116 The church urgently needed: Peterson, "Belmont's Blessing in Disguise."

116 Others had started a small consulting firm: Carol Pearson, "Two Celebrations: A Beginning—and an Ending," *The Boston Globe*, June 6, 1986.

117 A few years earlier: Peterson: "Belmont's Blessing in Disguise"; Frank Phillips and Don Aucoin, "Romney Quiet on Religious Beliefs," *The Boston Globe*, May 22, 1994.

117 "Some people in Belmont": Peterson, "Belmont's Blessing in Disguise."

117 Then something unexpected happened: Interview with Grant Bennett; Peterson, "Belmont's Blessing in Disguise."

117 "One of the things": Interview with Connie Eddington, 2011.

118 "There are still": Interview with Grant Bennett.

118 nearly three thousand: Peterson, "Belmont's Blessing in Disguise."

118 cartoon characters: David D. Kirkpatrick, "Romney, Searching and Earnest, Set His Path in '60s," *The New York Times*, November 15, 2007.

118 "I was a little surprised": Interview with Philip Barlow.

119 like a duty: Interview with Ken Hutchins.

119 Romney's church colleagues said: Stack, "Mitt and His Faith: Remembering When Candidate Romney was Bishop Romney."

119 "He was reasonable": Interview with Philip Barlow.

119 "His leadership has been": Interview with Douglas Anderson.

119 "He let the people": Interview with Ken Hutchins.

119 He put it: Interview with Philip Barlow.

119 "Next thing I know": Interview with David Gillette, 2011.

119 One Saturday: Interview with Grant Bennett.

120 weren't getting paid back: Interview with Helen Claire Sievers.

120 Romney has also upheld his obligation: E-mail from Eric Fehrnstrom, September 22, 2011.

120 On Super Bowl Sunday 1989: Interview with Douglas Anderson.

121   "Search diligently": Church of Jesus Christ of Latter-day Saints, *Doctrine and Covenants*, Section 90, Verse 24.

121   Outside on the steps: Interview with Douglas Anderson.

121   Romney helped lead: Interview with Joseph O'Donnell.

122   On one occasion: Interview with Grant Bennett.

122   In the spring of 1993: Interview with Helen Claire Sievers.

122   In the end: Stack, "Mitt and His Faith: Remembering When Candidate Romney was Bishop Romney"; interview with Helen Claire Sievers.

123   Tony Kimball said: Interview with Tony Kimball.

123   She felt it was about time: Interview with Helen Claire Sievers.

123   Ann Romney was not considered: Interview with member, 2011.

123   Mitt Romney showed flexibility: Interview with Tony Kimball.

123   "They feel needed": Julie A. Dockstader, "Enriching Theirs and Others' Lives," February 8, 1992, http://www.ldschurchnews.com/articles/22302/Enriching-theirs-and-others-lives.html.

124   Peggie Hayes had joined: Interview with Peggie Hayes, 2011.

125   the church encourages adoption: "Adoption," http://lds.org/study/topics/adoption?lang=eng.

125   Romney would later deny: Frank Phillips and Scot Lehigh, "Single Mother Tells of the Advice Romney Gave as Mormon Counselor," *The Boston Globe*, August 26, 1994.

126   In the fall of 1990: Anonymous, "Unheard," *Exponent II* 15, no. 4 (1990): 5.

126   Church leaders have said: "Abortion," http://lds.org/study/topics/abortion?lang=eng.

126   One day in the hospital: Jacquelynn Boyle, "A Positive Spirit Wins," *Detroit Free Press*, October 26, 1995.

126   Romney would later contend: Scot Lehigh and Frank Phillips, "Romney Admits Advice Against Abortion," *The Boston Globe*, October 20, 1994.

127   One woman: Interview with Judy Dushku, 2011.

128   The world map: Don L. Brugger, "Climate for Change," *Ensign*, September 1993.

128   living in the suburbs: Ibid.

128   "It has been a great challenge": Sheridan R. Sheffield, "Boston: Gospel Rolls Forward in One of Nation's Oldest Cities," September 28, 1991, www.ldschurchnews.com/articles/20681/Boston-Gospel-rolls-forward-in-one-of-nations-oldest-cities.html.

128   Romney and other Mormon officials: Brugger, "Climate for Change."

128   Missionaries worked: Sheffield, "Boston: Gospel Rolls Forward in One of Nation's Oldest Cities."

128   "Love those people": Interview with Keith Knighton, 2011.

128   David Gillette: Interview with David Gillette.

129   he was also able: Interview with Tony Kimball.

129   regaling missionaries: Interview with David Gillette.

129   "You feel humbled": Sheffield, "Boston: Gospel Rolls Forward in One of Nation's Oldest Cities."

129   "To understand my faith": Hewitt, *A Mormon in the White House?*, 96.

Chapter 6: The Moneymaker

130   "I remember him": Interview with Bill Bain, 2007.

131   "This guy is going to be": Ibid.

133   "light or flippant manner": Interview with Mitt Rómney, 2007.

133   "So," Bain explained: Interview with Bill Bain.

133   "I left a steady job": Romney announcement speech in Stratham, N.H., June 2, 2011.

134   If he took a briefcase home: Interview with Mitt Romney.

134   "wallow in the data": Mitt Romney, appearance at National Review Institute conference, Washington, D.C., January 27, 2007.

134   a decade younger: Interviews with Bain Capital partners, 2011.

135   "no sense": Robert Gay, speech, "Within and Beyond Ourselves: The Role of Conscience in Modern Business," delivered at Brigham Young University, April 26, 2002.

135   The most thorough analysis: Alex Brown, "Special Opportunities Fund: Investing in Funds Managed by Bain Capital," Deutsche Bank, 2000.

136   "expressed to me": Interview with Harry Strachan, 2011.

136   "We investigated": Mitchell Zuckoff and Ben Bradlee, Jr., "Romney's Business Record Gives Larger Picture," *The Boston Globe*, August 8, 1994.

137   "He was troubled": Interview with Coleman Andrews, 2007.

137   "punch him in the nose": Robert Gavin and Sacha Pfeiffer, "Study, Sweat, and Profit; Performance at Bain Capital Burnished Image, Fueled Critics," *The Boston Globe*, June 26, 2007; interview with Robert White, 2007.

137   "I always wondered": Interview with Bain partner, 2011.

138   "Mitt was struggling": Gavin and Pfeiffer, "Study, Sweat, and Profit; Performance at Bain Capital Burnished Image, Fueled Critics."

138   Key ran shuttle routes: Interview with Geoffrey Rehnert, 2011.

138   Another early deal: Interview with Robert White, 2011.

138   Holson Burnes: Interview with Geoffrey Rehnert, 2011.

139   "Look," Stemberg told Romney: Interviews with Thomas Stemberg, 2007 and 2011.

139   But after that, Romney took the lead: Ibid.

139   In all, it invested about $2.5 million: Gavin and Pfeiffer, "Study, Sweat, and Profit; Performance at Bain Capital Burnished Image, Fueled Critics."

140   At the initial public offering: Staples IPO prospectus, April 1989.

140   "a classic 'category killer' ": Steven Flax, "Perils of the Paper Clip Trade," *The New York Times*, June 11, 1989.

140   "helped to create tens of thousands": Interview with Mitt Romney, *On the Record*, CNN, September 6, 2011.

140   "That's why I'm always": Zuckoff and Bradlee, "Romney's Business Record Gives Larger Picture."

140   "What you really cannot do": Interview with Howard Anderson, 2011.

140   "Why not give him credit": Interview with Thomas Stemberg.

141   "the success of the enterprise": Interview with Mitt Romney.

142   Bain told management and unions: Frederick F. Reichheld and Thomas Teal, *The Loyalty Effect: The Hidden Force Behind Growth, Profits, and Lasting Value* (Boston: Harvard Business School Press, 1996), 172–173.

142   "loyalty effect": Ibid., 174.

142    $121 million: Ibid.

143   "I throw mine": Interview with Geoffrey Rehnert.

143   "led me to become": Romney announcement speech in Stratham, N.H., June 2, 2011.

143   "I never actually ran": Mitt Romney, *Turnaround: Crisis, Leadership, and the Olympic Games* (Washington, D.C.: Regnery Publishing, 2007), 16.

144   "Ivan the Terrible": "Investor 'Ivan the Terrible' Boesky," *Time*, December 1, 1986. Boesky struck a plea-bargain deal and was banned from securities trading for life.

144   "I am not a destroyer": Oliver Stone, director, *Wall Street*, 1987.

144–145 "incessantly destroying": Joseph Schumpeter, *Capitalism, Socialism and Democracy*, e-book, Taylor and Francis, 2003, 83.

145   "the problem with creative destruction": Alan Greenspan, testimony before Senate Committee on Banking, Housing and Urban Affairs, July 21, 2005.

145   "for governments to stand aside": Mitt Romney, *No Apology: Believe in America* (New York: St. Martin's Press, 2010), 110.

145   Indeed, he wrote: Mitt Romney, "Let Detroit Go Bankrupt," *The New York Times*, November 18, 2008.

146   "We were pretty happy": Interview with Josh Bekenstein, 2011.

146    $105 million: Interviews with Bain Capital partners, 2011.

147   "to Mitt's credit": Interview with Marc Wolpow, 2011.

147   Prosecutors who worked: Interviews with prosecutors of the Drexel case.

147   "We did not say": Zuckoff and Bradlee, "Romney's Business Record Gives Larger Picture."

148   Palais Royal: "Bain Acquires Apparel Chains," *The New York Times*, December 30, 1988.

148   "to make sure": Connie Bruck, *The Predators' Ball: The Inside Story of Drexel Burnham and the Rise of the Junk Bond Raiders* (New York: Penguin Books,

1989), 366–369. Bruck cited an affidavit by Joseph that mentioned Romney's phone call.

148   "By doing the deal": Interview with James T. Coffman, 2011.

148   Drexel pleaded guilty: "The Collapse of Drexel Burnham Lambert; Key Events for Drexel Burnham Lambert," *The New York Times*, February 14, 1990.

148   Milken eventually pleaded guilty: Stuart Pfeifer and Tom Petruno, "Michael Milken Is Still Seeking Redemption," *Los Angeles Times*, February 3, 2009.

148   By the following year: Monica Perin, "Clothing Retailer Slapped with Shareholder Suit," *Houston Business Journal*, April 11, 1999.

149   The department store company filed: Bain Capital partners said they weren't responsible for financial problems later encountered by the company. After reemerging from bankruptcy, the company said it had become profitable. Greg Hassell, "Bankruptcy Leaves Stage Feeling Fine; Retailer Has Strong Sales After Its Reorganization," *Houston Chronicle*, August 25, 2001.

149   "It was a terrible situation": Charles Stein, " 'Their Mission Is to Make Money'; but in Doing So, Romney Did More Building Than Slashing," *The Boston Globe*, October 9, 1994.

149   GST Steel: Robert Gavin, "As Bain Slashed Jobs, Romney Stayed to Side," *The Boston Globe*, January 27, 2008.

150   accumulating $1.6 billion in debt: Daniel G. Jacobs, "Coming off Life Support," *Smart Business*, May 2005.

150   When it merged with Behring Diagnostics: Gavin, "As Bain Slashed Jobs, Romney Stayed to Side."

150   At the same time: Josh Kosman, *The Buyout of America* (New York: Penguin, 2010), 107.

150   "It is one thing": David D. Kirkpatrick, "Romney's Fortunes Tied to Business Riches," *The New York Times*, June 4, 2007. Bain Capital partners said later that the Dade bankruptcy had occurred because the dollar had fallen against the euro and pushed the company's debt over a certain limit that the banks did not allow. They argue that it was debt investors who had pressured the company and that Bain had helped the company get back onto its feet.

150   In that time of immense success: Mitchell Zuckoff, "Romney Rescue of Bain & Co. a Study in Profit and Loss," *The Boston Globe*, October 25, 1994.

151   "going over a cliff": Hugh Hewitt, *A Mormon in the White House? 10 Things Every American Should Know About Mitt Romney* (Washington, D.C.: Regnery Publishing, 2007), 53.

151   "We have bad news": Ibid., 54–55.

151   The Massachusetts legal code: Massachusetts law has a maximum one-year punishment for the first violation of its wage law, which requires payment

within a week of the end of a pay period; see www.malegislature.gov/Laws/
GeneralLaws/PartI/TitleXXI/Chapter149/Section27C.

151   "It's a crime": Hewitt, *A Mormon in the White House?* 55–56.

152   At one point: Stephanie Strom, "A Fund for Distressed Companies Goes
Awry," *The New York Times*, September 25, 1996.

152   The discussion grew increasingly heated: Interview with Bain Capital partners, 2011. The blowup with a Goldman banker appears to have first been
reported in: David Snow, "The House That Mitt Built," *Private Equity International*, September 2007.

152   He convinced: Zuckoff, "Romney Rescue of Bain & Co. a Study in Profit
and Loss."

153   "He was willing": Interview with Harry Strachan, 2011.

153   Only one person left: Interview with Bain partners, 2011.

153   "If Bain & Co. went bankrupt": Interview with Geoffrey Rehnert, 2011.

154   the company had paid down: "Damon Corp., 5th Annual Globe 100/The Best
of Massachusetts Businesses-Profitability," *The Boston Globe*, June 8, 1993.

154   Romney personally reaped $473,000: Frank Phillips, "Romney Profited on
Firm Later Tied to Fraud," *The Boston Globe*, October 10, 2002. Romney
said his earnings included $103,000 he had made on the deal as well as a 5
percent share of Bain's profit, or $370,000, totaling $473,000.

154   The day after the merger: Meg Vaillancourt, "Romney-Aided Deal Closed
Damon Plant," *The Boston Globe*, October 9, 1994.

154   "corrective action": Phillips, "Romney Profited on Firm Later Tied to
Fraud."

154   paid $119 million in fines: Kimberly Blanton, "Needham Lab Fined $119M
for Fraud," *The Boston Globe*, October 10, 1996.

154   "There is a mess": Yvonne Abraham, "Candidates Spar over TV Ads," *The
Boston Globe*, October 22, 2002.

155   Lifelike: 2001 financial disclosure report, Mitt Romney, Massachusetts
State Ethics Commission.

156   Bain lost its money: Interview with Michael Goss, 2007.

156   Auto Palace/ADAP: Brown, "Special Opportunities Fund: Investing in
Funds Managed by Bain Capital."

156   Gartner would go on: Interview with Stephen Pagliuca, 2011.

156   "I don't think": Interview with Thomas Stemberg.

157   About forty-five minutes into: Interview with Mark Nunnelly, 2011.

157   In the fall of 1998: Chana R. Schoenberger, "Pie in the Sky; Is Domino's
Pizza Worth All the Trouble to Bain Capital?" *Forbes*, September 17, 2001.

157   "We're the biggest schmoes": Mitt Romney, appearance at National Review Institute conference.

157   Bain reaped more than $100 million: Domino's IPO filing, 2004.

157   earning a 500 percent return: "Domino Effect," *The Detroit News*, December 9, 2006.

158   "the Tin Man": Interviews with Bain partner, 2011.

158   "I don't care": Mitt Romney, "Searched," presidential campaign ad, December 17, 2007, http://abcnews.go.com/video/video?id=4026797&tab=9482931&section=2808950&page=1.

158   "Investment Firm Shuts": Shirley Leung, "Investment Firm Shuts to Help Find Girl," *The Boston Globe*, July 12, 1996.

158   "Shortly thereafter": Robert C. Gay, "Finding the Needle in a Haystack: Thoughts on Being Morally Courageous in Business," speech at Brigham Young University, 2003.

158–159 "It was a shocker": Peter Canellos, "Bain Capital Recalls NY Search," *The Boston Globe*, December 8, 1996.

159   "Mitt's done a lot of things": Romney, "Searched."

159   "more valuable than some": Canellos, "Bain Capital Recalls NY Search."

159   During Romney's fifteen years there: Brown, "Special Opportunities Fund: Investing in Funds Managed by Bain Capital"; interviews with Bain partner, 2011.

159   "the standard investment banker approach": Romney, *Turnaround*, 16.

159   The first deal: Interview with Scott Sperling, 2011.

160   Experian: Interviews with Bain Capital and Thomas H. Lee partners, 2011.

160   A mere seven weeks: James S. Hirsch and Matthew Rose, "Buyout Group Hits $500 Million Jackpot—Great Universal's Accord to Buy Experian Brings Fast Investor Bonanza," *The Wall Street Journal*, November 15, 1996.

160   "go down as one": Ibid. Partners involved in the deal said the seven-week turnaround time understated the time frame, explaining that Bain and Lee had had nearly six months of conditional ownership while the computer system was being tested.

160   "hit with the lucky stick": Interviews with Bain Capital partners, 2011.

160   "Gee, Mark, are you sure?": Interview with Mark Nunnelly.

161   "We stumbled on": Interview with Phil Cuneo, 2011.

161   "an Internet star": Ibid.

161   In just under three years: Stan Pace, "Rip the Band-Aid Off Quickly: Why 'Fast, Focused, and Simultaneous' Works Best in Corporate Transformations," *Strategy & Leadership* 30, no. 1 (2002): 4–9.

161   "wasn't like being": Interview with Bain Capital partner, 2011.

162   "The returns were just eye-popping": Interview with Geoffrey Rehnert.

162   "passbook" savings accounts: Mitt Romney, appearance at National Review Institute conference.

162   Romney's own wealth had increased exponentially: Zuckoff and Bradlee, "Romney's Business Record Gives Larger Picture."

162   "I'm not going to": Interview with Mitt Romney.

162   Under federal tax law: www.econlib.org/library/Enc/CapitalGainsTaxes
        .html. In 2011, the billionaire Warren Buffett drew the attention of Presi-
        dent Obama when he said that the tax rate differential contributed to the
        deficit and said it was unfair that he paid a lower tax rate than his secretary
        did.

163   "The objective is": Interview with Ross Gittell, 2007.

163   "It's the opposite": Interview with Marc Wolpow.

163   his claim is accurate: Interviews with Bain Capital partners, 2011.

164   "The goal of the investor": Interview with Howard Anderson.

165   "Work was never": Romney, *Turnaround*, 11–14.

165   "Do I really want to": Ibid., 14.

Chapter 7: Taking On an Icon

166   knew him as "Pops": Interview with Jim Davies, 2007; interview with
        Mickey Fedorko, 2011; e-mails from Davies and Fedorko, 2011.

166   "I'm so mad": Jack Thomas, "Ann Romney's Sweetheart Deal," *The Boston
        Globe*, October 20, 1994.

166   Decades earlier: E-mail from Jim Davies, 2011.

166   "He said, 'Ann, you've got' ": Thomas, "Ann Romney's Sweetheart Deal."

167   "I've been living here": *Campaign Almanac*, C-SPAN, 1994.

167   "Are we going to die": Ben Bradlee, Jr., "Romney Seeks New Chapter in
        Success," *The Boston Globe*, August 7, 1994.

167   "You can gripe": Ibid.

167   "No! No!": *Campaign Almanac*, C-SPAN, 1994.

167   what was left to do?: Mitt Romney, *Turnaround: Crisis, Leadership, and the
        Olympic Games* (Washington, D.C.: Regnery Publishing, 2004), 14.

167   That October: Bradlee, "Romney Seeks New Chapter in Success."

168   And Kennedy himself: Peter S. Canellos, ed., *Last Lion: The Fall and Rise of
        Ted Kennedy* (New York: Simon & Schuster, 2009), 261–281.

168   "I recognize my own shortcomings": Curtis Wilkie, "Kennedy Admits Per-
        sonal 'Faults' in Speech, Vows to 'Continue to Fight the Good Fight' for
        Liberal Causes," *The Boston Globe*, October 26, 1991.

168   "There was, from the beginning": Interview with former Kennedy staffer,
        2011.

169   His face was mottled: Sally Jacobs, "Bay State Again Takes Kennedy, Flaws
        and All," *The Boston Globe*, September 19, 1994.

169   "People said to me": Interview with Mitt Romney, 2007.

169   "This," Romney would later say: Anthony Flint and Andy Dabilis, "Sena-
        tor and His Challenger Both Exploit Romney Remark," *The Boston Globe*,
        October 31, 1994.

169   "He knew what the game was": Interview with GOP operative, 2011.

170   Even before all that: Interview with Joseph Malone, 2011.

170   "a very attractive": Interview with GOP operative.

170   "philosophically vacuous": Don Feder, "Romney: A Thin-Gruel Republican," *Boston Herald*, October 20, 1994.

171   "That was his entire focus": Interview with Seth Weinroth, 2011.

171   "tame the monster": Scot Lehigh and Frank Phillips, "GOP's Romney Declares, Says Kennedy Out of Date," *The Boston Globe*, February 3, 1994.

171   Early in 1994: Interview with John Lakian, 2011.

172   So the Romneys set out: Interviews with former campaign aides, 2011.

172   "We knew nobody": Ben Bradlee, Jr., and Daniel Golden, "Strategies Shaped an Epic Race," *The Boston Globe*, November 10, 1994.

172   a particularly popular attraction: Interviews with former campaign aides.

172   "He just made great strides": Interview with Rick Reed, 2011.

172   The campaign organized a team: Interviews with former campaign aides.

172   "Clearly everybody understood": Interview with Seth Weinroth.

172   "To this day": Interview with former staffer, 2011.

172   Ann assumed the role: Interview with former aide, 2011.

173   "empty suit": Gayle Fee and Laura Raposa, "Inside Track," *Boston Herald*, April 11, 1994.

173   "He doesn't sit": Interview with Michael Sununu, 2011.

173   "You kind of are": Interview with former adviser, 2011.

173   Romney's game plan: Interview with Seth Weinroth.

173   "Here was this": Ibid.

173   On at least one occasion: "Sins of the Father Visited on the Son," *The Boston Globe*, May 15, 1994.

174   *The Boston Globe* published a poll: Scot Lehigh and Frank Phillips, "Poll Sees Drop-off in Kennedy Support," *The Boston Globe*, May 14, 1994.

174   Romney used his convention speech: Frank Phillips and Peter J. Howe, "Romney Wins GOP Approval," *The Boston Globe*, May 15, 1994.

174   "He came out and said": Interview with John Lakian.

174   That evening: Interview with Seth Weinroth.

174   Some called on Lakian to drop out: Phillips and Howe, "Romney Wins GOP Approval."

174   Instead he cast: Scot Lehigh, "2 Senate GOP Hopefuls Clash over Abortion Funding," *The Boston Globe*, May 27, 1994.

175   "Ideologically, I'm not sure": Interview with John Lakian.

175   "There were a number of issues": Interview with Mitt Romney.

175   "a real opportunity": Romney for Senate videotape, 1994.

175   Kennedy and his advisers, meanwhile: Frank Phillips and Scot Lehigh,

"Kennedy Pressing Campaign Early On," *The Boston Globe*, June 22, 1994.

175 probe Romney's background: Frank Phillips, "Detectives Probe Romney for Kennedy Campaign," *The Boston Globe*, August 19, 1994.

175 "I'm looking out": Interview with former Kennedy aide, 2011.

176 an effective TV ad: Bradlee and Golden, "Strategies Shaped an Epic Race."

176 "polls do go up and down": P-I News Services, "Kennedy Has Tough Opponent," *Seattle Post-Intelligencer*, September 20, 1994.

176 Appearing before some five hundred: Geeta Anand and Frank Phillips, "Romney Savors Victory, Challenges Kennedy," *The Boston Globe*, September 21, 1994.

176 "I remember the feeling": Interview with former aide, 2011.

176 Romney said he told colleagues: Interview with Mitt Romney.

176 "There's just no way": Romney, *Turnaround*, 14.

176 "After the primary": Interview with Mitt Romney.

176 "You saw the flash of anger": Interview with former Kennedy aide, 2011.

177 Kennedy pollster Tom Kiley: Bradlee and Golden, "Strategies Shaped an Epic Race."

177 The room overflowed: Interview with former Kennedy adviser, 2011.

177 Shrum recommended: Bradlee and Golden, "Strategies Shaped an Epic Race."

177 "This is real": Interview with former Kennedy adviser.

177 For months, Kennedy researchers: Interviews with former Kennedy aides, 2011.

178 "the assets of SCM Office Supplies": Peter G. Gosselin, "Ind. Strikers Taking Fight to Romney," *The Boston Globe*, October 3, 1994.

178 One former Bain executive: Bob Drogin, "To Assess Romney, Look Beyond the Bottom Line," *Los Angeles Times*, December 17, 2007.

178 "It was devastating": Interview with former Kennedy staffer, 2011; Bradlee and Golden, "Strategies Shaped an Epic Race."

178 A union official called: Bradlee and Golden, "Strategies Shaped an Epic Race."

178 The ads featuring: Ibid.

178 Three days after the filming: Scot Lehigh and Frank Phillips, "Kennedy Raps Romney on Ind. Plant Strike," *The Boston Globe*, September 30, 1994.

179 "Wouldn't it be nice": Ibid.

179 "This is not fantasy land": Frank Phillips, "Romney Firm Tied to Labor Fight," *The Boston Globe*, September 23, 1994.

179 Romney did not air: Bradlee and Golden, "Strategies Shaped an Epic Race."

179 They distributed leaflets: Ibid.

179 On October 7: Frank Phillips, "Romney Agrees to Talk; Union Balks," *The Boston Globe*, October 8, 1994.

179   Two days later: Meg Vaillancourt and Sarah A. McNaught, "Romney Meets with Strikers; Workers Confront Candidate at Event," *The Boston Globe*, October 10, 1994.

180   Romney's poll numbers: Scot Lehigh and Sally Jacobs, "Romney Lashes Out at Kennedy over Television Attack Tactics," *The Boston Globe*, October 13, 1994.

180   "It's something we should have": Bradlee and Golden, "Strategies Shaped an Epic Race."

180   "You can tell": Interview with Joseph Malone.

180   "Letting go of that": Interview with former staffer, 2011.

181   "You'd get Republicans": Interview with former Romney aide, 2011.

181   It might have been: Sridhar Pappu, "The *Holy Cow!* Candidate," *Atlantic Monthly*, September 2005, 106.

181   "We've got this thing": Interview with Massachusetts Republican leader, 2011.

181   "socially innovative": P-I News Services, "Kennedy Has Tough Opponent."

181–182 In September: Interview with Richard Tafel, 2011.

182   "In Mitt's mind": Interview with Massachusetts Republican, 2011.

182   Romney was on the executive board: Peter G. Gosselin, "Kennedy, Romney Continue Trading Charges over Facts," *The Boston Globe*, October 27, 1994.

182   "I'm with you": Interview with Richard Tafel.

182   So not long afterward: Mitt Romney, letter to Log Cabin Republicans, October 6, 1994.

182   Romney had previously said: Scot Lehigh, "Kennedy, Romney Battle for the Middle," *The Boston Globe*, October 10, 1994.

183   Even after the group: Interview with Richard Tafel.

183   He was deeply touched: Richard Tafel, *Party Crasher: A Gay Republican Challenges Politics as Usual* (New York: Simon & Schuster, 1999), 224.

183   In fact, his professed views: Scot Lehigh, "Romney Takes More Heat on Abortion," *The Boston Globe*, September 10, 1994.

183   He later softened: Lehigh, "Kennedy, Romney Battle for the Middle."

183   appeared in June: Scot Lehigh, "Romney Admits Advice Against Abortion," *The Boston Globe*, October 20, 1994; Michael Levenson, "Romney Dismisses Photo Taken at Fund-raiser, Says He Is Firmly Antiabortion," *The Boston Globe*, December 19, 1997.

183   Ann Romney gave: Scott Helman, "Romney's Wife Made Contribution to Planned Parenthood," *The Boston Globe*, May 10, 2007.

183   October 1963 death: Ann Keenan, certificate of death, Michigan Department of Health, October 8, 1963.

183   Keenan's death: Interview with Jane Romney, 2007.

184   "championed a woman's right to choose": Romney for Governor campaign flyer, 2002.

184 In June 2005: Eileen McNamara, "Evolving History," *The Boston Globe*, June 26, 2005.

184 Abortion had been a defining issue: Interview with David Plawecki, 2011.

184 "I think we need": Noreen Murphy, "Lenore Speaks Out on Issues," *Owosso Argus-Press*, May 11, 1970.

184 "perverse": Scot Lehigh and Frank Phillips, "Romney Allegedly Faulted Gays in Talk to Mormons," *The Boston Globe*, July 15, 1994.

185 Romney maintained throughout: Lehigh, "Romney Admits Advice Against Abortion."

185 Further muddying his pitch: Bruce Mohl, "Mass. Antiabortion Group Backs Romney," *The Boston Globe*, September 8, 1994.

185 He supported raising: Interview with Mitt Romney on *This Week with David Brinkley*, ABC, October 16, 1994.

185 He rejected: Rod Dreher, "Kennedy Avoids Haymaker in Final Debate with Romney," *Washington Times*, October 28, 1994.

185 He backed two gun-control measures: Scott Helman, "Romney's Record on Guns Questioned," *The Boston Globe*, April 5, 2007; Scot Lehigh, "Lakian Vows to Fight Gun Control," *The Boston Globe*, August 19, 1994.

185 Romney also discouraged: Bradlee and Golden, "Strategies Shaped an Epic Race."

185 And he distanced himself: Dreher, "Kennedy Avoids Haymaker in Final Debate with Romney."

186 He expressed lukewarm support: Scot Lehigh and Frank Phillips, "Romney, Lakian Debate Crime, Tax and Abortion," *The Boston Globe*, September 7, 1994.

186 He was critical: Jeff Jacoby, "Lessons for Romney in Kennedy's Success," *The Boston Globe*, July 12, 1994.

186 Late in the race: John B. Judis, "Stormin' Mormon," *The New Republic*, November 7, 1994, 20.

186 A linchpin of Chafee's plan: Paul Starr, "What Happened to Health Care Reform?," *The American Prospect*, Winter 1995, 20–31.

186 "I told people": Interview with Mitt Romney.

186 Kennedy came out for "workfare": Peter G. Gosselin and Anthony Flint, "Liberties Taken with Some Facts," *The Boston Globe*, October 26, 1994.

186 And he tried: Lehigh, "Kennedy, Romney Battle for the Middle."

186 Everyone was there: Interviews with former campaign aides, 2011.

186 He later apologized: Frank Phillips, "Rep. Kennedy Apologizes to Romney on Mormon Issue," *The Boston Globe*, September 24, 1994.

187 "Where is Mr. Romney": Scot Lehigh, "Kennedy Believes Mormon-Racial Questions Proper," *The Boston Globe*, September 27, 1994.

187 "I do not speak": Scot Lehigh and Frank Phillips, "Romney Hits Kennedy on Faith Issue," *The Boston Globe*, September 28, 1994.

187   "George Romney literally dove": Interview with former Romney aide, 2011.

187   "I think it is absolutely wrong": Lehigh and Phillips, "Romney Hits Kennedy on Faith Issue."

187   "Every time": Bradlee and Golden, "Strategies Shaped an Epic Race."

187   "All my life": Ibid.

188   "I would ask Ann": Interview with former aide, 2011.

188   George's schedule was his own: Ibid.

188   "He's better": Hugh McDiarmid, "Ex-Governor Basks in Son's Strong Bid," *Detroit Free Press*, October 7, 1994.

189   A series of campaign appearances: *Campaign Almanac*, C-SPAN, 1994.

189   There was the time: Interview with former Romney aide, 2011.

189   Romney made a point: Interview with Rick Reed.

189   About a week before election day: Frank Phillips and Sally Jacobs, "Tracking Poll Shows Romney Gaining Ground," *The Boston Globe*, November 2, 1994.

189   Romney asked him: Interview with Ken Smith, 2011.

190   The day of the Boston debate: Bradlee and Golden, "Strategies Shaped an Epic Race."

190–191 "I saw little old ladies": Interview with former aide, 2011.

191   Kennedy basked: TV news coverage of October 1994 debate.

191   "You could stay overnight": Bradlee and Golden, "Strategies Shaped an Epic Race."

192   "I think the election is over": Interview with John Lakian.

192   "That question": Bradlee and Golden, "Strategies Shaped an Epic Race."

192   But Kennedy fared better: Canellos, *Last Lion*, 300.

192   With less than two weeks: Frank Phillips and Scot Lehigh, "Kennedy Opens Up 20-Point Lead in Poll," *The Boston Globe*, October 27, 1994.

193   Then, in his final debate: Bradlee and Golden, "Strategies Shaped an Epic Race."

193   "legacy of Kennedy country": Phillips and Jacobs, "Tracking Poll Shows Romney Gaining Ground."

193   "What do you expect": Frank Phillips and Scot Lehigh, "Poll: Kennedy Keeps Big Lead," *The Boston Globe*, October 29, 1994.

193   On the Southeast Expressway: Robert W. Trott, "Romney Angers Dorchester Residents with Remark About Urban Blight," Associated Press, October 28, 1994.

193   Ann Romney hadn't helped things: Thomas, "Ann Romney's Sweetheart Deal."

193   The campaign spent $100,000: Frank Phillips, "Romney Planning Half-Hour TV Spots," *The Boston Globe*, November 4, 1994.

193   TED KENNEDY OUT: Michael Cooper, "Romney Eclectic in Final Sprint," *The New York Times*, November 6, 1994.

193   In the days leading up: Interview with former Romney aide, 2011.

194   At about 8:45 a.m.: Anthony Flint, "Romney Run Comes to Bittersweet End," *The Boston Globe*, November 9, 1994.

194   That night, Romney put on: Ibid.

194   Ann Romney, worn out: Interview with former aide, 2011.

194   "You couldn't pay me": Bradlee and Golden, "Strategies Shaped an Epic Race."

194   But she also left an opening: Flint, "Romney Run Comes to Bittersweet End."

195   "Why couldn't he": Interview with former adviser, 2011.

195   He had sunk $3 million: Frank Phillips and Scot Lehigh, "GOP Captures Congress; Weld, Kennedy Reelected; Roosevelt, Romney Fall as State Bucks National Trend," *The Boston Globe*, November 9, 1994.

195   "I was surprised": Frank Phillips and Scot Lehigh, "Romney Says He'd Mull a Run Against Kerry," *The Boston Globe*, November 23, 1994.

195   "I think people were searching": Interview with Rick Reed.

195   "His main cause": Interview with longtime Republican, 2011.

195   Over dinner: Interview with Republican, 2011.

196   "I'm sure I ruffled a lot of feathers": Romney, *Turnaround*, 15.

Chapter 8: The Torch Is Lit

197   In their sixty-four years: Associated Press, "Lenore Romney, Widow of Former Governor, Dies," *Daily Globe* [Ironwood, Mich.], July 8, 1998.

197   At the dawn of their courtship: Marya Saunders and Bob Gains, "The Missionary Side of George Romney," *Family Weekly*, March 3, 1963, 5.

197   A little after 9 a.m: Ron Dzwonkowski and Angela Tuck, "Michigan Visionary; Colleagues Remember a Man of Uncommon Energy and Vision," *Detroit Free Press*, July 27, 1995; interview with former Romney aide, 2011.

198   On his birthday: Myrna Oliver, "George Romney; Governor Ran for President," *Los Angeles Times*, July 27, 1995.

198   "He got why I was": Interview of Jane Romney by Neil Swidey, 2006.

198   "His whole life": Interview with John Wright, 2011.

198   Some twelve hundred mourners: Malcolm Johnson, "Governor, Children Have Praise for George Romney," Associated Press, August 1, 1995.

198   "The experience of walking": Mitt Romney, *Turnaround: Crisis, Leadership, and the Olympic Games* (Washington, D.C.: Regnery Publishing, 2004), 16.

198   "I don't know": Interview with Mitt Romney, 2007.

199   The next five years: Romney, *Turnaround*, 16.

199   "I was in the investment business": Jo-Ann Barnas, "Romney Gives His All to Games; Michigan Native Saves Scandalous Olympics," *Detroit Free Press*, February 14, 2011.

199   In 1996, he was so concerned: David Warsh, "The Next Generation," *The Boston Globe*, February 21, 1999.

199   A couple years later: Interview with associate, 2011.

199   Romney also inquired: Interview with Republican, 2007.

200   Years earlier: Interviews with Grant Bennett and Ken Hutchins, 2011.

200   In the mid-1990s: Sridhar Pappu, "In Mitt Romney's Neighborhood, A Mormon Temple Casts a Shadow," *The Washington Post*, December 15, 2007.

200   After a visit: Interview with Grant Bennett.

200   "a reminder to the world": Michael Paulson, "A Visible Faith; for Mormon Church, Opening of Temple on Hill in Belmont Is Fulfillment of Dream, 'Reminder to the World,'" *The Boston Globe*, August 27, 2000.

200   "It makes you feel proud": Ibid.

200   But there was one glaring omission: Ibid.

200   One afternoon: Brian MacQuarrie, "Old Rivals Tour Mormon Temple," *The Boston Globe*, September 9, 2000.

201   The following spring: Michael Paulson and Caroline Louise Cole, "For Mormons, a Secular Victory; SJC Says Temple Can Have Its Steeple," *The Boston Globe*, May 17, 2001.

201   The more tests the doctor did: Interview with Mitt Romney.

201   "There weren't many": Ibid.

202   "She's failing test after test": Ibid.

202   "When I heard": Interview with Tagg Romney, 2007.

202   "It continued for Mitt": Neil Swidey and Stephanie Ebbert, "Journeys of a Shared Life," *The Boston Globe*, June 27, 2007.

202   The diagnosis also came: Niraj Warikoo, "Wife of Former Governor Had Opinions of Her Own," *Detroit Free Press*, July 8, 1998.

202   "Craig, I'm not going to die": Mike Melanson, "Ann Romney Tackles Multiple Sclerosis Head-On," *The Boston Globe*, September 13, 2003.

202   "I frankly would": Paula Parrish, "Mitt Romney Embraces Challenges, and This Might Be His Biggest," Scripps Howard News Service, February 4, 2002.

202   She was so weak: Annie Eldridge, "Dressage Makes Ann Romney's Soul Sing," *Chronicle of the Horse*, January 3, 2008.

203   She received intravenous: Shira Schoenberg, "Trail No Obstacle for Ann Romney; Uses Discipline to Manage MS," *The Boston Globe*, September 22, 2011.

203   "A lot of people": Interviews with John Wright and Grant Bennett.

203   Eventually, Ann hit upon: Michael Levenson, "Romney's Wife Opens Up on Campaign Issues," *The Boston Globe*, February 16, 2007.

203   "Riding exhilarated me": Eldridge, "Dressage Makes Ann Romney's Soul Sing."

203 "You don't have to be perfect": Interview with Tagg Romney.

203 "I am very strong right now": Matthew Rodriguez, "Romney Presses MS Awareness," *The Boston Globe*, May 12, 2004.

203 The lesson that she did not: Romney, *Turnaround*, 346.

204 So he went first to Ann: Ibid., 5.

204 Robert Garff, the chairman: Interview with Robert Garff, 2007.

205 "white knight": John Powers, "Hub's Romney Takes on Salt Lake City Games," *The Boston Globe*, February 12, 1999.

205 "Mitt had his father's": Interview with Robert Garff.

205 "I was looking for": Interview with Michael Leavitt, 2007.

205 With Romney emerging: Kristen Moulton, "Romney May Lead Salt Lake Committee," Associated Press, February 10, 1999.

205 "A search was never": Lisa Riley Roche and Alan Edwards, " 'White Knight' for SLOC Is Mitt Romney," *Deseret News*, February 11, 1999.

205 Huntsman would later assist: Thomas Burr, "Governor Supports McCain for President," *The Salt Lake Tribune*, July 20, 2006.

206 "Based on the fact": Interview with Michael Leavitt.

206 "Just think about it": Romney, *Turnaround*, 6.

206 One of his campaign consultants: Interview with Rick Reed, 2011.

206 "It was serendipitous": Interview with Robert Garff.

206 "It would have been a circus": Interview with Romney associate, 2011.

206 "We were facing a crucial event": Bob Gay, "Finding the Needle in a Haystack: Thoughts on Being Morally Courageous in Business," speech, 2003.

207 David D'Alessandro's office: Interview with David D'Alessandro, 2007.

207 "If this doesn't work": Ibid.

207 "I think he took it": John Powers, "Golden Opportunity," *The Boston Globe Magazine*, February 3, 2002.

207 "He loves emergencies": Ibid.

208 "We were in a psychological zombie-land": Barnas, "Romney Gives His All to Games."

208 "There is no justification": Salt Lake Organizing Committee news release, "W. Mitt Romney Acceptance Speech," February 11, 1999.

208 Whereas Nagano had plied: Mike Gorrell, "How the Games Were Won: Generosity or Bribery?," *The Salt Lake Tribune*, October 26, 2003.

208 Salt Lake City had handed out: Linda Fantin, "Top GOP Strategist Crafted 2002 Bid," *The Salt Lake Tribune*, July 3, 2000.

208–209 a stunning trove of booty: Special Collections on 2002 Olympic Winter Games, J. Willard Marriott Library, University of Utah.

209 T-shirts that depicted: Jerry Spangler, "Is SLOC Stifling Free Speech?," *Deseret News*, June 9, 1999.

209   "The best way": Interview with David D'Alessandro.

209   some people in Utah: Linda Fantin, "Huntsman Blasts [Olympic] Fund Raising," *The Salt Lake Tribune*, July 11, 1999.

209   In a state where nearly every: Church of Jesus Christ of Latter-day Saints, "Pres. Hinckley Speaks During Pioneer Devotional," July 28, 2001, http://www.ldschurchnews.com/articles/40269/Pres-Hinckley-speaks-during-Pioneer-devotional.html.

209   Documents in Garff's archives: Robert Heiner Garff Papers, J. Willard Marriott Library, Special Collections, University of Utah.

210   He had requested: Special Collections on 2002 Olympic Winter Games, J. Willard Marriott Library, University of Utah.

210   "We've got a chairman": Fantin, "Huntsman Blasts [Olympic] Fund Raising."

210   But church leaders: Interview with Richard Garff.

211   "was going to be": Powers, "Golden Opportunity"; interview with Fraser Bullock, 2007.

211   "We could be liquidated": Romney, *Turnaround*, 109.

211   After that he ordered pizza: Ibid., 55.

211   "Mitt was very adamant": Interview with Joan Guetschow, 2007.

211   "I ate more chili": Interview with David D'Alessandro.

211   When a local paper: Powers, "Golden Opportunity."

211   A review of archived records: Special Collections on 2002 Olympic Winter Games, J. Willard Marriott Library, University of Utah.

212   Direct federal aid: Interview with Robert Bennett, 2007.

212   "Most of the federal money": Ibid.

212   "Yes, we were out": Interview with Richard Garff.

212   But Fraser Bullock said: Interview with Fraser Bullock, 2007.

212   "The tsunami": Romney, *Turnaround*, 52.

212   "The I.O.C.'s sponsorships": Jeré Longman, "Potential Olympic Sponsors Said to Be Uneasy," *The New York Times*, January 21, 1999.

213   That cleared the way: Interview with Fraser Bullock.

213   His salesmanship: Special Collections on 2002 Olympic Winter Games, J. Willard Marriott Library, University of Utah.

213   The company had paid: "NBC Lands Sydney, Salt Lake Olympics," United Press International, August 7, 1995.

213   "the most powerful person": "The Sporting News: Most Powerful 100," *Sporting News*, December 30, 1996.

213   "I have no doubt": Interview with Dick Ebersol, 2007.

213   Yet Romney himself: Interview with Fraser Bullock.

213   "That's not a conflict": Interview with Mitt Romney.

213   In Garff's view: Interview with Richard Garff.

214   To be sure, Romney's entanglements: Mike Gorrell, "Romney Takes On Top Job at SLOC," *The Salt Lake Tribune*, February 12, 1999.

214   "They removed the people": Interview with Glenn Bailey, 2007.

214   Romney acted quickly: Romney, *Turnaround*, 29.

214   "We amassed significant": Interview with Max Wheeler, 2007.

214   "If you're going to fault": Interview with Zianibeth Shattuck-Owen, 2007.

215   "All we did": Interview with Richard Garff.

215   But Romney barred: Jennifer Toomer-Cook, "Wall of Honor Omits Welch and Johnson," *Deseret News*, December 8, 2001.

215   And he went so far: Interview with Tom Welch, 2007.

215   "Mitt called": Interview with Sydney Fonnesbeck, 2007.

215   "Tom was represented": Interview with Max Wheeler.

215   Yet even after the charges were dismissed: Romney, *Turnaround*, 23.

216   "Mitt's objective": Interview with Tom Welch.

216   The government's chief prosecutor: Interview with Richard Wiedis, 2007.

217   "He tried very hard": Interview with Ken Bullock, 2007.

217   Romney became the first: Debra Rosenberg, "Talk of the Town," *Newsweek*, February 8, 2002.

217   One pin depicted: Interview with Ken Bullock.

217   "You don't want me": Ibid.

217   a "very destructive" role: Interview with Rocky Anderson, 2007.

217   "Mitt saw him": Interview with Richard Garff.

217   "I wanted to know": Interview with Lillian Taylor, 2007.

218   Pace's group produced: Spangler, "Is SLOC Stifling Free Speech?"

218   "His first day in Utah": Interview with Stephen Pace, 2007.

218   Romney also impressed: Interview with Joan Guetschow.

219   "Both the Job Corps student": Interview with Terry Shaw, 2007.

219   Romney denied using: Interview with Mitt Romney.

219   "There were a lot of people": Interview with Peter Dawson, 2007.

219   The warm morning air: Yvonne Abraham, "True Believer; Aide Pushes Hard for Romney Agenda," *The Boston Globe*, March 30, 2003.

219   "like war": Romney, *Turnaround*, 302.

219   The blueprint had to be rewritten: Ibid., 303.

220   "I think Mitt wondered inwardly": Powers, "Golden Opportunity."

220   While he addressed: "Romney Rallies Olympic Staff," Associated Press, September 14, 2001.

220   "By the end": Interview with Zianibeth Shattuck-Owen.

220   "In the annals": Romney, *Turnaround*, 305.

220   The Salt Lake Games: "FBI Chat with Don Johnson, Special Agent in Charge of FBI Salt Lake City," Counterterrorism and Security Reports, September–October 2001.

220   "It was very easy": Interview with Robert Bennett.

221   Trouble began: Interview with and e-mails from A. J. Barto, 2007.

221   That made Romney: Stephanie Ebbert and Joanna Weiss, "Olympic Tickets Went to Legislators, Not Sept. 11 Families," *The Boston Globe*, April 12, 2002.

221   "I was outraged": Interview with A. J. Barto.

221   "It's been much more somber": Powers, "Golden Opportunity."

221   IOC officials had argued: Interview with Sandy Baldwin, 2007.

221   they ultimately relented: John Powers, "A Memory Relighted; at Winter Games, Symbols of Resolve and Renewal," *The Boston Globe*, February 9, 2002.

221   "Dwight Eisenhower": Ibid.

222   "gone through all the usual": Ibid.

222   winning thirty-four medals: Vicki Michaelis, "USA Basks in the Luster of Its Heavy Medal Push," *USA Today*, February 25, 2002; "World Viewership of Games Increases," *The New York Times*, June 18, 2002.

222   "The people who say": Interview with David D'Alessandro.

222   "It was an amazing thing": Schoenberg, "Trail No Obstacle for Ann Romney."

Chapter 9: The CEO Governor

224   A year before: Lisa Riley Roche and Bob Bernick, Jr., "Public Service for Romney," *Deseret News*, August 20, 2001.

224   "If politics is": John Powers, "Golden Opportunity," *The Boston Globe Magazine*, February 3, 2002.

224   Friends and his former chief: Ibid.

224   "I do not wish": Mitt Romney, letter to the editor, *The Salt Lake Tribune*, July 11, 2001.

225   And Massachusetts was clearly: LaVarr Webb and Ted Wilson, "Romney Has a Shot at Utah Governorship," *Deseret News*, December 30, 2001.

225   Romney had said: Powers, "Golden Opportunity."

225   "People are just sort of": Scott Helman, "GOP Activists Angling Toward Romney," *The Boston Globe*, March 15, 2002.

225   "I know you're really busy": Interview with Barbara Anderson, 2007.

225   He was noncommittal: Interview with Kerry Healey, 2007.

225   "The guy looks like": Dennis Romboy, "Mitt's Stock Is Sky-High," *Deseret News*, February 28, 2002.

225   "It was clear": Interview with Massachusetts Republican, 2007.

NOTES

[page 391]

NOTES

391

225 Romney never called: Interview with person with knowledge of the ballroom reservation, 2011.

226 "huge qualms": Glen Johnson, "Mitt Romney's Tough Decision," *The Boston Globe*, March 17, 2002.

226 a fresh poll indicating: Joe Battenfeld, "It's Mitt's Party—Mere 12% in GOP Stand with Swift If Romney Runs for Gov," *Boston Herald*, March 17, 2002.

226 The mother of three young children: Frank Phillips, "Swift Yields to Romney, Saying 'Something Had to Give,' Exits Race for Governor," *The Boston Globe*, March 20, 2002.

226 "Lest there be any doubt": Ibid.

226 "I learned in my race": "Cover Story," New England Cable News, October 31, 2002.

227 $6.3 million of his own money: Frank Phillips and Brian Mooney, "Governor's Race May Set a Record," *The Boston Globe*, June 10, 2006.

227 After evaluating a list: "Mitt Romney Running Mate Drops Middle Name for the Campaign," Associated Press, May 1, 2002.

227 the expedience of their alliance: Globe staff writers, "Romney Is Bedeviled by the Details During Radio Talk Show Appearance," *The Boston Globe*, April 14, 2002.

227 "Any effort": Frank Phillips and Rick Klein, "Democrats File to Halt Romney Bid," *The Boston Globe*, June 8, 2002.

228 But soon after: Ibid.

228 choosing George W. Bush over John McCain: Interview with Mitt Romney, 2007.

228 "credible in all respects": State Ballot Law Commission decision, June 25, 2002.

228 O'Brien herself had opposed: Stephanie Ebbert, "Major Abortion Rights Groups Give Nod to O'Brien," *The Boston Globe*, October 3, 2002.

229 "My position has not changed": Frank Phillips, "Romney Carries the Day with GOP," *The Boston Globe*, April 7, 2002.

229 "I respect": Ibid.

229 Ann Romney also tried: Ann and Mitt Romney, TV interview, www.youtube.com/watch?v=GKwVNUz52vo, 2007.

229 His answers: National Abortion Rights Action League questionnaire, April 8, 2002.

230 "We applaud your commitment": Republican Pro-Choice Coalition letter to Mitt Romney, October 2, 2002.

230 "promised to protect": Campaign flyers, Romney for Governor, 2002.

230 But Romney quickly distanced himself: Rick Klein, "Romney Kin Signed Petition to Ban Same-Sex Marriage," *The Boston Globe*, March 22, 2002.

230 Romney did not support: *Bay Windows* questionnaire, 2002.

230 Richard Babson, a board member: Interview with Richard Babson, 2007.

231   In addition, Romney said: "Don't Dismiss Romney, Gay Republicans Say,"
       Bay Windows, October 24, 2002.

231   his campaign distributed pink flyers: Campaign flyers, Romney for Gover-
       nor, 2002.

231   "protect already established rights": *Bay Windows* questionnaire.

231   Romney, according to one participant: Interviews with the meeting par-
       ticipant and another person familiar with the meeting, 2007.

231   "At a very young age": *Bay Windows* questionnaire.

232   "clean up the mess": Scott Helman, "Independents Key in Western Sub-
       urbs," *The Boston Globe*, November 14, 2002.

232   "microtargeting": Scott Helman, "Candidates Spend Heavily on Voter
       Lists," *The Boston Globe*, July 19, 2007.

232   He insisted: Rick Klein, "A Look at the Facts Behind the Exchanges," *The
       Boston Globe*, October 2, 2002.

232   "government by gimmickry": Scott Helman, "Romney Finds 'No New
       Taxes' Promise Suits Him After All," *The Boston Globe*, January 5, 2007.

232   That idea unnerved: Interview with former aide, 2011.

232   One of the first things: Interview with former Romney aide, 2011.

233   "Massachusetts," she said: Frank Phillips, "It's O'Brien for Democrats," *The
       Boston Globe*, September 18, 2002.

233   "I think it made him": Interview with Michael Murphy, 2007.

233   "Mike came and said": Brian C. Mooney, "Romney Guru Thrives in Politi-
       cal 'Show Business,'" *The Boston Globe*, June 12, 2005.

234   But Romney and his advisers: Scot Lehigh, "The Moment of Truth," *The
       Boston Globe*, November 6, 2002.

234   "That last debate": Ibid.

234   "as disgraceful": Rick Klein, "Ads Link O'Brien Spouse, Fund Loss," *The
       Boston Globe*, October 21, 2002.

234   "It was the Ted Kennedy punch": Interview with Michael Travaglini, 2007.

235   Romney had climbed back: Frank Phillips, "Poll Finds Romney and
       O'Brien in Dead Heat," *Boston Globe*, November 1, 2002.

235   "We took on": Frank Phillips, "Romney Sails to Victory," *The Boston Globe*,
       November 6, 2002.

235   Romney, while celebrating: David Guarino, "Romney Tops O'Brien for
       Corner Office," *Boston Herald*, November 6, 2002.

236   A few weeks after being sworn in: Interviews with current and former
       legislators, 2011.

236   "My usual approach": Interview with Andrea Nuciforo, 2011.

237   "in memory of Mitt Romney": Massachusetts Senate session transcript,
       State House News Service, July 17, 2003.

238   "We probably presented him": Interview with Robert Pozen, 2007.

238   "the most important": Anthony Flint, "Development's Impact to Test Next Governor," *The Boston Globe*, October 12, 2002.

238   "a decision maker": Interview with Eric Kriss, 2007.

239   The budget gap never became: Scott S. Greenberger, "Romney Often Casts Himself as Budget Hero but Speeches Omit Some Important Detail," *The Boston Globe*, October 24, 2005.

239   When Romney left office: Interview with Geoffrey Beckwith, executive director of the Massachusetts Municipal Association, 2007.

240   "Cities and towns": Interview with Eric Fehrnstrom, 2007.

240   "What people don't credit": Interview with Barbara Anderson, 2007.

240   "the most significant restructuring": Rick Klein, "Layoffs, New Fees in Romney Budget," *The Boston Globe*, February 27, 2003.

240   After four years: E-mail from Eric Fehrnstrom, April 2007, based on state Human Resources Division data.

241   Weld, for example: Frank Phillips and Peter J. Howe, "Hill Deal Hurts Weld's Outsider Image," *The Boston Globe*, December 11, 1994.

241   "Weld had a genuine curiosity": Interview with Thomas Finneran, 2007.

241   "You remember Richard Nixon": Interview with Democratic lawmaker, 2011.

241   "My program for creating jobs": Brian C. Mooney, Stephanie Ebbert, and Scott Helman, "Ambitions Grow and the Stances Shift," *The Boston Globe*, June 30, 2007.

241   The state lost about 200,000 jobs: U.S. Bureau of Labor Statistics, http:// www.bls.gov.

242   about a 1 percent increase: Ibid.

242   It was the fourth weakest: Interview with Mark Zandi, chief economist for Moody's Analytics, 2007.

242   Under Ranch C. Kimball: Ranch Kimball, PowerPoint presentation, 2006; intervieew with Kimball, 2007.

242   In 2006, Bristol-Myers Squibb: Massachusetts Executive Office of Housing and Economic Development data, August 2011; www.bms.com/sustainabil ity/worldwide_facilities/north_america/Pages/devens_massachusetts.aspx.

242   an unusual show of teamwork: Stephen Heuser, "Teamwork Landed Bristol-Myers," *The Boston Globe*, June 3, 2006.

242   "His theory of government": Interview with Salvatore DiMasi, 2007.

243   One was a bill: Scott S. Greenberger, "Lawmakers Toughen Drunken Driving Bill," *The Boston Globe*, October 28, 2005.

243   Another came when Romney refused: Raphael Lewis, "Lawmakers to Rescind Retroactive '02 Tax Bills," *The Boston Globe*, December 2, 2005.

243   "an embarrassment": Scott Helman, "Wiretap Mosques, Romney Suggests," *The Boston Globe*, September 15, 2005.

243   But Romney committed the state: Michael Levenson, "With Bay State Relief Effort, Romney Gets Moment to Shine," *The Boston Globe*, September 11, 2005.

244   "You're too big": Interview with witness, 2011.

244   As Romney and Amorello: Ibid.

244   "disappointment": Kimberly Atkins, "Pike Boss in Hot Water," *Boston Herald*, July 12, 2006.

244   "At a moment of crisis": Interview with David Luberoff, 2007.

244   Tony Kimball, Romney's former colleague: Interview with Tony Kimball, 2011.

245   Bulger had invoked: Shelley Murphy, "Bulger Refuses to Give Answers to House Panel," *The Boston Globe*, December 7, 2002.

245   In June 2003: Shelley Murphy, "Grilled by US Panel, Bulger Says He Did Not Aid Brother," *The Boston Globe*, June, 20, 2003.

245   The next day: Frank Phillips, "Governor Raps Bulger, Eyes Ouster," *The Boston Globe*, June 4, 2003; Shelley Murphy, "On Brother, Bulger's Memory Lapsed," *The Boston Globe*, December 4, 2002.

245   After trustees rebuffed Romney: Jenna Russell and Patrick Healy, "UMass Trustees Rebuff Governor on Bulger," *The Boston Globe*, June 27, 2003.

245   Lawyers negotiated: Jenna Russell, "Bulger Makes His Exit," *The Boston Globe*, August 7, 2003.

246   His hasty ouster: Frank Phillips, "New Head of Civil Service Resigns," *The Boston Globe*, August 29, 2003.

246   "Bill, my stomach is turning": Interview with William Monahan, 2007.

246   Citing the litigation: Interview with Mitt Romney.

247   "The first couple of years": Interview with James Vallee, 2007.

247   Romney would later contend: Interview with Mitt Romney.

247   "I never asked": Interview with Robert Travaglini, 2007

248   "He forced all of us": David Weber, "Democrats Take Sole Control of Statehouse as Romney Leaves Office," Associated Press, January 4, 2007.

248   "The review process was completely apolitical": Interview with Ralph Martin, 2007.

248   A July 2005 review: Raphael Lewis, "Romney Jurist Picks Not Tilted to GOP," *The Boston Globe*, July 25, 2005.

248   He similarly showed: Analysis of contribution records from Massachusetts Office of Campaign and Political Finance, 2011.

249   "I wanted to change": Interview with Mitt Romney.

249   reporters observed him: Sean Murphy and Connie Paige, "He Is the Director of the Department of Labor, the State Pays Him $108,000 a Year. But What Does Angelo Buonopane Do?" *The Boston Globe*, April 3, 2005.

249   Eric Fehrnstrom: Frank Phillips, "Romney Puts a Top Aide in Line for Pension," *The Boston Globe*, November 22, 2006.

249   "unwarranted political attacks": Frank Phillips, "Romney's Top Aide Leaves New Post," *The Boston Globe*, November 24, 2006.

249   Using the word "reform": Frank Phillips, "Address Opens Struggle with Democrats," *The Boston Globe*, January 16, 2004.

249   "Quite simply, reform": Mitt Romney, State of the State address, January 15, 2004, Massachusetts State Archives.

250   He began an aggressive recruiting drive: Raphael Lewis, "Romney Hosts Gala for Recruits," *The Boston Globe*, May 26, 2004.

250   Democrats were outraged: Raphael Lewis, "GOP Falls Short in State," *The Boston Globe*, November 3, 2004.

250   "We could have had": Interview with Massachusetts Republican, 2007.

250   With only 21 of 160 seats: Raphael Lewis, "GOP Has Its Fewest Seats Since 1867," *The Boston Globe*, November 4, 2004.

250   "He put his personal reputation": Lewis, "GOP Falls Short in State."

251   "From now on": Joan Vennochi, "See Mitt Run," *The Boston Globe*, January 4, 2005.

251   "The whole climate changed": Interview with Robert Travaglini.

251   That September: "A Big Stage for Romney," editorial in *The Boston Globe*, September 2, 2004.

251   ratcheted up the chatter: Rick Klein, "As Romney Rips Kerry, Iowans See an Audition," *The Boston Globe*, October 17, 2004.

251   Romney, in fact: Interviews with people familiar with the meetings, 2007.

251   The PAC went on: Scott Helman, "Romney Faces a Reckoning on '08," *The Boston Globe*, November 17, 2006.

252   "I didn't know": Interview with Mitt Romney.

252   Timing was everything: Interview with Republican, 2007.

252   Romney's furious preparations: Scott Helman and Michael Levenson, "Romney Camp Consulted with Mormon Leaders," *The Boston Globe*, October 19, 2006.

253   "We know Mitt": Thomas Burr, Joe Baird, and Peggy Fletcher Stack, "Romney Pal Takes Blame for Dust-Up," *The Salt Lake Tribune*, October 23, 2006.

253   a tally at the end of 2006 found: Brian C. Mooney, "Romney Left Mass. on 212 Days in '06," *The Boston Globe*, December 24, 2006.

253   On November 9, 2004: Scott Helman, "Romney's Journey to the Right," *The Boston Globe*, December 17, 2006.

254   On February 10, 2005: Pam Belluck, "Massachusetts Governor Opposes Stem Cell Work," *The New York Times*, February 10, 2005.

254   Some scientists wondered: Helman, "Romney's Journey to the Right."

254   "There's evidence": Scott S. Greenberger and Frank Phillips, "Romney Draws Fire on Stem Cells," *The Boston Globe*, February 11, 2005.

254   "Changing my position": Interview with Mitt Romney.

255   He turned Harvard: Helman, "Romney's Journey to the Right."

255   "convictions have evolved": Mitt Romney, "Why I Vetoed Contraception Bill," *The Boston Globe*, July 26, 2005.

255   Supporters of the bill: Helman, "Romney's Journey to the Right."

255   "an abortion pill": Pam Belluck, "Massachusetts Veto Seeks to Curb Morning-After Pill," *The New York Times*, July 26, 2005.

256   By 2006, Romney's aides: Maggie Mulvihill and Kimberly Atkins, "Romney Veers Right on Abortion, Gay Adoption," *Boston Herald*, March 1, 2006.

256   "He can come": Interview with former adviser, 2011.

256   Romney adopted just such a program: Helman, "Romney's Journey to the Right."

256   He backed out: Beth Daley, "Mass. Pulls Out of Agreement to Cut Power Plant Emissions," *The Boston Globe*, December 15, 2005.

257   Republican "extremists": Scott Helman, "Romney's '94 Remarks on Same-Sex Marriage Could Haunt Him," *The Boston Globe*, December 8, 2006.

257   He railed against "activist judges": Scott Helman, "Romney Rips SJC's Justices on Values," *The Boston Globe*, November 11, 2005.

257   He pressed Congress: Raphael Lewis, "Romney Urges Federal Ban on Gay-Marriage 'Experiment,'" *The Boston Globe*, June 23, 2004; Frank Phillips and Lisa Wangsness, "Same-Sex Marriage Ban Advances," *The Boston Globe*, January 3, 2007.

257   "the religion of secularism": Maria Cramer, "Broadcast Blasts Same-Sex Marriage," *The Boston Globe*, October 16, 2006.

257   "Some are actually": John J. Monahan, "Romney Talk Riles Gay Parents," *Worcester Telegram & Gazette*, February 26, 2005.

257   "San Francisco east": Peter Wallsten, "Activists Remember a Different Romney," *Los Angeles Times*, March 25, 2007.

257   He sought to amend: Jonathan Saltzman, "Romney Bill Seeks Adoption Exemption," *The Boston Globe*, March 16, 2006.

257   He backed away: Dan Balz and Shailagh Murray, "Mass. Governor's Rightward Shift Raises Questions," *The Washington Post*, December 21, 2006.

257   And, after a flap: "Romney Abolishes Governor's Commission on Gay and Lesbian Youth," Associated Press, July 21, 2006.

258   "He dealt a death blow": Interview with David LaFontaine, 2011.

258   "I stood at the center": Adam Nagourney, "Republican Hopefuls Seek Advantage on Social Issues," *The New York Times*, March 3, 2007.

258   "I know there are some": Interview with Mitt Romney.

258   The liberal House speaker: Scott Helman, "A Breakfast Best Served Cold," *The Boston Globe*, March 20, 2006.

259   "It's almost as though": Interview with Richard Tisei, 2007.

259 A majority of voters: Lisa Wangsness, "Voters Voice Regard, Regret over Romney," *The Boston Globe*, December 26, 2006.

Chapter 10: Health Care Revolutionary

261 It was a sunny October afternoon: Interview with Richard Whitney, 2011. Unless otherwise noted, all 2011 interviews for this chapter were conducted by Brian C. Mooney for a two-part *Boston Globe* series on Mitt Romney's push for universal health care.

263 a "doubting Thomas": Interview with Thomas Stemberg, 2011.

263 "It struck me": Interview with Mitt Romney, 2011.

263 "You can find a way": Mitt Romney, "Health Care for Everyone? We Found a Way," *The Wall Street Journal*, April 11, 2006.

264 "The next thing I know": Interview with Thomas Stemberg.

264 "In the same bipartisan fashion": Mitt Romney, "My Plan for Massachusetts Health Insurance Reform," *The Boston Globe*, November 21, 2004.

264 "This administration hasn't been willing": Benjamin Gedan, "Romney Plan on Health Insurance Questioned," *The Boston Globe*, November 22, 2004.

264 "We're basically stalemated": Benjamin Gedan and Scott S. Greenberger, "Kennedy Lauds Romney for Airing Plan," *The Boston Globe*, November 24, 2004.

265 Romney had provided: Interview with Stacey Sachs, 2011.

265 His first secretary: Ronald Preston, "Health Care Reform: Covering the Commonwealth's Uninsured Residents," May 3, 2004.

265 "You were talking": Interview with Christine Ferguson, 2011.

265 And the picture that emerged: "Commonwealth Care," PowerPoint presentation, Massachusetts State Archives, February 28, 2005.

266 Gruber had a computer model: Interview with Amy Lischko, 2011.

266 "made enormous sense": Interview with Jonathan Gruber, 2011.

266 "They were making the realpolitik argument": Ibid.

266 "We discussed the need": Interview with Amy Lischko.

267 "Everybody in the room": Interviews with Timothy Murphy, 2007.

267 In the early 1990s: Stuart M. Butler, "The Heritage Consumer Choice Health Plan," Heritage Foundation, March 5, 1992; Robert E. Moffit, "Personal Freedom, Responsibility, and Mandates," *Health Affairs*, January 1994.

267 One of the early GOP advocates: Health Equity and Access Reform Today Act, introduced November 23, 1993, by John Chafee and twenty cosponsors.

268 "What about the individual requirement?": Interviews with Andrew Dreyfus and Timothy Murphy, 2011.

268 "No more free riding": Scott S. Greenberger, "Romney Eyes Penalties for Those Lacking Insurance," *The Boston Globe*, June 22, 2005.

269 "eureka moment": Interview with Timothy Murphy, 2011.

269   a higher percentage of employers offered: Massachusetts Division of Health Care Finance and Policy, "Massachusetts Employer Survey 2007."

269   Massachusetts, too: Randall R. Bovbjerg, Alison Evans Cuellar, and John Holahan, "Market Competition and Uncompensated Care Pools," Urban Institute, March 2000, http://www.urban.org/publications/309525.html.

269   It was also one: Leigh Wachenheim and Hans Leida, "The Impact of Guaranteed Issue and Community Rating Reform on Individual Insurance Markets," America's Health Insurance Plans, July 10, 2007.

269   But by the time: Dolores Kong, "Beacon Hill Reopens Health Care Debate," *The Boston Globe*, May 1, 2005; interview with Bruce Bullen, 2011.

269   Leading the way: Interviews with Jack Connors, 2011.

270   But he faced an obstacle: Interview with John Sasso, 2011.

270   O'Donnell warned Sasso: Interviews with Joseph O'Donnell and John Sasso, 2011.

270   "I've got to admit": Mitt Romney, "Ted's Attack for Success," *The Boston Globe*, January 30, 1995.

270   When the three sat down: Interviews with Joseph O'Donnell and John Sasso, 2011.

270   "This was talking about policy": Interview with Mitt Romney.

271   On January 14, 2005: Alice Dembner and Rick Klein, "Mass., US Reach Deal on Funding; Averts Major Cut in Medicaid Match," *The Boston Globe*, January 15, 2005.

271   "I was never absolutely sure": Interview with Edward Kennedy, 2007.

271   "People kept coming down": Interview with Stacey Sachs.

271   Neither Kennedy nor his staff: Ibid.

272   "It didn't go as smooth": Interview with Mitt Romney.

272   "Romney knew his best chance": Interview with Alan Macdonald, 2011.

272   In early December 2005: Scott S. Greenberger, "Kennedy Joins Mass. Healthcare Push," *The Boston Globe*, December 5, 2005.

273   The state missed: Interview with Timothy Murphy, 2011.

273   "The plane is circling": Ibid.

273   "How often does the governor": Interview with Robert Travaglini, 2007.

273   About two weeks later: Interview with Salvatore DiMasi, 2007.

273   The group of business leaders: Christopher Rowland, "Health Executives Emerge as State's New Power Players," *The Boston Globe*, March 13, 2006.

273   The next night: Interview with Robert Travaglini, 2011.

274   "Today, Massachusetts has set itself apart": Scott Helman, "Mass. Bill Requires Health Insurance," *The Boston Globe*, April 4, 2006.

274   "We are where we'd hoped": Mitt Romney, press conference, March 3, 2006.

275      $45 million was the estimate: Helman, "Mass. Bill Requires Health Insurance."

276   "It didn't bother me": Interview with Robert Travaglini by author, 2011.

276   An audience of several hundred VIPs: "Signing of an Act Providing Access to Affordable, Quality, Accountable Health Care," video, City of Boston Cable Office video library.

276   Privately, Romney had been uneasy: Interview with former Romney aide by author, 2011.

277   "I have to admit": Mitt Romney, press availability, April 12, 2006.

277   "We carried out an experiment": Interview with Mitt Romney.

277   "I don't know": Interview with Michael Leavitt, 2007.

277   A detailed examination: Brian C. Mooney, " 'RomneyCare'—A Revolution That Basically Worked," *The Boston Globe*, June 26, 2011.

278   Those are the lowest: "Health Insurance Coverage in Massachusetts: Results from 2008–2010 Massachusetts Health Insurance Surveys," Massachusetts Division of Health Care Finance and Policy, December 2010, http://www.mass.gov/Eeohhs2/docs/dhcfp/r/pubs/10/mhis_report_12-2010.pdf.

278   Recent U.S. Census data: U.S. Census Bureau, Health Insurance Historical Tables, www.census.gov/hhes/www/hlthins/data/historical/HIB_tables.html.

278   Many more businesses: "Massachusetts Employer Survey 2010," Massachusetts Division of Health Care Finance and Policy, July 2011, http://www.mass.gov/Eeohhs2/docs/dhcfp/r/pubs/11/mes_results_2010.pdf.

278   And support for the requirement: Kay Lazar, "Support for Health Law Rises; Residents Split on Coverage Mandate," *The Boston Globe*, June 5, 2011.

279   "I was a little concerned": Mitt Romney, appearance at National Review Institute conference, Washington, D.C., January 27, 2007.

279   (Romney later contended): Interview with Mitt Romney, 2007.

279   He later blamed: David S. Bernstein, "Romney Rewrites, Sampler," *Boston Phoenix*, February 10, 2011, http://thephoenix.com/Blogs/talkingpolitics.

279   "A lot of pundits": Matt Viser, "Romney Says He Stands by Mass. Law; Offers a Plan Enabling States on Health Care," *The Boston Globe*, May 13, 2011.

279   He recounted an instance: Interview with Mitt Romney.

279   "Significant successes": Interview with Brian Gilmore, 2007.

279   "I'm not naive enough": Scott Helman, "Romney Exits with Pomp, Ambition; Mixed Legacy: Health Plan, Jobs, Fee Hikes," *The Boston Globe*, January 4, 2007.

280   "The truth is": Interview with Mitt Romney, 2007.

280   On the evening of January 3, 2007: Andrea Estes and Scott Helman, "Romney Exits with Pomp, Ambition; Ends Term, Takes 1st Formal Step for White House Bid," *The Boston Globe*, January 4, 2007.

Chapter 11: A Right Turn on the Presidential Trail

281   "I'm not changing my religion": Interviews with Doug Gross, Richard
         Schwarm, and third individual at Ritz-Carlton meeting, 2011.

282   The things that troubled Gross: According to Romney's spokesman, Eric
         Fehrnstrom, Romney would not discuss what happened in the failed presi-
         dential campaign, saying, "I can't imagine a more unproductive use of our
         time." Interview with Eric Fehrnstrom, 2011.

283   perception—phony: 2006 PowerPoint presentation for Mitt Romney cam-
         paign.

284   "optimistic, conservative leader": Ibid.

284   "There is a perception": Harvard Institute of Politics, *Campaign for President:
         The Campaign Managers Speak* (Lanham, Maryland: Rowman & Littlefield,
         2009), 17. See also "Dole Institute of Politics Post Election Conference,"
         www.doleinstitute.org/documents/Post-Election08-Transcript1.pdf.

284   A poll: *Los Angeles Times*/Bloomberg Poll, published in *The Hotline*, July
         5, 2006; Gallup Poll, October 3, 2006, Gallup News Service. (For com-
         parison, the Gallup Poll said that 40 percent of Americans believed the
         country was not ready for an African-American president.)

285   "You can't pay me": Interview with Mark DeMoss, 2011.

286   "You do understand": Interview with Richard Land, 2011.

286   Romney promised to think about it: Kennedy's speech is included as an
         appendix in a book written by Land, which Land presented to Romney.
         The book is Richard Land, *The Divided States of America: What Liberals and
         Conservatives Get Wrong About Faith and Politics* (Nashville, Tenn.: Thomas
         Nelson, 2011).

286   "There is always": Interviews with Richard Land and Mark DeMoss.

286   "dive right": Interview with Romney aide, 2011.

287   "Svengali," "mastermind," "alter ego": Interviews with Romney aides, 2011.

287   It was Murphy who: Brian C. Mooney, "Romney Guru Thrives in Political
         'Show Business,'" *The Boston Globe*, June 12, 2005.

288   "sheer brilliance": Mitt Romney, *Turnaround: Crisis, Leadership, and the
         Olympic Games* (Washington, D.C.: Regnery Publishing, 2007), 381.

288   "a pro-life Mormon": John J. Miller, "Matinee Mitt—The Governor of
         Massachusetts May Soon Be Appearing in a (Political) Theater Near You,"
         The National Review, June 20, 2005.

288   I'm out: Interviews with Romney aide, 2011.

288   "Mike Murphy grew up": Interview with Doug Gross.

288   "Beth said, 'I don't want to be'": Interview with Romney aide, 2011.

289   "Win early": Interviews with Romney aides. Myers declined comment.

289   Romney ended the day: Scott Helman, "On to a New Campaign Trail—
         Romney Event Nets More than $6.5M," *The Boston Globe*, January 9, 2007.

289   "akin to a nightmare": Michael Levenson, "Top Spender Romney Could Soon Run Short," *The Boston Globe*, October 15, 2007.

289   "A loan's a loan": Scott Helman, "Romney Tops GOP in Race for Funds," *The Boston Globe*, April 3, 2007.

289   In fact, Romney eventually loaned: Michael Kranish, "Romney Not Getting His $45M Back, Says He Won't Seek Gifts to Repay Campaign Loans," *The Boston Globe*, July 17, 2008.

289   Bruce Keough: Interview with Bruce Keough, 2011.

290   "the nation's second-least-churchgoing state": Frank Newport, "Mississippians Go to Church the Most; Vermonters, Least," www.gallup.com/poll/125999/mississippians-go-church-most-vermonters-least.aspx.

291   "That was an argument": Interview with Richard Schwarm.

291   The memo noted: Romney campaign strategy memo by South Carolina advisers.

291   "didn't want to deal with it": Interview with Romney campaign aides, 2011.

292   In his speech he emphasized: Mitt Romney, announcement speech, February 14, 2007.

292   He felt as if everyone: Interview with Romney campaign aide, 2011. ABC-TV/*Washington Post* poll, cited in Associated Press, "National Presidential Preference Poll Results," February 28, 2007.

293   "effectively prochoice": Romney-O'Brien gubernatorial debate, October 30, 2002. Romney frequently acknowledged that he had been "effectively prochoice" including during an interview on Fox News Channel on January 6, 2008.

294   An analysis of advertising: Nielsen Co., analysis of 2008 presidential campaign commercials, spreadsheet prepared at request of *The Boston Globe*.

294   "He's not a very notional leader": Thomas Beaumont, " 'CEO' Takes Businesslike Approach," *Des Moines Register*, December 26, 2007.

295   "we had a lot of data": Interview with Doug Gross; interviews with Huckabee aides, 2011.

295   "Wait, he's not dead yet": Harvard Institute of Politics, *Campaign for President*, 65.

295   It showed Romney saying: John Dickerson, "A Sneak Preview of John McCain's Secret Anti-Romney Ad," *Slate*, www.slate.com/id/2181005/.

296   "we were bleeding": interview with Romney aides, 2011; Interview with Alex Castellanos, 2011.

297   "the Bain way": Interviews with Romney aides, 2011. (Stuart Stevens and Russ Schriefer declined comment.)

297   "he was the turnaround guy": Interview with Mandy Fletcher, 2011.

297   "The glaring deficiency": Interview with Warren Tompkins, 2011.

297   "There is definitely a creative dynamic": Jonathan Martin, "Dissension Hits Romney Campaign," November 14, 2007, www.politico.com/news/stories/1107/6870.html.

298   But Romney was committed: Lisa Wangsness, "Romney Pulls Out All the Stops for Iowa GOP Straw Poll Today; Top Rivals Opt Out, Casting Doubt on Event Significance," *The Boston Globe*, August 11, 2007; Thomas Beaumont, "Huckabee Faces Pressure to Lure Voters After Win," *Des Moines Register*, August 13, 2007. Huckabee actually came in second place, but the *Register*'s reference to his "win" demonstrated how the perception of Huckabee's result may have meant more than Romney's actual win.

299   "Given the nature": Steve Deace podcast, 2007, WHO radio.

299   "You are waging": Steve Deace, August 9, 2007, WHO radio.

299   "If you look at concentric circles": Interview with Eric Woolson, 2011.

299   "I don't believe": Interview with Steve Deace, 2011.

300   Mickelson used most of the interview: Jan Mickelson, interview with Romney, August 2, 2007. The full interview and off-air segments were posted online at numerous sites.

300   "cost Romney central Iowa": Interview with Jan Mickelson, 2011.

301   "I can't buy you": Lisa Wangsness, "Romney Trounces GOP Field in Iowa Straw Poll; Ex-governor Tops 2d-tier Candidates," *The Boston Globe*, August 12, 2007.

301   Why not declare victory: Interviews with Romney aides, 2011.

301   "If you can't compete": Mitt Romney on *Fox News Sunday*, August 12, 2007.

302   "a bit hollow": David Yepsen, "Republicans Seem Lethargic About Choices," *Des Moines Register*, August 12, 2007.

302   "As a Christian": Campaign notebook, "Politics Overshadow Pulpits in Endorsement," *The Boston Globe*, October 17, 2007.

302   "a mistake": Interview with Richard Land.

303   "serve no one religion": Mitt Romney, speech on religion, Houston, Tex., December 6, 2007.

304   A few days after Romney's speech: "Religion Takes Center Stage Among Republican Rivals," *The New York Times*, December 12, 2007.

304   "I think attacking someone's religion": Glen Johnson, "Romney: Attacks on Religion Go Too Far," Associated Press, December 12, 2007.

304   "Why the sudden focus on Huckabee?": E-mail from Stuart Stevens to other Romney campaign aides, obtained by *The Boston Globe*.

305   "Mike Huckabee granted": Proposed ad produced by Romney campaign team; it never ran.

305   Concerned that the attack: Interviews with Romney aides, 2011.

305   "Choice: The Record": Michael Levenson and Sasha Issenberg, "Romney Draws Contrast with Huckabee," *The Boston Globe*, December 10, 2007; Maria Cramer and Maria Sacchetti, "More Immigrant Woes for Romney," *The Boston Globe*, December 5, 2007.

305   By the time the Iowa campaign was over: Interview with Romney aide, 2011.

The $10 million figure is an aide's estimate; campaign finance records do not accurately show how much is spent in a state because many of the advertising and consulting expenses are funneled through businesses in other states.

305   Indeed, a Huckabee aide: Interview with Huckabee aide, 2011.

306   "TrustHuckabee.com": Interview with Patrick Davis, 2011. Davis arranged for a coauthor to listen in on one of the telephone calls made during the Iowa campaign.

306   "We were waiting": Interview with Brian Kennedy, 2011.

306   Romney raced back to the Granite State: Michael Kranish, "While Most Zig to Iowa, Romney Zags to New Hampshire," *The Boston Globe*, December 26, 2007.

306   But Romney was under attack: Ibid.

307   "Granite Staters want": "Conservatives Are Coming Home," *Union Leader* (New Hampshire), December 26, 2007.

307   "I know something": Kranish, "While Most Zig to Iowa, Romney Zags to New Hampshire."

307   "We're going up against": Byron York, "Inside Huckabee's Victory," *The National Review*, January 4, 2008.

308   "Eric's quote just shows": Interview with Chip Saltsman, 2011.

308   "There are a lot of Iowans": Interview with Richard Schwarm.

308   "If we had known": Harvard Institute of Politics, *Campaign for President*, 73–78; Hugh Winebrenner and Dennis J. Goldford, *The Iowa Precinct Caucuses: The Making of a Media Event*, 3rd ed. (Iowa City: University of Iowa Press, 2010), 328.

309   "The pollsters were absolutely sure": Interview with Judd Gregg, 2011. Gregg later endorsed Romney's 2012 presidential bid.

310   "His preference in traveling": Interview with Bruce Keough.

310   "Keough fired off a memo": Ibid.

310   "rocket sled process": Romney campaign PowerPoint presentation, 2006.

311   "I was surprised": Dole Institute of Politics, Post Election Conference Transcript, www.doleinstitute.org/documents/Post-Election08-Transcript1.pdf, 15.

311   "We are going to be": Michael Kranish, "Huckabee Hires Former Reagan Adviser; 'Bare Knuckles' Brawler, Rollins Eyes a N.H. Win," *The Boston Globe*, December 15, 2007.

311   But Giuliani was waging: Brian C. Mooney, "Skills at Ready When Crisis Struck Giuliani," *The Boston Globe*, November 7, 2008.

312   "When Rudy stopped advertising": Harvard Institute of Politics, *Campaign for President*, 61.

312   "The President is not an expert": Michael Kranish, "Romney Says Qualified Advisers Are Key to a Sound Foreign Policy," *The Boston Globe*, December 28, 2007; Michael Levenson, "Romney Never Saw Father on King March; Rom-

ney Says He Never Saw His Father March with King; Defends Figurative Words; Evidence Contradicts Story," *The Boston Globe*, December 21, 2007.

312   "Washington is broken": Interview with Romney aide, 2011.

312   "I agree, you are": Republican presidential debate, January 5, 2008.

313   "money could buy": Interview with Bruce Keough.

313   That night, election returns: Scott Helman and Susan Milligan, "Clinton Edges Obama in N.H.; McCain Topples Romney," *The Boston Globe*, January 9, 2008.

313   "Alex. Well, change was it": E-mail from Mitt Romney to Alex Castellanos, January 8, 2008.

314   "I don't have lobbyists": Romney's interchange with Glen Johnson was captured by videographers at the Staples store in South Carolina; see www .youtube.com/watch?v=upGhWD4Bny0.

315   "a tough job to do": Jennifer Park, "Mitt Keeps It Classy on Leno," Political Punch, ABC News, http://abcnews.go.com/blogs/politics/2008/01/mitt -keeps-it-c/. Romney appeared on *The Tonight Show* on January 19, 2008.

315   "the truth of the matter": Harvard Institute of Politics, *Campaign for President*, 83.

316   The state staff prepared: "All Roads Lead to Florida," PowerPoint presentation, Florida Romney campaign team.

317   "This was a major debate": Interview with Mandy Fletcher.

317   "I really cared": Interview with Sally Bradshaw, 2011.

317   "If you had to boil": Interview with Mandy Fletcher.

318   "Governor Romney wanted to set": Sasha Issenberg and Michael Levenson, "McCain Accuses Romney of Wanting to Set Deadline for Iraq Withdrawal," *The Boston Globe*, January 27, 2008.

318   "We played Romney like a fiddle": Interview with McCain aide.

318   McCain beat Romney: Michael Cooper and Megan Thee, "McCain Defeats Romney in Florida Vote," *The New York Times*, January 29, 2008.

318   "I was really tired": Interview with John McCain, 2011.

319   "We're going to keep on": Michael Levenson and Sasha Issenberg, "Romney Wins Mass.; Huckabee Racks Up Victories in the South," *The Boston Globe*, February 6, 2008.

Chapter 12: Back into the Fire

322   "I had every confidence": Interview with John McCain, 2011.

323   "And so, in May 2008, a few months after": Ibid.; details of the visit and luncheon were also described by McCain aides.

324   "honored": *Hannity & Colmes*, Fox News Channel, March 11, 2008.

324   "This is a Massachusetts flip-flopper": Interviews with McCain aides, 2011.

324   Romney was thoroughly vetted: Interviews with McCain aide, 2011.

324   "wrong track": Harvard University Institute of Politics, *Campaign for President* (Lanham, Maryland: Rowman and Littlefield, 2009), 89–90.

324   "If we don't do anything": Ibid., 90.

325   But Land's key message: Interview with Richard Land, 2011.

326   "I never had a strategist": Interviews with Romney adviser, 2011.

327   "I think that one of the things": Glen Johnson, "Romney Says He Should Have Stuck to Economy in '08," Associated Press, March 24, 2010.

328   "agenda for a free and strong America": Mitt Romney, *No Apology: Believe in America* (New York: St. Martin's Press, 2010), 265, 269, 301–305.

328   "It's always a great interest": Sasha Issenberg, "In Book, Romney Styles Himself Wonk, Not Warrior," *The Boston Globe*, March 2, 2010.

328   With their help: Dan Eggen and T. W. Farnam, "Romney Has Eye on One Prize Right Now: Money," *The Washington Post*, May 11, 2011.

329   "The populism I'm referring to": Issenberg, "In Book, Romney Styles Himself Wonk, Not Warrior."

329   "I do as I'm commanded": Matt Viser, "Once a Suit-and-Tie Guy, Romney Opts for Jeans," *The Boston Globe*, June 8, 2011.

329   "Just got a Trim at Tommy's": Matt Viser, "Romney Drops by Atlanta for Business Meetings, Haircut," *The Boston Globe*, March 2, 2011.

330   "Corporations are people": www.c-span.org/Events/GOP-Presidential-Hopefuls-Campaign-in-Iowa/10737423434–6/; Ashley Parker, "'Corporations Are People,' Romney Tells Iowa Hecklers Angry over His Tax Policy," *The New York Times*, August 11, 2011.

330   "Light the Fire Within": Mitt Romney, *Turnaround: Crisis, Leadership, and the Olympic Games* (Washington, D.C.: Regnery Publishing, 2004), xxii.

331   "He was the same passionate person": Interview with Dane McBride, 2011.

Afterword

333   new "American century": Philip Rucker, "Romney Calls for New 'American century,' with Muscular Foreign Policy," *The Washington Post*, October 7, 2011.

333   modeled after the Pantheon: The Franklin Institute, accessed April 2012, http://www2.fi.edu/exhibits/permanent/franklin_national_memorial.php.

334   "watching Mitt Romney try to connect": Callum Borchers, "Economic Uncertainty Will Play a Key Role in Presidential Tilt, Analysts Say," *The Boston Globe*, April 11, 2012.

335   His state had also created jobs: "Texas-Size Recovery," Factcheck.org, August 31, 2011, http://www.factcheck.org/2011/08/texas-size-recovery.

335   "Father, our heart breaks": Thomas Beaumont, "Perry's Prayer Rally Draws a Big Crowd – And Criticism," Associated Press, August 7, 2011; Manny Fernandez, "Perry Leads Prayer Rally for 'Nation in Crisis,'" *The New York Times*, August 6, 2011.

336   he began courting: "Texas Governor Seeks Advice on a GOP Bid," Associated Press, July 12, 2011.

336   "I'll work every day": Glen Johnson, "Perry Launches Presidential Bid in
        S.C., N.H., Slamming Obama," *The Boston Globe,* August 14, 2011.

336   his poll numbers were sinking: 2012 National GOP Primary polls, *The
        Huffington Post,* http://elections.huffingtonpost.com/pollster/2012-national-
        gop-primary#!.

336   Romney himself: 2012 National GOP Primary polls, *The Huffington Post.*

336   the 9-9-9 plan: Jackie Kucinich, "Is Herman Cain's 9-9-9 Tax Plan Fair?"
        USA Today, October 10, 2001, http://www.usatoday.com/money/perfi/
        taxes/story/2011-10-10/herman-cain-9-9-9-tax-plan/50723976/1.

336   overtaking Romney: 2012 National GOP Primary polls, *The Huffington
        Post.*

336   foolishly combative performance: Michael Kranish, "Gingrich's Surge
        Gives Romney Camp Pause; Ignoring His Rivals Could Be Hurting," *The
        Boston Globe*, December 2, 2011.

337   "trying to feed a dog a pill": James Carville, *Anderson Cooper 360*, February
        10, 2012.

337   Romney's camp started believing: Ashley Parker and Jeff Zeleny, "Romney
        Will Compete in Iowa, if a Bit Cautiously," *The New York Times*, October
        21, 2011.

337   The rallies got bigger: Matt Viser and Michael Levenson, "It's Romney
        by 8 Votes, Santorum Close 2d in Iowa; Paul Is 3d; Perry May Quit," *The
        Boston Globe*, January 4, 2012.

337   It was a cold winter day: Interviews with Romney aides and Restore Our
        Future officials, 2012. Mitt and Ann Romney contributed $150,000 to-
        ward his presidential bid in May 2012, after he became the presumptive
        nominee, and did not rule out giving more later.

338   "We knew what needed to be done": Interview with Larry McCarthy, 2012.

339   "in any way": Mitt Romney on MSNBC's *Morning Joe,* December 20, 2011.
        A public interest group, Democracy 21, asked the Justice Department in
        March 2012 to investigate what it called "illegal coordination" between
        the campaign and the Super PAC. Romney's appearance at a fundraiser for
        Restore Our Future was reported by Peter Stone of the Center for Public
        Integrity. See: Peter Stone, "Democrats and Republicans alike are exploit-
        ing new fundraising loophole," *iWatch News*, July 27, 2011, http://www
        .iwatchnews.org/2011/07/27/5409/democrats-and-republicans-alike-are-
        exploiting-new-fundraising-loophole. Charles Spies, the lawyer for Restore
        Our Future, is quoted in this *Politico* story saying that Romney had at-
        tended "fundraising events" for the Super PAC: Alexander Burns, "Mitt
        Romney Addressing Super PAC Fundraisers," *Politico*, July 28, 2011, http://
        www.politico.com/news/stories/0711/60143.html.

339   A public interest group: The Democracy 21 complaint is at http://www
        .democracy21.org/index.asp?Type=B_PR&SEC={35BD37FA-E6F9-
        41ED-928E-884DAE00EAEC}&DE={93FE85F3-1A6D-415A-A069-
        2A10C47BB3B6}.

339      $51 million: PAC Profile: Restore Our Future, *iWatch News*, Center for Public Integrity, accessed April 2012, http://www.iwatchnews.org/node/7977.

339   Bob Perry: Michael Levenson, "'Swift Boat' Donor Throws Big Money Behind Romney," *The Boston Globe*, April 13, 2012. Perry had given $4.5 million to Swift Boat Veterans for Truth in 2004.

340   "My goodness": Matt Viser, "Romney Hits Stump in N.H. with McCain; Encouraged by Win, Gets Ready for Tuesday," *The Boston Globe*, January 5, 2012.

340   At 2:30 a.m.: Viser and Levenson, "It's Romney by 8 Votes, Santorum Close 2d in Iowa; Paul Is 3d; Perry May Quit."

341   built a sizable lead: RealClearPolitics compilation of South Carolina polls, accessed May 2012, http://www.realclearpolitics.com/epolls/2012/president/sc/south_carolina_republican_presidential_primary-1590.html#polls.

341   "The first few minutes": Michael Kranish, "Gingrich Plans to Stay in Race, Says Republican Contest Only in 'First Few Minutes,'" *The Boston Globe*, January 8, 2012.

341   Romney arrived in Florence: Matt Viser, "Mitt Romney Says He Has Been Paying Taxes Near 15 percent Rate," *The Boston Globe*, January 17, 2012.

341   in a January 19 debate: A transcript of the debate can be found at http://www.washingtonpost.com/wp-srv/politics/2012-presidential-debates/republican-primary-debate-january-19-2012/.

342   "bitter politics of envy": Jeff Zeleny, Trip Gabriel, Nicholas Confessore, Dalia Sussman, and Allison Kopicki, "Romney Is Winner in New Hampshire, Blunting Attacks," *The New York Times*, January 11, 2012.

342   his effective tax rate: Ashley Parker, "Romney Says His Effective Tax Rate Is About 15 Percent," *The New York Times*, January 17, 2012, http://thecaucus.blogs.nytimes.com/2012/01/17/romney-says-his-effective-tax-rate-is-about-15-percent/.

342   13.9 percent: Nicholas Confessore and David Kocieniewski, "For Romneys, Friendly Code Reduces Taxes," *The New York Times*, January 24, 2012, http://www.nytimes.com/2012/01/25/us/politics/romneys-tax-returns-show-21-6-million-income-in-10.html?pagewanted=all.

342   "It was a pained look": Interviews with Romney adviser, 2012.

342   "What I would say": Tim Mak, "Chris Christie: Mitt Romney Should Release Tax Returns," *Politico*, January 18, 2012, http://www.politico.com/news/stories/0112/71590.html.

343   a series of political missteps: Matt Viser, "In Detroit, Romney Says Wife Has 2 Cadillacs," *The Boston Globe*, February 25, 2012.

343   while at the Daytona 500: Jeff Zeleny, Jim Rutenberg, Michael Barbaro, and Katharine Q. Seelye, "Michigan a Testing Ground for Doubt on Romney," *The New York Times*, February 28, 2012.

343   Iowa Republicans had revised: Jennifer Jacobs, "2012 GOP Caucus Count Unresolved," *Des Moines Register*, January 19, 2012.

343   overtaking Romney: RealClearPolitics compilation of South Carolina polls.

344   "Everybody got knocked off": Interview with Romney adviser, 2012.

344   Polls in the wake of the South Carolina vote: RealClearPolitics compilation of Florida polls, accessed May 2012, http://www.realclearpolitics.com/epolls/2012/president/fl/florida_republican_presidential_primary-1597.html#polls.

344   "less than 0.1 percent": Jonathan Karl, "The Statistic of the Campaign: Romney's Single Positive Ad in Florida," *ABC News*, January 31, 2012. http://abcnews.go.com/blogs/politics/2012/01/the-statistic-of-the-campaign-romneys-single-positive-ad-in-florida/. The analysis of advertisements cited by ABC was done by Kantar Media's CMAG.

345   This was especially evident: Matt Viser, "Romney Still Has the Math in His Favor," *The Boston Globe*, March 15, 2012.

345   Romney had lost every state: Mike Allen and Evan Thomas, *Inside the Circus: Romney, Santorum and the GOP Race* (New York: Random House, 2012), iPhone edition, 313.

345   In early February: David A. Fahrenthold, "Mitt Romney Was 'Severely Conservative' Governor, He Tells CPAC," *The Washington Post*, February 10, 2012.

345   Those words weren't: Interview with Romney adviser, 2012.

345   She had been instrumental: Katie Leslie, "Ann Romney Campaigns in Georgia," *Atlanta Journal Constitution*, March 1, 2012; interview with Romney aide, 2012.

346   "I was tired": Ann Romney interview on *Entertainment Tonight*. Lisa Hirsch, "Ann Romney on Recent MS Scare on Campaign Trail," April 25, 2012. http://www.etonline.com/news/121147_Ann_Romney_on_Recent_MS_Scare_on_Campaign_Trail/index.html.

347   "He's the guy": Allen and Thomas, *Inside the Circus*, 267.

347   Back when it began: Glen Johnson, "Live Blog of Romney Presidential Announcement," Political Intelligence blog, *The Boston Globe*, June 2, 2011, http://www.boston.com/news/politics/politicalintelligence/2011/06/live_blog_of_ro.html.

348   some two dozen times: Fahrenthold, "Mitt Romney Was 'Severely Conservative' Governor, He Tells CPAC."

348   Longtime adviser Eric Fehrnstrom: Matt Viser, "Words Boost, Bedevil Romney; Jeb Bush Gives a Coveted Backing; Aide Likens Campaign to an Etch A Sketch," *The Boston Globe*, March 22, 2012.

349   He emerged: Allen and Thomas, *Inside the Circus*, 245.

INDEX